MARPOL 73/78
Consolidated Edition, 2002

Articles, Protocols, Annexes, Unified Interpretations
of the International Convention for the Prevention
of Pollution from Ships, 1973, as modified by
the Protocol of 1978 relating thereto

IMO
London, 2002

Published in 2002
by the INTERNATIONAL MARITIME ORGANIZATION
4 Albert Embankment, London SE1 7SR

Consolidated edition, 1991
Consolidated edition, 1997
Consolidated edition, 2002

Printed in the United Kingdom by William Clowes Ltd, Beccles, Suffolk

ISBN 92-801-5125-8

2 4 6 8 10 9 7 5 3 1

IMO PUBLICATION
Sales number: IMO-520E

Introduction

The International Convention for the Prevention of Pollution from Ships, 1973, was adopted by the International Conference on Marine Pollution convened by IMO from 8 October to 2 November 1973. Protocols I (Provisions concerning Reports on Incidents involving Harmful Substances) and II (Arbitration) were adopted at the same Conference. This Convention was subsequently modified by the Protocol of 1978 relating thereto, which was adopted by the International Conference on Tanker Safety and Pollution Prevention (TSPP Conference) convened by IMO from 6 to 17 February 1978. The Convention, as modified by the 1978 Protocol, is known as the "International Convention for the Prevention of Pollution from Ships, 1973, as modified by the Protocol of 1978 relating thereto", or, in short form, "MARPOL 73/78". Regulations covering the various sources of ship-generated pollution are contained in the five Annexes of the Convention. The Convention has also been modified by the Protocol of 1997, whereby a sixth Annex was adopted, but this Protocol has not yet been accepted by sufficient States for it to enter into force.

The Marine Environment Protection Committee (MEPC), since its inception in 1974, has reviewed various provisions of MARPOL 73/78 that have been found to require clarification or have given rise to difficulties in implementation. In order to resolve such ambiguities and difficulties in a uniform manner, the MEPC agreed that it was desirable to develop unified interpretations. In certain cases, the MEPC recognized that there was a need to amend existing regulations or to introduce new regulations with the aim of reducing even further operational and accidental pollution from ships. These activities by the MEPC have resulted in a number of unified interpretations and amendments to the Convention.

The purpose of this publication is to provide an easy reference to the up-to-date provisions and unified interpretations of the articles, protocols and Annexes of MARPOL 73/78, including the incorporation of all of the amendments that have been adopted by the MEPC and have entered into force, up to and including the 2000 amendments (as adopted by resolution MEPC.89(45)). It should be noted, however, that the Secretariat has no intention of changing the authentic texts editorially or otherwise. For legal purposes, the authentic texts of the provisions of MARPOL 73/78 should always be consulted.

An exception to the above is the amendments to regulation 13G of Annex I and to the Supplement to the IOPP Certificate (as adopted on 16 May 2001 by resolution MEPC.95(46)). The date for tacit acceptance of these important amendments is 1 March 2002, and if they are accepted on that date they will enter into force on 1 September 2002. As of the date of publication of this edition, the criteria for entry into force of these amendments have not been met. It was felt, however, that the amendments might enter into force before the next revision of the present consolidated edition of MARPOL 73/78. Therefore, the text of resolution MEPC.95(46) is reproduced as item 7 of the Additional Information section. An associated text (the Condition Assessment Scheme) that was adopted by resolution MEPC.94(46) is item 8 of the same section.

In addition to incorporating the applicable amendments into the texts of Protocol I and Annexes I to V to MARPOL 73/78, the Secretariat has updated the 1997 Consolidated Edition by adding the text of the Protocol of 1997 and of Annex VI. A unified interpretation for hydrostatic balance loading, relating to regulation 13G of Annex I, as approved by the MEPC, has also been added to the appendices to the unified interpretations of Annex I. Resolution MEPC.88(44), relating to the adoption of a revised Annex IV, and the text of the revised Annex are included as items 5 and 6 of the Additional Information section.

For consistency in providing information, guidelines which are not made mandatory by the applicable Annex, and which are contained in another IMO publication, are omitted from the 2002 Consolidated Edition.

Protocol I – Provisions concerning Reports on Incidents involving Harmful Substances

This Protocol was adopted on 2 November 1973 and subsequently amended by:

- 1985 amendments (resolution MEPC.21(22)) by which the Protocol was replaced by a revised text: entered into force on 6 April 1987; and

- 1996 amendments (resolution MEPC.68(38)) on amendments to article II(1): entered into force on 1 January 1998.

Annex I – Regulations for the Prevention of Pollution by Oil

Annex I entered into force on 2 October 1983 and, as between the Parties to MARPOL 73/78, supersedes the International Convention for the Prevention of Pollution of the Sea by Oil, 1954, as amended in 1962 and 1969, which was then in force. A number of amendments to Annex I have been adopted by the MEPC and have entered into force as summarized below:

- 1984 amendments (resolution MEPC.14(20)) on control of discharge of oil; retention of oil on board; pumping, piping and discharge arrangements of oil tankers; subdivision and stability: entered into force on 7 January 1986;

- 1987 amendments (resolution MEPC.29(25)) on designation of the Gulf of Aden as a special area: entered into force on 1 April 1989;

- 1990 amendments (resolution MEPC.39(29)) on the introduction of the harmonized system of survey and certification: entered into force on 3 February 2000;

- 1990 amendments (resolution MEPC.42(30)) on designation of the Antarctic area as a special area: entered into force on 17 March 1992;

- 1991 amendments (resolution MEPC.47(31)) on new regulation 26, Shipboard Oil Pollution Emergency Plan, and other amendments to Annex I: entered into force on 4 April 1993;

- 1992 amendments (resolution MEPC.51(32)) on discharge criteria of Annex I: entered into force on 6 July 1993;

- 1992 amendments (resolution MEPC.52(32)) on new regulations 13F and 13G and related amendments to Annex I: entered into force on 6 July 1993;

- 1994 amendments (resolution 1 adopted on 2 November 1994 by the Conference of Parties to MARPOL 73/78) on port State control on operational requirements: entered into force on 3 March 1996;

- 1997 amendments (resolution MEPC.75(40)) on designation of North West European waters as a special area and new regulation 25A; entered into force on 1 February 1999;

- 1999 amendments (resolution MEPC.78(43)) on amendments to regulations 13G and 26 and the IOPP Certificate: entered into force on 1 January 2001; and

- 2001 amendments (resolution MEPC.95(46)) on amendments to regulation 13G: if they are accepted on 1 March 2002 they will enter into force on 1 September 2002.

Annex II – Regulations for the Control of Pollution by Noxious Liquid Substances in Bulk

To facilitate implementation of the Annex, the original text underwent amendments in 1985, by resolution MEPC.16(22), in respect of pumping, piping and control requirements. At its twenty-second session, the MEPC also decided that, in accordance with article II of the 1978 Protocol, "Parties shall be bound by the provisions of Annex II of MARPOL 73/78 as amended from 6 April 1987" (resolution MEPC.17(22)). Subsequent amendments have been adopted by the MEPC and have entered into force as summarized below:

- 1989 amendments (resolution MEPC.34(27)), which updated appendices II and III to make them compatible with chapters 17/VI and 18/VII of the IBC Code and BCH Code, respectively: entered into force on 13 October 1990;

- 1990 amendments (resolution MEPC.39(29)) on the introduction of the harmonized system of survey and certification: entered into force on 3 February 2000;

- 1992 amendments (resolution MEPC.57(33)) on designation of the Antarctic area as a special area and lists of liquid substances in appendices to Annex II: entered into force on 1 July 1994;

- 1994 amendments (resolution 1 adopted on 2 November 1994 by the Conference of Parties to MARPOL 73/78) on port State control on operational requirements: entered into force on 3 March 1996; and

- 1999 amendments (resolution MEPC.78(43)) on addition of new regulation 16: entered into force on 1 January 2001.

Annex III – Regulations for the Prevention of Pollution by Harmful Substances Carried by Sea in Packaged Form

Annex III entered into force on 1 July 1992. However, long before this entry into force date, the MEPC, with the concurrence of the Maritime Safety Committee (MSC), agreed that the Annex should be implemented through the IMDG Code. The IMDG Code had amendments covering marine pollution prepared by the MSC (Amendment 25-89) and these amendments were implemented from 1 January 1991. Subsequent amendments have been adopted by the MEPC and have entered into force as summarized below:

- 1992 amendments (resolution MEPC.58(33)), which totally revised Annex III as a clarification of the requirements in the original version of Annex III rather than a change of substance, and incorporated the reference to the IMDG Code: entered into force on 28 February 1994;

- 1994 amendments (resolution 2 adopted on 2 November 1994 by the Conference of Parties to MARPOL 73/78) on port State control on operational requirements: entered into force on 3 March 1996; and

- 2000 amendments (MEPC.84(44)), deleting a clause relating to tainting of seafood: entered into force on 1 January 2002.

Annex IV – Regulations for the Prevention of Pollution by Sewage from Ships

Annex IV is not yet in force. As of 21 September 2001, the Annex had been ratified by 81 States, the combined merchant fleet of which represented approximately 46% of the gross tonnage of the world's merchant fleet. Therefore, ratification by States covering an additional 4% of the gross tonnage of the world's merchant fleet was required before the entry-into-force requirements of article 16(2)(f) of the Convention were satisfied. A resolution (MEPC.88(44)) was adopted in March 2000 by which a revised text of Annex IV would be considered for adoption at the same time as the conditions for entry into force of the original Annex are met, and this revised text was adopted by the MEPC on the same date as the resolution. The resolution and the revised text are included as items 5 and 6 of the Additional Information section.

Annex V – Regulations for the Prevention of Pollution by Garbage from Ships

Annex V entered into force on 31 December 1988. Subsequent amendments have been adopted by the MEPC and have entered into force as summarized below:

– 1989 amendments (resolution MEPC.36(28)) on designation of the North Sea as a special area and amendment of regulation 6, Exceptions: entered into force on 18 February 1991;

– 1990 amendments (resolution MEPC.42(30)) on designation of the Antarctic area as a special area: entered into force on 17 March 1992;

– 1991 amendments (resolution MEPC.48(31)) on designation of the Wider Caribbean area as a special area: entered into force on 4 April 1993;

– 1994 amendments (resolution 3 adopted on 2 November 1994 by the Conference of Parties to MARPOL 73/78) on port State control on operational requirements: entered into force on 3 March 1996;

– 1995 amendments (resolution MEPC.65(37)) on amendment of regulation 2 and the addition of a new regulation 9 of Annex V: entered into force on 1 July 1997; and

– 2000 amendments (resolution MEPC.89(45)) on amendments to regulations 1, 3, 5 and 9 and to the Record of Garbage Discharge: will enter into force on 1 March 2002.

Annex VI – Regulations for the Prevention of Air Pollution from Ships

Annex VI is appended to the Protocol of 1997 to amend the International Convention for the Prevention of Pollution from Ships, 1973, as modified by the Protocol of 1978 relating thereto, which was adopted by the International Conference of Parties to MARPOL 73/78 in September 1997. In accordance with article 6, this Protocol will enter into force twelve months after the date on which not less than fifteen States, the combined merchant fleets of which constitute not less than 50% of the gross tonnage of the world's merchant shipping, have expressed their consent to be bound by it. As of 21 September 2001, there were three Contracting States.

Contents

Annex V of MARPOL 73/78: *Regulations for the Prevention of Pollution by Garbage from Ships*

Appendix to Annex V

Annex VI of MARPOL 73/78: *Regulations for the Prevention of Air Pollution from Ships*

Chapter I – *General*

Chapter II – *Survey, certification and means of control*

International Convention for the Prevention of Pollution from Ships, 1973

International Convention for the Prevention of Pollution from Ships, 1973

THE PARTIES TO THE CONVENTION,

BEING CONSCIOUS of the need to preserve the human environment in general and the marine environment in particular,

RECOGNIZING that deliberate, negligent or accidental release of oil and other harmful substances from ships constitutes a serious source of pollution,

RECOGNIZING ALSO the importance of the International Convention for the Prevention of Pollution of the Sea by Oil, 1954, as being the first multilateral instrument to be concluded with the prime objective of protecting the environment, and appreciating the significant contribution which that Convention has made in preserving the seas and coastal environment from pollution,

DESIRING to achieve the complete elimination of intentional pollution of the marine environment by oil and other harmful substances and the minimization of accidental discharge of such substances,

CONSIDERING that this object may best be achieved by establishing rules not limited to oil pollution having a universal purport,

HAVE AGREED as follows:

Article 1
General obligations under the Convention

(1) The Parties to the Convention undertake to give effect to the provisions of the present Convention and those Annexes thereto by which they are bound, in order to prevent the pollution of the marine environment by the discharge of harmful substances or effluents containing such substances in contravention of the Convention.

(2) Unless expressly provided otherwise, a reference to the present Convention constitutes at the same time a reference to its Protocols and to the Annexes.

Article 2

Definitions

For the purposes of the present Convention, unless expressly provided otherwise:

(1) *Regulation* means the regulations contained in the Annexes to the present Convention.

(2) *Harmful substance* means any substance which, if introduced into the sea, is liable to create hazards to human health, to harm living resources and marine life, to damage amenities or to interfere with other legitimate uses of the sea, and includes any substance subject to control by the present Convention.

(3) (a) *Discharge*, in relation to harmful substances or effluents containing such substances, means any release howsoever caused from a ship and includes any escape, disposal, spilling, leaking, pumping, emitting or emptying;

    (b) *Discharge* does not include:

        (i) dumping within the meaning of the Convention on the Prevention of Marine Pollution by Dumping of Wastes and Other Matter, done at London on 13 November 1972; or

        (ii) release of harmful substances directly arising from the exploration, exploitation and associated offshore processing of sea-bed mineral resources; or

        (iii) release of harmful substances for purposes of legitimate scientific research into pollution abatement or control.

(4) *Ship* means a vessel of any type whatsoever operating in the marine environment and includes hydrofoil boats, air-cushion vehicles, submersibles, floating craft and fixed or floating platforms.

(5) *Administration* means the Government of the State under whose authority the ship is operating. With respect to a ship entitled to fly a flag of any State, the Administration is the Government of that State. With respect to fixed or floating platforms engaged in exploration and exploitation of the sea-bed and subsoil thereof adjacent to the coast over which the coastal State exercises sovereign rights for the purposes of exploration and exploitation of their natural resources, the Administration is the Government of the coastal State concerned.

(6) *Incident* means an event involving the actual or probable discharge into the sea of a harmful substance, or effluents containing such a substance.

4

(7)   *Organization* means the Inter-Governmental Maritime Consultative Organization.*

Article 3
Application

(1)   The present Convention shall apply to:

    (a)   ships entitled to fly the flag of a Party to the Convention; and

    (b)   ships not entitled to fly the flag of a Party but which operate under the authority of a Party.

(2)   Nothing in the present article shall be construed as derogating from or extending the sovereign rights of the Parties under international law over the sea-bed and subsoil thereof adjacent to their coasts for the purposes of exploration and exploitation of their natural resources.

(3)   The present Convention shall not apply to any warship, naval auxiliary or other ship owned or operated by a State and used, for the time being, only on government non-commercial service. However, each Party shall ensure by the adoption of appropriate measures not impairing the operations or operational capabilities of such ships owned or operated by it, that such ships act in a manner consistent, so far as is reasonable and practicable, with the present Convention.

Article 4
Violation

(1)   Any violation of the requirements of the present Convention shall be prohibited and sanctions shall be established therefor under the law of the Administration of the ship concerned wherever the violation occurs. If the Administration is informed of such a violation and is satisfied that sufficient evidence is available to enable proceedings to be brought in respect of the alleged violation, it shall cause such proceedings to be taken as soon as possible, in accordance with its law.

(2)   Any violation of the requirements of the present Convention within the jurisdiction of any Party to the Convention shall be prohibited and sanctions shall be established therefor under the law of that Party. Whenever such a violation occurs, that Party shall either:

    (a)   cause proceedings to be taken in accordance with its law; or

---

* The name of the Organization was changed to "International Maritime Organization" by virtue of amendments to the Organization's Convention which entered into force on 22 May 1982.

(b) furnish to the Administration of the ship such information and evidence as may be in its possession that a violation has occurred.

(3) Where information or evidence with respect to any violation of the present Convention by a ship is furnished to the Administration of that ship, the Administration shall promptly inform the Party which has furnished the information or evidence, and the Organization, of the action taken.

(4) The penalties specified under the law of a Party pursuant to the present article shall be adequate in severity to discourage violations of the present Convention and shall be equally severe irrespective of where the violations occur.

Article 5
Certificates and special rules on inspection of ships

(1) Subject to the provisions of paragraph (2) of the present article a certificate issued under the authority of a Party to the Convention in accordance with the provisions of the regulations shall be accepted by the other Parties and regarded for all purposes covered by the present Convention as having the same validity as a certificate issued by them.

(2) A ship required to hold a certificate in accordance with the provisions of the regulations is subject, while in the ports or offshore terminals under the jurisdiction of a Party, to inspection by officers duly authorized by that Party. Any such inspection shall be limited to verifying that there is on board a valid certificate, unless there are clear grounds for believing that the condition of the ship or its equipment does not correspond substantially with the particulars of that certificate. In that case, or if the ship does not carry a valid certificate, the Party carrying out the inspection shall take such steps as will ensure that the ship shall not sail until it can proceed to sea without presenting an unreasonable threat of harm to the marine environment. That Party may, however, grant such a ship permission to leave the port or offshore terminal for the purpose of proceeding to the nearest appropriate repair yard available.

(3) If a Party denies a foreign ship entry to the ports or offshore terminals under its jurisdiction or takes any action against such a ship for the reason that the ship does not comply with the provisions of the present Convention, the Party shall immediately inform the consul or diplomatic representative of the Party whose flag the ship is entitled to fly, or if this is not possible, the Administration of the ship concerned. Before denying entry or taking such action the Party may request consultation with the Administration of the ship concerned.

Information shall also be given to the Administration when a ship does not carry a valid certificate in accordance with the provisions of the regulations.

(4)   With respect to the ship of non-Parties to the Convention, Parties shall apply the requirements of the present Convention as may be necessary to ensure that no more favourable treatment is given to such ships.

Article 6
Detection of violations and enforcement of the Convention

(1)   Parties to the Convention shall co-operate in the detection of violations and the enforcement of the provisions of the present Convention, using all appropriate and practicable measures of detection and environmental monitoring, adequate procedures for reporting and accumulation of evidence.

(2)   A ship to which the present Convention applies may, in any port or offshore terminal of a Party, be subject to inspection by officers appointed or authorized by that Party for the purpose of verifying whether the ship has discharged any harmful substances in violation of the provisions of the regulations. If an inspection indicates a violation of the Convention, a report shall be forwarded to the Administration for any appropriate action.

(3)   Any Party shall furnish to the Administration evidence, if any, that the ship has discharged harmful substances or effluents containing such substances in violation of the provisions of the regulations. If it is practicable to do so, the competent authority of the former Party shall notify the master of the ship of the alleged violation.

(4)   Upon receiving such evidence, the Administration so informed shall investigate the matter, and may request the other Party to furnish further or better evidence of the alleged contravention. If the Administration is satisfied that sufficient evidence is available to enable proceedings to be brought in respect of the alleged violation, it shall cause such proceedings to be taken in accordance with its law as soon as possible. The Administration shall promptly inform the Party which has reported the alleged violation, as well as the Organization, of the action taken.

(5)   A Party may also inspect a ship to which the present Convention applies when it enters the ports or offshore terminals under its jurisdiction, if a request for an investigation is received from any Party together with sufficient evidence that the ship has discharged harmful substances or effluents containing such substances in any place. The

report of such investigation shall be sent to the Party requesting it and to the Administration so that the appropriate action may be taken under the present Convention.

Article 7
Undue delay to ships

(1) All possible efforts shall be made to avoid a ship being unduly detained or delayed under articles 4, 5 or 6 of the present Convention.

(2) When a ship is unduly detained or delayed under articles 4, 5 or 6 of the present Convention, it shall be entitled to compensation for any loss or damage suffered.

Article 8
Reports on incidents involving harmful substances

(1) A report of an incident shall be made without delay to the fullest extent possible in accordance with the provisions of Protocol I to the present Convention.

(2) Each Party to the Convention shall:

    (a) make all arrangements necessary for an appropriate officer or agency to receive and process all reports on incidents; and

    (b) notify the Organization with complete details of such arrangements for circulation to other Parties and Member States of the Organization.

(3) Whenever a Party receives a report under the provisions of the present article, that Party shall relay the report without delay to:

    (a) the Administration of the ship involved; and

    (b) any other State which may be affected.

(4) Each Party to the Convention undertakes to issue instructions to its maritime inspection vessels and aircraft and to other appropriate services, to report to its authorities any incident referred to in Protocol I to the present Convention. That Party shall, if it considers it appropriate, report accordingly to the Organization and to any other Party concerned.

Article 9
Other treaties and interpretation

(1) Upon its entry into force, the present Convention supersedes the International Convention for the Prevention of Pollution of the Sea by Oil, 1954, as amended, as between Parties to that Convention.

(2) Nothing in the present Convention shall prejudice the codification and development of the law of the sea by the United Nations Conference on the Law of the Sea convened pursuant to resolution 2750 C(XXV) of the General Assembly of the United Nations nor the present or future claims and legal views of any State concerning the law of the sea and the nature and extent of coastal and flag State jurisdiction.

(3) The term "jurisdiction" in the present Convention shall be construed in the light of international law in force at the time of application or interpretation of the present Convention.

Article 10
Settlement of disputes

Any dispute between two or more Parties to the Convention concerning the interpretation or application of the present Convention shall, if settlement by negotiation between the Parties involved has not been possible, and if these Parties do not otherwise agree, be submitted upon request of any of them to arbitration as set out in Protocol II to the present Convention.

Article 11
Communication of information

(1) The Parties to the Convention undertake to communicate to the Organization:

(a) the text of laws, orders, decrees and regulations and other instruments which have been promulgated on the various matters within the scope of the present Convention;

(b) a list of non-governmental agencies which are authorized to act on their behalf in matters relating to the design, construction and equipment of ships carrying harmful substances in accordance with the provisions of the regulations;*

(c) a sufficient number of specimens of their certificates issued under the provisions of the regulations;

(d) a list of reception facilities including their location, capacity and available facilities and other characteristics;

(e) official reports or summaries of official reports in so far as they show the results of the application of the present Convention; and

---

* The text of this subparagraph is replaced by that contained in article III of the 1978 Protocol.

(f) an annual statistical report, in a form standardized by the Organization, of penalties actually imposed for infringement of the present Convention.

(2) The Organization shall notify Parties of the receipt of any communications under the present article and circulate to all Parties any information communicated to it under subparagraphs (1)(b) to (f) of the present article.

Article 12
Casualties to ships

(1) Each Administration undertakes to conduct an investigation of any casualty occurring to any of its ships subject to the provisions of the regulations if such casualty has produced a major deleterious effect upon the marine environment.

(2) Each Party to the Convention undertakes to supply the Organization with information concerning the findings of such investigation, when it judges that such information may assist in determining what changes in the present Convention might be desirable.

Article 13
Signature, ratification, acceptance, approval
and accession

(1) The present Convention shall remain open for signature at the Headquarters of the Organization from 15 January 1974 until 31 December 1974 and shall thereafter remain open for accession. States may become Parties to the present Convention by:

(a) signature without reservation as to ratification, acceptance or approval; or

(b) signature subject to ratification, acceptance or approval, followed by ratification, acceptance or approval; or

(c) accession.

(2) Ratification, acceptance, approval or accession shall be effected by the deposit of an instrument to that effect with the Secretary-General of the Organization.

(3) The Secretary-General of the Organization shall inform all States which have signed the present Convention or acceded to it of any signature or of the deposit of any new instrument of ratification, acceptance, approval or accession and the date of its deposit.

Article 14
Optional Annexes

(1) A State may at the time of signing, ratifying, accepting, approving or acceding to the present Convention declare that it does not accept any one or all of Annexes III, IV and V (hereinafter referred to as "Optional Annexes") of the present Convention. Subject to the above, Parties to the Convention shall be bound by any Annex in its entirety.

(2) A State which has declared that it is not bound by an Optional Annex may at any time accept such Annex by depositing with the Organization an instrument of the kind referred to in article 13(2).

(3) A State which makes a declaration under paragraph (1) of the present article in respect of an Optional Annex and which has not subsequently accepted that Annex in accordance with paragraph (2) of the present article shall not be under any obligation nor entitled to claim any privileges under the present Convention in respect of matters related to such Annex and all references to Parties in the present Convention shall not include that State in so far as matters related to such Annex are concerned.

(4) The Organization shall inform the States which have signed or acceded to the present Convention of any declaration under the present article as well as the receipt of any instrument deposited in accordance with the provisions of paragraph (2) of the present article.

Article 15
Entry in force

(1) The present Convention shall enter into force 12 months after the date on which not less than 15 States, the combined merchant fleets of which constitute not less than 50 per cent of the gross tonnage of the world's merchant shipping, have become parties to it in accordance with article 13.

(2) An Optional Annex shall enter into force 12 months after the date on which the conditions stipulated in paragraph (1) of the present article have been satisfied in relation to that Annex.

(3) The Organization shall inform the States which have signed the present Convention or acceded to it of the date on which it enters into force and of the date on which an Optional Annex enters into force in accordance with paragraph (2) of the present article.

(4) For States which have deposited an instrument of ratification, acceptance, approval or accession in respect of the present Convention

or any Optional Annex after the requirements for entry into force thereof have been met but prior to the date of entry into force, the ratification, acceptance, approval or accession shall take effect on the date of entry into force of the Convention or such Annex or three months after the date of deposit of the instrument whichever is the later date.

(5) For States which have deposited an instrument of ratification, acceptance, approval or accession after the date on which the Convention or an Optional Annex entered into force, the Convention or the Optional Annex shall become effective three months after the date of deposit of the instrument.

(6) After the date on which all the conditions required under article 16 to bring an amendment to the present Convention or an Optional Annex into force have been fulfilled, any instrument of ratification, acceptance, approval or accession deposited shall apply to the Convention or Annex as amended.

Article 16
Amendments

(1) The present Convention may be amended by any of the procedures specified in the following paragraphs.

(2) Amendments after consideration by the Organization:

    (a) any amendment proposed by a Party to the Convention shall be submitted to the Organization and circulated by its Secretary-General to all Members of the Organization and all Parties at least six months prior to its consideration;

    (b) any amendment proposed and circulated as above shall be submitted to an appropriate body by the Organization for consideration;

    (c) Parties to the Convention, whether or not Members of the Organization, shall be entitled to participate in the proceedings of the appropriate body;

    (d) amendments shall be adopted by a two-thirds majority of only the Parties to the Convention present and voting;

    (e) if adopted in accordance with subparagraph (d) above, amendments shall be communicated by the Secretary-General of the Organization to all the Parties to the Convention for acceptance;

    (f) an amendment shall be deemed to have been accepted in the following circumstances:

(i)   an amendment to an article of the Convention shall be deemed to have been accepted on the date on which it is accepted by two thirds of the Parties, the combined merchant fleets of which constitute not less than 50 per cent of the gross tonnage of the world's merchant fleet;

(ii)   an amendment to an Annex to the Convention shall be deemed to have been accepted in accordance with the procedure specified in subparagraph (f)(iii) unless the appropriate body, at the time of its adoption, determines that the amendment shall be deemed to have been accepted on the date on which it is accepted by two thirds of the Parties, the combined merchant fleets of which constitute not less than 50 per cent of the gross tonnage of the world's merchant fleet. Nevertheless, at any time before the entry into force of an amendment to an Annex to the Convention, a Party may notify the Secretary-General of the Organization that its express approval will be necessary before the amendment enters into force for it. The latter shall bring such notification and the date of its receipt to the notice of Parties;

(iii)   an amendment to an appendix to an Annex to the Convention shall be deemed to have been accepted at the end of a period to be determined by the appropriate body at the time of its adoption, which period shall be not less than ten months, unless within that period an objection is communicated to the Organization by not less than one third of the Parties or by the Parties the combined merchant fleets of which constitute not less than 50 per cent of the gross tonnage of the world's merchant fleet whichever condition is fulfilled;

(iv)   an amendment to Protocol I to the Convention shall be subject to the same procedures as for the amendments to the Annexes to the Convention, as provided for in subparagraphs (f)(ii) or (f)(iii) above;

(v)   an amendment to Protocol II to the Convention shall be subject to the same procedures as for the amendments to an article of the Convention, as provided for in subparagraph (f)(i) above;

(g)   the amendment shall enter into force under the following conditions:

(i)   in the case of an amendment to an article of the Convention, to Protocol II, or to Protocol I or to an Annex to the Convention not under the procedure specified in subparagraph (f)(iii), the amendment accepted in conformity with the foregoing provisions shall enter into force six months after the date of its acceptance with respect to the Parties which have declared that they have accepted it;

(ii) in the case of an amendment to Protocol I, to an appendix to an Annex or to an Annex to the Convention under the procedure specified in subparagraph (f)(iii), the amendment deemed to have been accepted in accordance with the foregoing conditions shall enter into force six months after its acceptance for all the Parties with the exception of those which, before that date, have made a declaration that they do not accept it or a declaration under subparagraph (f)(ii), that their express approval is necessary.

(3) Amendment by a Conference:

(a) Upon the request of a Party, concurred in by at least one third of the Parties, the Organization shall convene a Conference of Parties to the Convention to consider amendments to the present Convention.

(b) Every amendment adopted by such a Conference by a two-thirds majority of those present and voting of the Parties shall be communicated by the Secretary-General of the Organization to all Contracting Parties for their acceptance.

(c) Unless the Conference decides otherwise, the amendment shall be deemed to have been accepted and to have entered into force in accordance with the procedures specified for that purpose in paragraph (2)(f) and (g) above.

(4) (a) In the case of an amendment to an Optional Annex, a reference in the present article to a "Party to the Convention" shall be deemed to mean a reference to a Party bound by that Annex.

(b) Any Party which has declined to accept an amendment to an Annex shall be treated as a non-Party only for the purpose of application of that amendment.

(5) The adoption and entry into force of a new Annex shall be subject to the same procedures as for the adoption and entry into force of an amendment to an article of the Convention.

(6) Unless expressly provided otherwise, any amendment to the present Convention made under this article, which relates to the structure of a ship, shall apply only to ships for which the building contract is placed, or in the absence of a building contract, the keel of which is laid, on or after the date on which the amendment comes into force.

(7)   Any amendment to a Protocol or to an Annex shall relate to the substance of that Protocol or Annex and shall be consistent with the articles of the present Convention.

(8)   The Secretary-General of the Organization shall inform all Parties of any amendments which enter into force under the present article, together with the date on which each such amendment enters into force.

(9)   Any declaration of acceptance or of objection to an amendment under the present article shall be notified in writing to the Secretary-General of the Organization. The latter shall bring such notification and the date of its receipt to the notice of the Parties to the Convention.

Article 17
Promotion of technical co-operation

The Parties to the Convention shall promote, in consultation with the Organization and other international bodies, with assistance and co-ordination by the Executive Director of the United Nations Environment Programme, support for those Parties which request technical assistance for:

(a)   the training of scientific and technical personnel;

(b)   the supply of necessary equipment and facilities for reception and monitoring;

(c)   the facilitation of other measures and arrangements to prevent or mitigate pollution of the marine environment by ships; and

(d)   the encouragement of research;

preferably within the countries concerned, so furthering the aims and purposes of the present Convention.

Article 18
Denunciation

(1)   The present Convention or any Optional Annex may be denounced by any Parties to the Convention at any time after the expiry of five years from the date on which the Convention or such Annex enters into force for that Party.

(2)   Denunciation shall be effected by notification in writing to the Secretary-General of the Organization who shall inform all the other Parties of any such notification received and of the date of its receipt as well as the date on which such denunciation takes effect.

(3)   A denunciation shall take effect 12 months after receipt of the notification of denunciation by the Secretary-General of the

Organization or after the expiry of any other longer period which may be indicated in the notification.

Article 19
Deposit and registration

(1) The present Convention shall be deposited with the Secretary-General of the Organization who shall transmit certified true copies thereof to all States which have signed the present Convention or acceded to it.

(2) As soon as the present Convention enters into force, the text shall be transmitted by the Secretary-General of the Organization to the Secretary-General of the United Nations for registration and publication, in accordance with Article 102 of the Charter of the United Nations.

Article 20
Languages

The present Convention is established in a single copy in the English, French, Russian and Spanish languages, each text being equally authentic. Official translations in the Arabic, German, Italian and Japanese languages shall be prepared and deposited with the signed original.

IN WITNESS WHEREOF the undersigned* being duly authorized by their respective Governments for that purpose have signed the present Convention.

DONE AT LONDON this second day of November, one thousand nine hundred and seventy-three.

---

* Signatures omitted.

Protocol of 1978 relating to the International Convention for the Prevention of Pollution from Ships, 1973

Protocol of 1978 relating to the International Convention for the Prevention of Pollution from Ships, 1973

THE PARTIES TO THE PRESENT PROTOCOL,

RECOGNIZING the significant contribution which can be made by the International Convention for the Prevention of Pollution from Ships, 1973, to the protection of the marine environment from pollution from ships,

RECOGNIZING ALSO the need to improve further the prevention and control of marine pollution from ships, particularly oil tankers,

RECOGNIZING FURTHER the need for implementing the Regulations for the Prevention of Pollution by Oil contained in Annex I of that Convention as early and as widely as possible,

ACKNOWLEDGING HOWEVER the need to defer the application of Annex II of that Convention until certain technical problems have been satisfactorily resolved,

CONSIDERING that these objectives may best be achieved by the conclusion of a Protocol relating to the International Convention for the Prevention of Pollution from Ships, 1973,

HAVE AGREED as follows:

Article I
General obligations

1   The Parties to the present Protocol undertake to give effect to the provisions of:

    (a)   the present Protocol and the Annex hereto which shall constitute an integral part of the present Protocol; and

    (b)   the International Convention for the Prevention of Pollution from Ships, 1973 (hereinafter referred to as "the Convention"), subject to the modifications and additions set out in the present Protocol.

2   The provisions of the Convention and the present Protocol shall be read and interpreted together as one single instrument.

3   Every reference to the present Protocol constitutes at the same time a reference to the Annex hereto.

Article II
Implementation of Annex II of the Convention

1   Notwithstanding the provisions of article 14(1) of the Convention, the Parties to the present Protocol agree that they shall not be bound by the provisions of Annex II of the Convention for a period of three years from the date of entry into force of the present Protocol or for such longer period as may be decided by a two-thirds majority of the Parties to the present Protocol in the Marine Environment Protection Committee (hereinafter referred to as "the Committee") of the Inter-Governmental Maritime Consultative Organization (hereinafter referred to as "the Organization").[*]

2   During the period specified in paragraph 1 of this article, the Parties to the present Protocol shall not be under any obligations nor entitled to claim any privileges under the Convention in respect of matters relating to Annex II of the Convention and all reference to Parties in the Convention shall not include the Parties to the present Protocol in so far as matters relating to that Annex are concerned.

Article III
Communication of information

The text of article 11(1)(b) of the Convention is replaced by the following:

> "a list of nominated surveyors or recognized organizations which are authorized to act on their behalf in the administration of matters relating to the design, construction, equipment and operation of ships carrying harmful substances in accordance with the provisions of the regulations for circulation to the Parties for information of their officers. The Administration shall therefore notify the Organization of the specific responsibilities and conditions of the authority delegated to nominated surveyors or recognized organizations."

---

[*] The name of the Organization was changed to "International Maritime Organization" by virtue of amendments to the Organization's Convention which entered into force on 22 May 1982.

Article IV
Signature, ratification, acceptance, approval and accession

1   The present Protocol shall be open for signature at the Headquarters of the Organization from 1 June 1978 to 31 May 1979 and shall thereafter remain open for accession. States may become Parties to the present Protocol by:

(a)   signature without reservation as to ratification, acceptance or approval; or

(b)   signature, subject to ratification, acceptance or approval, followed by ratification, acceptance or approval; or

(c)   accession.

2   Ratification, acceptance, approval or accession shall be effected by the deposit of an instrument to that effect with the Secretary-General of the Organization.

Article V
Entry into force

1   The present Protocol shall enter into force 12 months after the date on which not less than 15 States, the combined merchant fleets of which constitute not less than 50 per cent of the gross tonnage of the world's merchant shipping, have become Parties to it in accordance with article IV of the present Protocol.

2   Any instrument of ratification, acceptance, approval or accession deposited after the date on which the present Protocol enters into force shall take effect three months after the date of deposit.

3   After the date on which an amendment to the present Protocol is deemed to have been accepted in accordance with article 16 of the Convention, any instrument of ratification, acceptance, approval or accession deposited shall apply to the present Protocol as amended.

Article VI
Amendments

The procedures set out in article 16 of the Convention in respect of amendments to the articles, an Annex and an appendix to an Annex of the Convention shall apply respectively to amendments to the articles, the Annex and an appendix to the Annex of the present Protocol.

Protocol of 1978

Article VII
Denunciation

1   The present Protocol may be denounced by any Party to the present Protocol at any time after the expiry of five years from the date on which the Protocol enters into force for that Party.

2   Denunciation shall be effected by the deposit of an instrument of denunciation with the Secretary-General of the Organization.

3   A denunciation shall take effect 12 months after receipt of the notification by the Secretary-General of the Organization or after the expiry of any other longer period which may be indicated in the notification.

Article VIII
Depositary

1   The present Protocol shall be deposited with the Secretary-General of the Organization (hereinafter referred to as "the Depositary").

2   The Depositary shall:

(a)   inform all States which have signed the present Protocol or acceded thereto of:

(i)   each new signature or deposit of an instrument of ratification, acceptance, approval or accession, together with the date thereof;

(ii)   the date of entry into force of the present Protocol;

(iii)   the deposit of any instrument of denunciation of the present Protocol together with the date on which it was received and the date on which the denunciation takes effect;

(iv)   any decision made in accordance with article II(1) of the present Protocol;

(b)   transmit certified true copies of the present Protocol to all States which have signed the present Protocol or acceded thereto.

3   As soon as the present Protocol enters into force, a certified true copy thereof shall be transmitted by the Depositary to the Secretariat of the United Nations for registration and publication in accordance with Article 102 of the Charter of the United Nations.

Article IX
Languages

The present Protocol is established in a single original in the English, French, Russian and Spanish languages, each text being equally authentic. Official translations in the Arabic, German, Italian and Japanese languages shall be prepared and deposited with the signed original.

IN WITNESS WHEREOF the undersigned* being duly authorized by their respective Governments for that purpose have signed the present Protocol.

DONE AT LONDON this seventeenth day of February one thousand nine hundred and seventy-eight.

Protocol of 1978

---

* Signatures omitted.

Protocol I
(including amendments)

Provisions concerning Reports on Incidents Involving Harmful Substances

Protocol I
(including amendments)

Provisions concerning Reports on Incidents Involving Harmful Substances
(in accordance with article 8 of the Convention)

Article I
Duty to report

(1) The master or other person having charge of any ship involved in an incident referred to in article II of this Protocol shall report the particulars of such incident without delay and to the fullest extent possible in accordance with the provisions of this Protocol.

(2) In the event of the ship referred to in paragraph (1) of this article being abandoned, or in the event of a report from such a ship being incomplete or unobtainable, the owner, charterer, manager or operator of the ship, or their agent shall, to the fullest extent possible, assume the obligations placed upon the master under the provisions of this Protocol.

Article II
When to make reports

(1) The report shall be made when an incident involves:

(a) a discharge above the permitted level or probable discharge of oil or of noxious liquid substances for whatever reason including those for the purpose of securing the safety of the ship or for saving life at sea; or

(b) a discharge or probable discharge of harmful substances in packaged form, including those in freight containers, portable tanks, road and rail vehicles and shipborne barges; or

    (c)   damage, failure or breakdown of a ship of 15 metres in length or above which:

        (i)   affects the safety of the ship; including but not limited to collision, grounding, fire, explosion, structural failure, flooding and cargo shifting; or

        (ii)   results in impairment of the safety of navigation; including but not limited to, failure or breakdown of steering gear, propulsion plant, electrical generating system, and essential shipborne navigational aids; or

    (d)   a discharge during the operation of the ship of oil or noxious liquid substances in excess of the quantity or instantaneous rate permitted under the present Convention.

(2)   For the purposes of this Protocol:

    (a)   *Oil* referred to in subparagraph 1(a) of this article means oil as defined in regulation 1(1) of Annex I of the Convention.

    (b)   *Noxious liquid substances* referred to in subparagraph 1(a) of this article means noxious liquid substances as defined in regulation 1(6) of Annex II of the Convention.

    (c)   *Harmful substances* in packaged form referred to in subparagraph 1(b) of this article means substances which are identified as marine pollutants in the International Maritime Dangerous Goods Code (IMDG Code).

Article III
Contents of report

Reports shall in any case include:

    (a)   identity of ships involved;

    (b)   time, type and location of incident;

    (c)   quantity and type of harmful substance involved;

    (d)   assistance and salvage measures.

Article IV
Supplementary report

Any person who is obliged under the provisions of this Protocol to send a report shall, when possible:

    (a)   supplement the initial report, as necessary, and provide information concerning further developments; and

    (b)   comply as fully as possible with requests from affected States for additional information.

Article V
Reporting procedures

(1)   Reports shall be made by the fastest telecommunications channels available with the highest possible priority to the nearest coastal State.

(2)   In order to implement the provisions of this Protocol, Parties to the present Convention shall issue, or cause to be issued, regulations or instructions on the procedures to be followed in reporting incidents involving harmful substances, based on guidelines developed by the Organization.*

* Refer to the General Principles for Ship Reporting Systems and Ship Reporting Requirements, including Guidelines for Reporting Incidents Involving Dangerous Goods, Harmful Substances and/or Marine Pollutants adopted by the Organization by resolution A.851(20); see IMO sales publication IMO-516E.

Protocol I

Protocol II

Arbitration

Protocol II

Arbitration

(in accordance with article 10 of the Convention)

Article I

Arbitration procedure, unless the Parties to the dispute decide otherwise, shall be in accordance with the rules set out in this Protocol.

Article II

(1) An Arbitration Tribunal shall be established upon the request of one Party to the Convention addressed to another in application of article 10 of the present Convention. The request for arbitration shall consist of a statement of the case together with any supporting documents.

(2) The requesting Party shall inform the Secretary-General of the Organization of the fact that it has applied for the establishment of a Tribunal, of the names of the Parties to the dispute, and of the articles of the Convention or Regulations over which there is in its opinion disagreement concerning their interpretation or application. The Secretary-General shall transmit this information to all Parties.

Article III

The Tribunal shall consist of three members: one Arbitrator nominated by each Party to the dispute and a third Arbitrator who shall be nominated by agreement between the two first named, and shall act as its Chairman.

Article IV

(1) If, at the end of a period of 60 days from the nomination of the second Arbitrator, the Chairman of the Tribunal shall not have been nominated, the Secretary-General of the Organization upon request of either Party shall within a further period of 60 days proceed to such nomination, selecting him from a list of qualified persons previously drawn up by the Council of the Organization.

(2) If, within a period of 60 days from the date of the receipt of the request, one of the Parties shall not have nominated the member of

the Tribunal for whose designation it is responsible, the other Party may directly inform the Secretary-General of the Organization who shall nominate the Chairman of the Tribunal within a period of 60 days, selecting him from the list prescribed in paragraph (1) of the present article.

(3)   The Chairman of the Tribunal shall, upon nomination, request the Party which has not provided an Arbitrator, to do so in the same manner and under the same conditions. If the Party does not make the required nomination, the Chairman of the Tribunal shall request the Secretary-General of the Organization to make the nomination in the form and conditions prescribed in the preceding paragraph.

(4)   The Chairman of the Tribunal, if nominated under the provisions of the present article, shall not be or have been a national of one of the Parties concerned, except with the consent of the other Party.

(5)   In the case of the decease or default of an Arbitrator for whose nomination one of the Parties is responsible, the said Party shall nominate a replacement within a period of 60 days from the date of decease or default. Should the said Party not make the nomination, the arbitration shall proceed under the remaining Arbitrators. In case of the decease or default of the Chairman of the Tribunal, a replacement shall be nominated in accordance with the provisions of article III above, or in the absence of agreement between the members of the Tribunal within a period of 60 days of the decease or default, according to the provisions of the present article.

Article V

The Tribunal may hear and determine counter-claims arising directly out of the subject matter of the dispute.

Article VI

Each Party shall be responsible for the remuneration of its Arbitrator and connected costs and for the costs entailed by the preparation of its own case. The remuneration of the Chairman of the Tribunal and of all general expenses incurred by the Arbitration shall be borne equally by the Parties. The Tribunal shall keep a record of all its expenses and shall furnish a final statement thereof.

Article VII

Any Party to the Convention which has an interest of a legal nature and which may be affected by the decision in the case may, after giving written notice to the Parties which have originally initiated the procedure, join in the arbitration procedure with the consent of the Tribunal.

Article VIII

Any Arbitration Tribunal established under the provisions of the present Protocol shall decide its own rules of procedure.

Article IX

(1) Decisions of the Tribunal both as to its procedure and its place of meeting and as to any question laid before it, shall be taken by majority votes of its members; the absence or abstention of one of the members of the Tribunal for whose nomination the Parties were responsible, shall not constitute an impediment to the Tribunal reaching a decision. In cases of equal voting, the vote of the Chairman shall be decisive.

(2) The Parties shall facilitate the work of the Tribunal and in particular, in accordance with their legislation, and using all means at their disposal:

(a) provide the Tribunal with the necessary documents and information;

(b) enable the Tribunal to enter their territory, to hear witnesses or experts, and to visit the scene.

(3) Absence or default of one Party shall not constitute an impediment to the procedure.

Article X

(1) The Tribunal shall render its award within a period of five months from the time it is established unless it decides, in the case of necessity, to extend the time limit for a further period not exceeding three months. The award of the Tribunal shall be accompanied by a statement of reasons. It shall be final and without appeal and shall be communicated to the Secretary-General of the Organization. The Parties shall immediately comply with the award.

(2) Any controversy which may arise between the Parties as regards interpretation or execution of the award may be submitted by either Party for judgment to the Tribunal which made the award, or, if it is not available to another Tribunal constituted for this purpose, in the same manner as the original Tribunal.

Protocol II

Protocol of 1997
to amend the
International Convention for the
Prevention of Pollution
from Ships, 1973,
as modified by the
Protocol of 1978
relating thereto

Protocol of 1997 to amend the International Convention for the Prevention of Pollution from Ships, 1973, as modified by the Protocol of 1978 relating thereto

THE PARTIES TO THE PRESENT PROTOCOL,

BEING Parties to the Protocol of 1978 relating to the International Convention for the Prevention of Pollution from Ships, 1973,

RECOGNIZING the need to prevent and control air pollution from ships,

RECALLING Principle 15 of the Rio Declaration on Environment and Development which calls for the application of a precautionary approach,

CONSIDERING that this objective could best be achieved by the conclusion of a Protocol of 1997 to amend the International Convention for the Prevention of Pollution from Ships, 1973, as modified by the Protocol of 1978 relating thereto,

HAVE AGREED as follows:

Article 1
Instrument to be amended

The instrument which the present Protocol amends is the International Convention for the Prevention of Pollution from Ships, 1973, as modified by the Protocol of 1978 relating thereto (hereinafter referred to as the "Convention").

Article 2
Addition of Annex VI to the Convention

Annex VI entitled Regulations for the Prevention of Air Pollution from Ships, the text of which is set out in the annex to the present Protocol, is added.

Article 3
General obligations

1   The Convention and the present Protocol shall, as between the Parties to the present Protocol, be read and interpreted together as one single instrument.

2   Every reference to the present Protocol constitutes at the same time a reference to the annex hereto.

Article 4
Amendment procedure

In applying article 16 of the Convention to an amendment to Annex VI and its appendices, the reference to "a Party to the Convention" shall be deemed to mean the reference to a Party bound by that Annex.

FINAL CLAUSES

Article 5
Signature, ratification, acceptance, approval and accession

1   The present Protocol shall be open for signature at the Headquarters of the International Maritime Organization (hereinafter referred to as the "Organization") from 1 January 1998 until 31 December 1998 and shall thereafter remain open for accession. Only Contracting States to the Protocol of 1978 relating to the International Convention for the Prevention of Pollution from Ships, 1973 (hereinafter referred to as the "1978 Protocol") may become Parties to the present Protocol by:

   (a)   signature without reservation as to ratification, acceptance or approval; or

   (b)   signature, subject to ratification, acceptance or approval, followed by ratification, acceptance or approval; or

   (c)   accession.

2   Ratification, acceptance, approval or accession shall be effected by the deposit of an instrument to that effect with the Secretary-General of the Organization (hereinafter referred to as the "Secretary-General").

Protocol of 1997

Article 6
Entry into force

1   The present Protocol shall enter into force twelve months after the date on which not less than fifteen States, the combined merchant fleets of which constitute not less than 50 per cent of the gross tonnage of the world's merchant shipping, have become Parties to it in accordance with article 5 of the present Protocol.

2   Any instrument of ratification, acceptance, approval or accession deposited after the date on which the present Protocol enters into force shall take effect three months after the date of deposit.

3   After the date on which an amendment to the present Protocol is deemed to have been accepted in accordance with article 16 of the Convention, any instrument of ratification, acceptance, approval or accession deposited shall apply to the present Protocol as amended.

Article 7
Denunciation

1   The present Protocol may be denounced by any Party to the present Protocol at any time after the expiry of five years from the date on which the Protocol enters into force for that Party.

2   Denunciation shall be effected by the deposit of an instrument of denunciation with the Secretary-General.

3   A denunciation shall take effect twelve months after receipt of the notification by the Secretary-General or after the expiry of any other longer period which may be indicated in the notification.

4   A denunciation of the 1978 Protocol in accordance with article VII thereof shall be deemed to include a denunciation of the present Protocol in accordance with this article. Such denunciation shall take effect on the date on which denunciation of the 1978 Protocol takes effect in accordance with article VII of that Protocol.

Article 8
Depositary

1   The present Protocol shall be deposited with the Secretary-General (hereinafter referred to as the "Depositary").

Protocol of 1997

2   The Depositary shall:

    (a)   inform all States which have signed the present Protocol or acceded thereto of:

        (i)   each new signature or deposit of an instrument of ratification, acceptance, approval or accession, together with the date thereof;

        (ii)   the date of entry into force of the present Protocol; and

        (iii)   the deposit of any instrument of denunciation of the present Protocol, together with the date on which it was received and the date on which the denunciation takes effect; and

    (b)   transmit certified true copies of the present Protocol to all States which have signed the present Protocol or acceded thereto.

3   As soon as the present Protocol enters into force, a certified true copy thereof shall be transmitted by the Depositary to the Secretariat of the United Nations for registration and publication in accordance with Article 102 of the Charter of the United Nations.

Article 9

Languages

The present Protocol is established in a single copy in the Arabic, Chinese, English, French, Russian and Spanish languages, each text being equally authentic.

IN WITNESS WHEREOF the undersigned, being duly authorized by their respective Governments for that purpose, have signed* the present Protocol.

DONE AT LONDON this twenty-sixth day of September, one thousand nine hundred and ninety-seven.

---

* Signatures omitted.

Annex I of MARPOL 73/78
(including amendments)

Regulations for the Prevention of Pollution by Oil

Annex I of MARPOL 73/78
(including amendments)

Regulations for the Prevention of Pollution by Oil

Chapter I – General

Regulation 1
Definitions

For the purposes of this Annex:

(1) *Oil* means petroleum in any form including crude oil, fuel oil, sludge, oil refuse and refined products (other than petrochemicals which are subject to the provisions of Annex II of the present Convention) and, without limiting the generality of the foregoing, includes the substances listed in appendix I to this Annex.

SEE INTERPRETATION 1A.0

---

(2) *Oily mixture* means a mixture with any oil content.

(3) *Oil fuel* means any oil used as fuel in connection with the propulsion and auxiliary machinery of the ship in which such oil is carried.

(4) *Oil tanker* means a ship constructed or adapted primarily to carry oil in bulk in its cargo spaces and includes combination carriers and any "chemical tanker" as defined in Annex II of the present Convention when it is carrying a cargo or part cargo of oil in bulk.

SEE INTERPRETATIONS 1.0 AND 6.1

---

(5) *Combination carrier* means a ship designed to carry either oil or solid cargoes in bulk.

(6) *New ship* means a ship:

    (a) for which the building contract is placed after 31 December 1975; or

(b)   in the absence of a building contract, the keel of which is laid or which is at a similar stage of construction after 30 June 1976; or

(c)   the delivery of which is after 31 December 1979; or

(d)   which has undergone a major conversion:

    (i)   for which the contract is placed after 31 December 1975; or

    (ii)   in the absence of a contract, the construction work of which is begun after 30 June 1976; or

    (iii)   which is completed after 31 December 1979.

SEE INTERPRETATIONS 1.1 AND 1.2

---

(7)   *Existing ship* means a ship which is not a new ship.

(8)   (a)   *Major conversion* means a conversion of an existing ship:

    (i)   which substantially alters the dimensions or carrying capacity of the ship; or

    (ii)   which changes the type of the ship; or

    (iii)   the intent of which in the opinion of the Administration is substantially to prolong its life; or

    (iv)   which otherwise so alters the ship that, if it were a new ship, it would become subject to relevant provisions of the present Convention not applicable to it as an existing ship.

SEE INTERPRETATION 1.3

(b)   Notwithstanding the provisions of subparagraph (a) of this paragraph, conversion of an existing oil tanker of 20,000 tons deadweight and above to meet the requirements of regulation 13 of this Annex shall not be deemed to constitute a major conversion for the purposes of this Annex.

(c)   Notwithstanding the provisions of subparagraph (a) of this paragraph, conversion of an existing oil tanker to meet the requirements of regulation 13F or 13G of this Annex shall not be deemed to constitute a major conversion for the purpose of this Annex.

(9)   *Nearest land.* The term "from the nearest land" means from the baseline from which the territorial sea of the territory in question is established in accordance with international law, except that, for the purposes of the present Convention "from the nearest land" off the north-eastern coast of Australia shall mean from a line drawn from a point on the coast of Australia in

latitude 11°00′ S, longitude 142°08′ E
to a point in latitude 10°35′ S, longitude 141°55′ E,
thence to a point latitude 10°00′ S, longitude 142°00′ E,
thence to a point latitude 09°10′ S, longitude 143°52′ E,
thence to a point latitude 09°00′ S, longitude 144°30′ E,
thence to a point latitude 10°41′ S, longitude 145°00′ E,
thence to a point latitude 13°00′ S, longitude 145°00′ E,
thence to a point latitude 15°00′ S, longitude 146°00′ E,
thence to a point latitude 17°30′ S, longitude 147°00′ E,
thence to a point latitude 21°00′ S, longitude 152°55′ E,
thence to a point latitude 24°30′ S, longitude 154°00′ E,
thence to a point on the coast of Australia
in latitude 24°42′ S, longitude 153°15′ E.

(10) *Special area* means a sea area where for recognized technical reasons in relation to its oceanographical and ecological condition and to the particular character of its traffic the adoption of special mandatory methods for the prevention of sea pollution by oil is required. Special areas shall include those listed in regulation 10 of this Annex.

(11) *Instantaneous rate of discharge of oil content* means the rate of discharge of oil in litres per hour at any instant divided by the speed of the ship in knots at the same instant.

(12) *Tank* means an enclosed space which is formed by the permanent structure of a ship and which is designed for the carriage of liquid in bulk.

(13) *Wing tank* means any tank adjacent to the side shell plating.

(14) *Centre tank* means any tank inboard of a longitudinal bulkhead.

(15) *Slop tank* means a tank specifically designated for the collection of tank drainings, tank washings and other oily mixtures.

(16) *Clean ballast* means the ballast in a tank which since oil was last carried therein, has been so cleaned that effluent therefrom if it were discharged from a ship which is stationary into clean calm water on a clear day would not produce visible traces of oil on the surface of the water or on adjoining shorelines or cause a sludge or emulsion to be deposited beneath the surface of the water or upon adjoining shorelines. If the ballast is discharged through an oil discharge monitoring and control system approved by the Administration, evidence based on such a system to the effect that the oil content of the effluent did not exceed 15 parts per million shall be determinative that the ballast was clean, notwithstanding the presence of visible traces.

(17) *Segregated ballast* means the ballast water introduced into a tank which is completely separated from the cargo oil and oil fuel system and which is permanently allocated to the carriage of ballast or to the

carriage of ballast or cargoes other than oil or noxious substances as variously defined in the Annexes of the present Convention.

SEE INTERPRETATION 1.4

(18) *Length (L)* means 96% of the total length on a waterline at 85% of the least moulded depth measured from the top of the keel, or the length from the foreside of the stem to the axis of the rudder stock on that waterline, if that be greater. In ships designed with a rake of keel the waterline on which this length is measured shall be parallel to the designed waterline. The length *(L)* shall be measured in metres.

(19) *Forward and after perpendiculars* shall be taken at the forward and after ends of the length *(L)*. The forward perpendicular shall coincide with the foreside of the stem on the waterline on which the length is measured.

(20) *Amidships* is at the middle of the length *(L)*.

(21) *Breadth (B)* means the maximum breadth of the ship, measured amidships to the moulded line of the frame in a ship with a metal shell and to the outer surface of the hull in a ship with a shell of any other material. The breadth *(B)* shall be measured in metres.

(22) *Deadweight* (DW) means the difference in metric tons between the displacement of a ship in water of a specify gravity of 1.025 at the load waterline corresponding to the assigned summer freeboard and the lightweight of the ship.

(23) *Lightweight* means the displacement of a ship in metric tons without cargo, fuel, lubricating oil, ballast water, fresh water and feed water in tanks, consumable stores, and passengers and crew and their effects.

(24) *Permeability* of a space means the ratio of the volume within that space which is assumed to be occupied by water to the total volume of that space.

(25) *Volumes* and *areas* in a ship shall be calculated in all cases to moulded lines.

(26) Notwithstanding the provisions of paragraph (6) of this regulation, for the purposes of regulations 13, 13B, 13E and 18(4) of this Annex, *new oil tanker* means an oil tanker:

   (a)   for which the building contract is placed after 1 June 1979; or

   (b)   in the absence of a building contract, the keel of which is laid or which is at a similar stage of construction after 1 January 1980; or

   (c)   the delivery of which is after 1 June 1982; or

(d)   which has undergone a major conversion:

   (i)   for which the contract is placed after 1 June 1979; or

   (ii)   in the absence of a contract, the construction work of which is begun after 1 January 1980; or

   (iii)   which is completed after 1 June 1982;

except that, for oil tankers of 70,000 tons deadweight and above, the definition in paragraph (6) of this regulation shall apply for the purposes of regulation 13(1) of this Annex.

SEE INTERPRETATIONS 1.1 AND 1.2

(27)   Notwithstanding the provisions of paragraph (7) of this regulation, for the purposes of regulations 13, 13A, 13B, 13C, 13D, 18(5) and 18(6)(c) of this Annex, *existing oil tanker* means an oil tanker which is not a new oil tanker as defined in paragraph (26) of this regulation.

(28)   *Crude oil* means any liquid hydrocarbon mixture occurring naturally in the earth whether or not treated to render it suitable for transportation and includes:

   (a)   crude oil from which certain distillate fractions may have been removed; and

   (b)   crude oil to which certain distillate fractions may have been added.

(29)   *Crude oil tanker* means an oil tanker engaged in the trade of carrying crude oil.

(30)   *Product carrier* means an oil tanker engaged in the trade of carrying oil other than crude oil.

(31)   *Anniversary date* means the day and the month of each year which will correspond to the date of expiry of the International Oil Pollution Prevention Certificate.

Regulation 2
Application

(1)   Unless expressly provided otherwise, the provisions of this Annex shall apply to all ships.

(2)   In ships other than oil tankers fitted with cargo spaces which are constructed and utilized to carry oil in bulk of an aggregate capacity of 200 cubic metres or more, the requirements of regulations 9, 10, 14, 15(1), (2) and (3), 18, 20 and 24(4) of this Annex for oil tankers shall also apply to the construction and operation of those spaces, except that where such aggregate capacity is less than 1,000 cubic metres the requirements of regulation 15(4) of this Annex may apply in lieu of regulation 15(1), (2) and (3).

(3) Where a cargo subject to the provisions of Annex II of the present Convention is carried in a cargo space of an oil tanker, the appropriate requirements of Annex II of the present Convention shall also apply.

(4) (a) Any hydrofoil, air-cushion vehicle and other new type of vessel (near-surface craft, submarine craft, etc.) whose constructional features are such as to render the application of any of the provisions of chapters II and III of this Annex relating to construction and equipment unreasonable or impracticable may be exempted by the Administration from such provisions, provided that the construction and equipment of that ship provides equivalent protection against pollution by oil, having regard to the service for which it is intended.

(b) Particulars of any such exemption granted by the Administration shall be indicated in the Certificate referred to in regulation 5 of this Annex.

(c) The Administration which allows any such exemption shall, as soon as possible, but not more than 90 days thereafter, communicate to the Organization particulars of same and the reasons therefor, which the Organization shall circulate to the Parties to the Convention for their information and appropriate action, if any.

Regulation 3
Equivalents

SEE INTERPRETATION 1.5

(1) The Administration may allow any fitting, material, appliance or apparatus to be fitted in a ship as an alternative to that required by this Annex if such fitting, material, appliance or apparatus is at least as effective as that required by this Annex. This authority of the Administration shall not extend to substitution of operational methods to effect the control of discharge of oil as equivalent to those design and construction features which are prescribed by regulations in this Annex.

(2) The Administration which allows a fitting, material, appliance or apparatus, as an alternative to that required by this Annex shall communicate to the Organization for circulation to the Parties to the Convention particulars thereof, for their information and appropriate action, if any.

Regulation 4
Surveys

(1) Every oil tanker of 150 tons gross tonnage and above, and every other ship of 400 tons gross tonnage and above shall be subject to the surveys specified below:

    (a) An initial survey before the ship is put in service or before the Certificate required under regulation 5 of this Annex is issued for the first time, which shall include a complete survey of its structure, equipment, systems, fittings, arrangements and material in so far as the ship is covered by this Annex. This survey shall be such as to ensure that the structure, equipment, systems, fittings, arrangements and material fully comply with the applicable requirements of this Annex.

    (b) A renewal survey at intervals specified by the Administration, but not exceeding five years, except where regulation 8(2), 8(5), 8(6) or 8(7) of this Annex is applicable. The renewal survey shall be such as to ensure that the structure, equipment, systems, fittings, arrangements and material fully comply with applicable requirements of this Annex.

    (c) An intermediate survey within three months before or after the second anniversary date or within three months before or after the third anniversary date of the Certificate which shall take the place of one of the annual surveys specified in paragraph (1)(d) of this regulation. The intermediate survey shall be such as to ensure that the equipment and associated pump and piping systems, including oil discharge monitoring and control systems, crude oil washing systems, oily-water separating equipment and oil filtering systems, fully comply with the applicable requirements of this Annex and are in good working order. Such intermediate surveys shall be endorsed on the Certificate issued under regulation 5 or 6 of this Annex.

SEE INTERPRETATION 1A.1

    (d) An annual survey within three months before or after each anniversary date of the Certificate, including a general inspection of the structure, equipment, systems, fittings, arrangements and material referred to in paragraph (1)(a) of this regulation to ensure that they have been maintained in accordance with paragraph (4) of this regulation and that they remain satisfactory for the service for which the ship is intended. Such annual surveys shall be endorsed on the Certificate issued under regulation 5 or 6 of this Annex.

(e) An additional survey either general or partial, according to the circumstances, shall be made after a repair resulting from investigations prescribed in paragraph (4) of this regulation, or whenever any important repairs or renewals are made. The survey shall be such as to ensure that the necessary repairs or renewals have been effectively made, that the material and workmanship of such repairs or renewals are in all respects satisfactory and that the ship complies in all respects with the requirements of this Annex.

(2) The Administration shall establish appropriate measures for ships which are not subject to the provisions of paragraph (1) of this regulation in order to ensure that the applicable provisions of this Annex are complied with.

(3) (a) Surveys of ships as regards the enforcement of the provisions of this Annex shall be carried out by officers of the Administration. The Administration may, however, entrust the surveys either to surveyors nominated for the purpose or to organizations recognized by it.

(b) An Administration nominating surveyors or recognizing organizations to conduct surveys as set forth in subparagraph (a) of this paragraph shall, as a minimum, empower any nominated surveyor or recognized organization to:

(i) require repairs to a ship; and

(ii) carry out surveys, if requested by the appropriate authorities of a port State.

The Administration shall notify the Organization of the specific responsibilities and conditions of the authority delegated to the nominated surveyors or recognized organizations, for circulation to Parties to the present Protocol for the information of their officers.

(c) When a nominated surveyor or recognized organization determines that the condition of the ship or its equipment does not correspond substantially with the particulars of the Certificate or is such that the ship is not fit to proceed to sea without presenting an unreasonable threat of harm to the marine environment, such surveyor or organization shall immediately ensure that corrective action is taken and shall in due course notify the Administration. If such corrective action is not taken the Certificate should be withdrawn and the Administration shall be notified immediately; and if the ship is in a port of another Party, the appropriate authorities of the port State shall also be notified immediately. When an officer of the Administration, a nominated surveyor or a recognized organization has notified the appropriate authorities of the port State, the Government of the port State concerned shall give such officer, surveyor or organization any necessary assistance to carry out their obligations under this regulation. When applicable, the Government of the port State concerned shall take such steps as will ensure that the ship shall not sail until it can proceed to sea or leave the port for the purpose of proceeding to the nearest appropriate repair yard available without presenting an unreasonable threat of harm to the marine environment.

(d)   In every case, the Administration concerned shall fully guarantee the completeness and efficiency of the survey and shall undertake to ensure the necessary arrangements to satisfy this obligation.

(4) (a)   The condition of the ship and its equipment shall be maintained to conform with the provisions of the present Convention to ensure that the ship in all respects will remain fit to proceed to sea without presenting an unreasonable threat of harm to the marine environment.

(b)   After any survey of the ship under paragraph (1) of this regulation has been completed, no change shall be made in the structure, equipment, systems, fittings, arrangements or material covered by the survey, without the sanction of the Administration, except the direct replacement of such equipment and fittings.

(c)   Whenever an accident occurs to a ship or a defect is discovered which substantially affects the integrity of the ship or the efficiency or completeness of its equipment covered by this Annex the master or owner of the ship shall report at the earliest opportunity to the Administration, the recognized organization or the nominated surveyor responsible for issuing the relevant certificate, who shall cause investigations to be initiated to determine whether a survey as required by paragraph (1) of this regulation is necessary. If the ship is in a port of another Party, the master or owner shall also report immediately to the appropriate authorities of the port State and the nominated surveyor or recognized organization shall ascertain that such report has been made.

Regulation 5
Issue or endorsement of Certificate

SEE INTERPRETATIONS 2.0 and 2.1

(1)   An International Oil Pollution Prevention Certificate shall be issued, after an initial or renewal survey in accordance with the provisions of regulation 4 of this Annex, to any oil tanker of 150 tons gross tonnage and above and any other ships of 400 tons gross tonnage and above

which are engaged in voyages to ports or offshore terminals under the jurisdiction of other Parties to the Convention.

SEE INTERPRETATIONS 2.2, 2.3, 2.4

(2)   Such Certificate shall be issued or endorsed either by the Administration or by any persons or organization duly authorized by it. In every case the Administration assumes full responsibility for the Certificate.

(3)   Notwithstanding any other provisions of the amendments to this Annex adopted by the Marine Environment Protection Committee (MEPC) by resolution MEPC.39(29), any International Oil Pollution Prevention Certificate, which is current when these amendments enter into force, shall remain valid until it expires under the terms of this Annex prior to the amendments entering into force.

Regulation 6
Issue or endorsement of a Certificate
by another Government

(1)   The Government of a Party to the Convention may, at the request of the Administration, cause a ship to be surveyed and, if satisfied that the provisions of this Annex are complied with, shall issue or authorize the issue of an International Oil Pollution Prevention Certificate to the ship, and where appropriate, endorse or authorize the endorsement of that Certificate on the ship, in accordance with this Annex.

(2)   A copy of the Certificate and a copy of the survey report shall be transmitted as soon as possible to the requesting Administration.

(3)   A Certificate so issued shall contain a statement to the effect that it has been issued at the request of the Administration and it shall have the same force and receive the same recognition as the Certificate issued under regulation 5 of this Annex.

(4)   No International Oil Pollution Prevention Certificate shall be issued to a ship which is entitled to fly the flag of a State which is not a Party.

Regulation 7
Form of Certificate

SEE INTERPRETATION 2.4A

The International Oil Pollution Prevention Certificate shall be drawn up in a form corresponding to the model given in appendix II to this Annex. If the language used is neither English nor French, the text shall include a translation into one of these languages.

Regulation 8
Duration and validity of Certificate

SEE INTERPRETATION 2.5

(1)   An International Oil Pollution Prevention Certificate shall be issued for a period specified by the Administration, which shall not exceed five years.

(2) (a)   Notwithstanding the requirements of paragraph (1) of this regulation, when the renewal survey is completed within three months before the expiry date of the existing Certificate, the new Certificate shall be valid from the date of completion of the renewal survey to a date not exceeding five years from the date of expiry of the existing Certificate.

     (b)   When the renewal survey is completed after the expiry date of the existing Certificate, the new Certificate shall be valid from the date of completion of the renewal survey to a date not exceeding five years from the date of expiry of the existing Certificate.

     (c)   When the renewal survey is completed more than three months before the expiry date of the existing Certificate, the new Certificate shall be valid from the date of completion of the renewal survey to a date not exceeding five years from the date of completion of the renewal survey.

(3)   If a Certificate is issued for a period of less than five years, the Administration may extend the validity of the Certificate beyond the expiry date to the maximum period specified in paragraph (1) of this regulation, provided that the surveys referred to in regulation 4(1)(c) and 4(1)(d) of this Annex applicable when a Certificate is issued for a period of five years are carried out as appropriate.

(4)   If a renewal survey has been completed and a new Certificate cannot be issued or placed on board the ship before the expiry date of the existing Certificate, the person or organization authorized by the Administration may endorse the existing Certificate and such a Certificate shall be accepted as valid for a further period which shall not exceed five months from the expiry date.

(5)   If a ship at the time when a Certificate expires is not in a port in which it is to be surveyed, the Administration may extend the period of validity of the Certificate but this extension shall be granted only for

the purpose of allowing the ship to complete its voyage to the port in which it is to be surveyed, and then only in cases where it appears proper and reasonable to do so. No Certificate shall be extended for a period longer than three months, and a ship to which an extension is granted shall not, on its arrival in the port in which it is to be surveyed, be entitled by virtue of such extension to leave that port without having a new Certificate. When the renewal survey is completed, the new Certificate shall be valid to a date not exceeding five years from the date of expiry of the existing Certificate before the extension was granted.

(6)   A Certificate issued to a ship engaged on short voyages which has not been extended under the foregoing provisions of this regulation may be extended by the Administration for a period of grace of up to one month from the date of expiry stated on it. When the renewal survey is completed, the new Certificate shall be valid to a date not exceeding five years from the date of expiry of the existing Certificate before the extension was granted.

(7)   In special circumstances, as determined by the Administration, a new Certificate need not be dated from the date of expiry of the existing Certificate as required by paragraph (2)(b), (5) or (6) of this regulation. In these special circumstances, the new Certificate shall be valid to a date not exceeding five years from the date of completion of the renewal survey.

(8)   If an annual or intermediate survey is completed before the period specified in regulation 4 of this Annex, then:

(a)   the anniversary date shown on the Certificate shall be amended by endorsement to a date which shall not be more than three months later than the date on which the survey was completed;

(b)   the subsequent annual or intermediate survey required by regulation 4 of this Annex shall be completed at the intervals prescribed by that regulation using the new anniversary date;

(c)   the expiry date may remain unchanged provided one or more annual or intermediate surveys, as appropriate, are carried out so that the maximum intervals between the surveys prescribed by regulation 4 of this Annex are not exceeded.

(9)   A Certificate issued under regulation 5 or 6 of this Annex shall cease to be valid in any of the following cases:

(a)   if the relevant surveys are not completed within the periods specified under regulation 4(1) of this Annex;

(b)   if the Certificate is not endorsed in accordance with regulation 4(1)(c) or 4(1)(d) of this Annex.

(c) Upon transfer of the ship to the flag of another State. A new Certificate shall only be issued when the Government issuing the new Certificate is fully satisfied that the ship is in compliance with the requirements of regulation 4(4)(a) and 4(4)(b) of this Annex. In the case of a transfer between Parties, if requested within three months after the transfer has taken place, the Government of the Party whose flag the ship was formerly entitled to fly shall, as soon as possible, transmit to the Administration copies of the Certificate carried by the ship before the transfer and, if available, copies of the relevant survey reports.

Regulation 8A
*Port State control on operational requirements**

(1) A ship when in a port or an offshore terminal of another Party is subject to inspection by officers duly authorized by such Party concerning operational requirements under this Annex, where there are clear grounds for believing that the master or crew are not familiar with essential shipboard procedures relating to the prevention of pollution by oil.

(2) In the circumstances given in paragraph (1) of this regulation, the Party shall take such steps as will ensure that the ship shall not sail until the situation has been brought to order in accordance with the requirements of this Annex.

(3) Procedures relating to the port State control prescribed in article 5 of the present Convention shall apply to this regulation.

(4) Nothing in this regulation shall be construed to limit the rights and obligations of a Party carrying out control over operational requirements specifically provided for in the present Convention.

---

* Refer to the procedures for port State control adopted by the Organization by resolution A.787(19) and amended by A.882(21); see IMO sales publication IMO-650E.

Chapter II – Requirements for control of operational pollution

Regulation 9
Control of discharge of oil

(1) Subject to the provisions of regulations 10 and 11 of this Annex and paragraph (2) of this regulation, any discharge into the sea of oil or oily mixtures from ships to which this Annex applies shall be prohibited except when all the following conditions are satisfied:

   (a) for an oil tanker, except as provided for in subparagraph (b) of this paragraph:

   (i) the tanker is not within a special area;

   (ii) the tanker is more than 50 nautical miles from the nearest land;

   (iii) the tanker is proceeding *en route*;

   (iv) the instantaneous rate of discharge of oil content does not exceed 30 litres per nautical mile;

   (v) the total quantity of oil discharged into the sea does not exceed for existing tankers 1/15,000 of the total quantity of the particular cargo of which the residue formed a part, and for new tankers 1/30,000 of the total quantity of the particular cargo of which the residue formed a part; and

SEE INTERPRETATION 3.2

   (vi) the tanker has in operation an oil discharge monitoring and control system and a slop tank arrangement as required by regulation 15 of this Annex.

   (b) from a ship of 400 tons gross tonnage and above other than an oil tanker and from machinery space bilges excluding cargo pump-room bilges of an oil tanker unless mixed with oil cargo residue:

   (i) the ship is not within a special area;

   (ii) the ship is proceeding *en route*;

   (iii) the oil content of the effluent without dilution does not exceed 15 parts per million; and

   (iv) the ship has in operation equipment as required by regulation 16 of this Annex.

SEE INTERPRETATION 3.1

(2) In the case of a ship of less than 400 tons gross tonnage other than an oil tanker whilst outside the special area, the Administration shall ensure that it is equipped as far as practicable and reasonable with installations to ensure the storage of oil residues on board and their discharge to reception facilities or into the sea in compliance with the requirements of paragraph (1)(b) of this regulation.

(3) Whenever visible traces of oil are observed on or below the surface of the water in the immediate vicinity of a ship or its wake, Governments of Parties to the Convention should, to the extent they are reasonably able to do so, promptly investigate the facts bearing on the issue of whether there has been a violation of the provisions of this regulation or regulation 10 of this Annex. The investigation should include, in particular, the wind and sea conditions, the track and speed of the ship, other possible sources of the visible traces in the vicinity, and any relevant oil discharge records.

(4) The provisions of paragraph (1) of this regulation shall not apply to the discharge of clean or segregated ballast or unprocessed oily mixtures which without dilution have an oil content not exceeding 15 parts per million and which do not originate from cargo pump-room bilges and are not mixed with oil cargo residues.

(5) No discharge into the sea shall contain chemicals or other substances in quantities or concentrations which are hazardous to the marine environment or chemicals or other substances introduced for the purpose of circumventing the conditions of discharge specified in this regulation.

(6) The oil residues which cannot be discharged into the sea in compliance with paragraphs (1), (2) and (4) of this regulation shall be retained on board or discharged to reception facilities.

(7) In the case of a ship, referred to in regulation 16(6) of this Annex, not fitted with equipment as required by regulation 16(1) or 16(2) of this Annex, the provisions of paragraph (1)(b) of this regulation will not apply until 6 July 1998 or the date on which the ship is fitted with such equipment, whichever is the earlier. Until this date any discharge from machinery space bilges into the sea of oil or oily mixtures from such a ship shall be prohibited except when all the following conditions are satisfied:

    (a) the oily mixture does not originate from the cargo pump-room bilges;

    (b) the oily mixture is not mixed with oil cargo residues;

    (c) the ship is not within a special area;

    (d) the ship is more than 12 nautical miles from the nearest land;

    (e) the ship is proceeding *en route*;

Annex I

(f)   the oil content of the effluent is less than 100 parts per million; and

(g)   the ship has in operation oily-water separating equipment of a design approved by the Administration, taking into account the specification recommended by the Organization.[*]

Regulation 10
Methods for the prevention of oil pollution from ships while operating in special areas

(1)   For the purpose of this Annex, the special areas are the Mediterranean Sea area, the Baltic Sea area, the Black Sea area, the Red Sea area, the "Gulfs area", the Gulf of Aden area, the Antarctic area and the North-West European waters, which are defined as follows:

(a)   The *Mediterranean Sea area* means the Mediterranean Sea proper including the gulfs and seas therein with the boundary between the Mediterranean and the Black Sea constituted by the 41° N parallel and bounded to the west by the Straits of Gibraltar at the meridian of 5°36′ W.

(b)   The *Baltic Sea area* means the Baltic Sea proper with the Gulf of Bothnia, the Gulf of Finland and the entrance to the Baltic Sea bounded by the parallel of the Skaw in the Skagerrak at 57°44.8′ N.

(c)   The *Black Sea area* means the Black Sea proper with the boundary between the Mediterranean and the Black Sea constituted by the parallel 41° N.

(d)   The *Red Sea area* means the Red Sea proper including the Gulfs of Suez and Aqaba bounded at the south by the rhumb line between Ras si Ane (12°28.5′ N, 43°19.6′ E) and Husn Murad (12°40.4′ N, 43°30.2′ E).

(e)   The *Gulfs area* means the sea area located north-west of the rhumb line between Ras al Hadd (22°30′ N, 59°48′ E) and Ras al Fasteh (25°04′ N, 61°25′ E).

(f)   The *Gulf of Aden area* means that part of the Gulf of Aden between the Red Sea and the Arabian Sea bounded to the west by the rhumb line between Ras si Ane (12°28.5′ N, 43°19.6′ E) and Husn Murad (12°40.4′ N, 43°30.2′ E) and to the east by the rhumb line between Ras Asir (11°50′ N, 51°16.9′ E) and Ras Fartak (15°35′ N, 52°13.8′ E).

---

[*] Refer to the Guidelines and specifications for pollution prevention equipment for machinery space bilges of ships adopted by the Marine Environment Protection Committee of the Organization by resolution MEPC.60(33); see IMO sales publication IMO-646E.

(g)   The *Antarctic area* means the sea area south of latitude 60° S.

(h)   The *North-West European waters* include the North Sea and its approaches, the Irish Sea and its approaches, the Celtic Sea, the English Channel and its approaches and part of the North-East Atlantic immediately to the west of Ireland. The area is bounded by lines joining the following points:

   (i)   48°27′ N on the French coast;

   (ii)   48°27′ N, 6°25′ W;

   (iii)   49°52′ N, 7°44′ W;

   (iv)   50°30′ N, 12° W;

   (v)   56°30′ N, 12° W;

   (vi)   62° N, 3° W;

   (vii) 62° N on the Norwegian coast;

   (viii) 57°44.8′ N on the Danish and Swedish coasts.

(2)   Subject to the provisions of regulation 11 of this Annex:

(a)   Any discharge into the sea of oil or oily mixture from any oil tanker and any ship of 400 tons gross tonnage and above other than an oil tanker shall be prohibited while in a special area. In respect of the Antarctic area, any discharge into the sea of oil or oily mixture from any ship shall be prohibited.

(b)   Except as provided for in respect of the Antarctic area under subparagraph 2(a) of this regulation, any discharge into the sea of oil or oily mixture from a ship of less than 400 tons gross tonnage, other than an oil tanker, shall be prohibited while in a special area, except when the oil content of the effluent without dilution does not exceed 15 parts per million.

(3)   (a)   The provisions of paragraph (2) of this regulation shall not apply to the discharge of clean or segregated ballast.

(b)   The provisions of subparagraph (2)(a) of this regulation shall not apply to the discharge of processed bilge water from machinery spaces, provided that all of the following conditions are satisfied:

   (i)   the bilge water does not originate from cargo pump-room bilges;

   (ii)   the bilge water is not mixed with oil cargo residues;

   (iii)   the ship is proceeding *en route*;

   (iv)   the oil content of the effluent without dilution does not exceed 15 parts per million;

   (v)   the ship has in operation oil filtering equipment complying with regulation 16(5) of this Annex; and

(vi) the filtering system is equipped with a stopping device which will ensure that the discharge is automatically stopped when the oil content of the effluent exceeds 15 parts per million.

SEE INTERPRETATION 3.4

(4) (a) No discharge into the sea shall contain chemicals or other substances in quantities or concentrations which are hazardous to the marine environment or chemicals or other substances introduced for the purpose of circumventing the conditions of discharge specified in this regulation.

(b) The oil residues which cannot be discharged into the sea in compliance with paragraph (2) or (3) of this regulation shall be retained on board or discharged to reception facilities.

(5) Nothing in this regulation shall prohibit a ship on a voyage only part of which is in a special area from discharging outside the special area in accordance with regulation 9 of this Annex.

(6) Whenever visible traces of oil are observed on or below the surface of the water in the immediate vicinity of a ship or its wake, the Governments of Parties to the Convention should, to the extent they are reasonably able to do so, promptly investigate the facts bearing on the issue of whether there has been a violation of the provisions of this regulation or regulation 9 of this Annex. The investigation should include, in particular, the wind and sea conditions, the track and speed of the ship, other possible sources of the visible traces in the vicinity, and any relevant oil discharge records.

(7) Reception facilities within special areas:

(a) Mediterranean Sea, Black Sea and Baltic Sea areas:

(i) The Government of each Party to the Convention the coastline of which borders on any given special area undertakes to ensure that not later than 1 January 1977 all oil loading terminals and repair ports within the special area are provided with facilities adequate for the reception and treatment of all the dirty ballast and tank washing water from oil tankers. In addition all ports within the special area shall be provided with adequate reception facilities for other residues and oily mixtures from all ships. Such facilities shall have adequate capacity to meet the needs of the ships using them without causing undue delay.

(ii) The Government of each Party having under its jurisdiction entrances to seawater courses with low depth contour which might require a reduction of draught by the discharge of ballast undertakes to ensure the provision of the facilities referred to in subparagraph (a)(i) of this paragraph but with the proviso that ships required to discharge slops or dirty ballast could be subject to some delay.

(iii) During the period between the entry into force of the present Convention (if earlier than 1 January 1977) and 1 January 1977 ships while navigating in the special areas shall comply with the requirements of regulation 9 of this Annex. However, the Governments of Parties the coastlines of which border any of the special areas under this subparagraph may establish a date earlier than 1 January 1977, but after the date of entry in force of the present Convention, from which the requirements of this regulation in respect of the special areas in question shall take effect:

    (1) if all the reception facilities required have been provided by the date so established; and

    (2) provided that the Parties concerned notify the Organization of the date so established at least six months in advance, for circulation to other Parties.

(iv) After 1 January 1977, or the date established in accordance with subparagraph (a)(iii) of this paragraph if earlier, each Party shall notify the Organization for transmission to the Contracting Governments concerned of all cases where the facilities are alleged to be inadequate.

(b) Red Sea area, Gulfs area, Gulf of Aden area and North-West European waters:

(i) The Government of each Party the coastline of which borders on the special areas undertakes to ensure that as soon as possible all oil loading terminals and repair ports within these special areas are provided with facilities adequate for the reception and treatment of all the dirty ballast and tank washing water from tankers. In addition all ports within the special area shall be provided with adequate reception facilities for other residues and oily mixtures from all ships. Such facilities shall have adequate capacity to meet the needs of the ships using them without causing undue delay.

(ii)   The Government of each Party having under its jurisdiction entrances to seawater courses with low depth contour which might require a reduction of draught by the discharge of ballast shall undertake to ensure the provision of the facilities referred to in subparagraph (b)(i) of this paragraph but with the proviso that ships required to discharge slops or dirty ballast could be subject to some delay.

(iii)   Each Party concerned shall notify the Organization of the measures taken pursuant to provisions of subparagraph (b)(i) and (ii) of this paragraph. Upon receipt of sufficient notifications the Organization shall establish a date from which the requirements of this regulation in respect of the area in question shall take effect. The Organization shall notify all Parties of the date so established no less than twelve months in advance of that date.

(iv)   During the period between the entry into force of the present Convention and the date so established, ships while navigating in the special area shall comply with the requirements of regulation 9 of this Annex.

(v)   After such date oil tankers loading in ports in these special areas where such facilities are not yet available shall also fully comply with the requirements of this regulation. However, oil tankers entering these special areas for the purpose of loading shall make every effort to enter the area with only clean ballast on board.

(vi)   After the date on which the requirements for the special area in question take effect, each Party shall notify the Organization for transmission to the Parties concerned of all cases where the facilities are alleged to be inadequate.

(vii)   At least the reception facilities as prescribed in regulation 12 of this Annex shall be provided by 1 January 1977 or one year after the date of entry into force of the present Convention, whichever occurs later.

(8)   Notwithstanding paragraph (7) of this regulation, the following rules apply to the Antarctic area:

(a)   The Government of each Party to the Convention at whose ports ships depart *en route* to or arrive from the Antarctic area undertakes to ensure that as soon as practicable adequate facilities are provided for the reception of all sludge, dirty ballast, tank washing water, and other oily residues and mixtures from all ships, without causing undue delay, and according to the needs of the ships using them.

(b) The Government of each Party to the Convention shall ensure that all ships entitled to fly its flag, before entering the Antarctic area, are fitted with a tank or tanks of sufficient capacity on board for the retention of all sludge, dirty ballast, tank washing water and other oily residues and mixtures while operating in the area and have concluded arrangements to discharge such oily residues at a reception facility after leaving the area.

Regulation 11
Exceptions

Regulations 9 and 10 of this Annex shall not apply to:

(a) the discharge into the sea of oil or oily mixture necessary for the purpose of securing the safety of a ship or saving life at sea; or

(b) the discharge into the sea of oil or oily mixture resulting from damage to a ship or its equipment:

(i) provided that all reasonable precautions have been taken after the occurrence of the damage or discovery of the discharge for the purpose of preventing or minimizing the discharge; and

(ii) except if the owner or the master acted either with intent to cause damage, or recklessly and with knowledge that damage would probably result; or

(c) the discharge into the sea of substances containing oil, approved by the Administration, when being used for the purpose of combating specific pollution incidents in order to minimize the damage from pollution. Any such discharge shall be subject to the approval of any Government in whose jurisdiction it is contemplated the discharge will occur.

Regulation 12
Reception facilities

(1) Subject to the provisions of regulation 10 of this Annex, the Government of each Party undertakes to ensure the provision at oil loading terminals, repair ports, and in other ports in which ships have oily residues to discharge, of facilities for the reception of such residues and oily mixtures as remain from oil tankers and other ships adequate to meet the needs of the ships using them without causing undue delay to ships.

(2) Reception facilities in accordance with paragraph (1) of this regulation shall be provided in:

(a)   all ports and terminals in which crude oil is loaded into oil tankers where such tankers have immediately prior to arrival completed a ballast voyage of not more than 72 hours or not more than 1,200 nautical miles;

(b)   all ports and terminals in which oil other than crude oil in bulk is loaded at an average quantity of more than 1,000 metric tons per day;

(c)   all ports having ship repair yards or tank cleaning facilities;

(d)   all ports and terminals which handle ships provided with the sludge tank(s) required by regulation 17 of this Annex;

(e)   all ports in respect of oily bilge waters and other residues, which cannot be discharged in accordance with regulation 9 of this Annex; and

(f)   all loading ports for bulk cargoes in respect of oil residues from combination carriers which cannot be discharged in accordance with regulation 9 of this Annex.

SEE INTERPRETATION 3.5

(3)   The capacity for the reception facilities shall be as follows:

(a)   Crude oil loading terminals shall have sufficient reception facilities to receive oil and oily mixtures which cannot be discharged in accordance with the provisions of regulation 9(1)(a) of this Annex from all oil tankers on voyages as described in paragraph (2)(a) of this regulation.

(b)   Loading ports and terminals referred to in paragraph (2)(b) of this regulation shall have sufficient reception facilities to receive oil and oily mixtures which cannot be discharged in accordance with the provisions of regulation 9(1)(a) of this Annex from oil tankers which load oil other than crude oil in bulk.

(c)   All ports having ship repair yards or tank cleaning facilities shall have sufficient reception facilities to receive all residues and oily mixtures which remain on board for disposal from ships prior to entering such yards or facilities.

(d)   All facilities provided in ports and terminals under paragraph (2)(d) of this regulation shall be sufficient to receive all residues retained according to regulation 17 of this Annex from all ships that may reasonably be expected to call at such ports and terminals.

(e)   All facilities provided in ports and terminals under this regulation shall be sufficient to receive oily bilge waters and other residues which cannot be discharged in accordance with regulation 9 of this Annex.

(f)   The facilities provided in loading ports for bulk cargoes shall take into account the special problems of combination carriers as appropriate.

(4)   The reception facilities prescribed in paragraphs (2) and (3) of this regulation shall be made available no later than one year from the date of entry into force of the present Convention or by 1 January 1977, whichever occurs later.

(5)   Each Party shall notify the Organization for transmission to the Parties concerned of all cases where the facilities provided under this regulation are alleged to be inadequate.

Regulation 13
Segregated ballast tanks, dedicated clean ballast tanks and crude oil washing

---
SEE INTERPRETATIONS 2.1 AND 4.6

Subject to the provisions of regulations 13C and 13D of this Annex, oil tankers shall comply with the requirements of this regulation.

New oil tankers of 20,000 tons deadweight and above

(1)   Every new crude oil tanker of 20,000 tons deadweight and above and every new product carrier of 30,000 tons deadweight and above shall be provided with segregated ballast tanks and shall comply with paragraphs (2), (3) and (4), or paragraph (5) as appropriate, of this regulation.

(2)   The capacity of the segregated ballast tanks shall be so determined that the ship may operate safely on ballast voyages without recourse to the use of cargo tanks for water ballast except as provided for in paragraphs (3) or (4) of this regulation. In all cases, however, the capacity of segregated ballast tanks shall be at least such that, in any ballast condition at any part of the voyage, including the conditions consisting of lightweight plus segregated ballast only, the ship's draughts and trim can meet each of the following requirements:

(a)   the moulded draught amidships (d_m) in metres (without taking into account any ship's deformation) shall not be less than:

$$d_m = 2.0 + 0.02L;$$

(b)   the draughts at the forward and after perpendiculars shall correspond to those determined by the draught amidships (d_m) as specified in subparagraph (a) of this paragraph, in association with the trim by the stern of not greater than $0.015L$; and

(c) in any case the draught at the after perpendicular shall not be less than that which is necessary to obtain full immersion of the propeller(s).

(3) In no case shall ballast water be carried in cargo tanks, except:

(a) on those rare voyages when weather conditions are so severe that, in the opinion of the master, it is necessary to carry additional ballast water in cargo tanks for the safety of the ship; and

(b) in exceptional cases where the particular character of the operation of an oil tanker renders it necessary to carry ballast water in excess of the quantity required under paragraph (2) of this regulation, provided that such operation of the oil tanker falls under the category of exceptional cases as established by the Organization.

Such additional ballast water shall be processed and discharged in compliance with regulation 9 of this Annex and in accordance with the requirements of regulation 15 of this Annex and an entry shall be made in the Oil Record Book referred to in regulation 20 of this Annex.

SEE INTERPRETATION 4.1

(4) In the case of new crude oil tankers, the additional ballast permitted in paragraph (3) of this regulation shall be carried in cargo tanks only if such tanks have been crude oil washed in accordance with regulation 13B of this Annex before departure from an oil unloading port or terminal.

SEE INTERPRETATION 4.2

(5) Notwithstanding the provisions of paragraph (2) of this regulation, the segregated ballast conditions for oil tankers less than 150 metres in length shall be to the satisfaction of the Administration.

SEE INTERPRETATION 4.3

(6) Every new crude oil tanker of 20,000 tons deadweight and above shall be fitted with a cargo tank cleaning system using crude oil washing. The Administration shall undertake to ensure that the system fully complies with the requirements of regulation 13B of this Annex within one year after the tanker was first engaged in the trade of carrying crude oil or by the end of the third voyage carrying crude oil suitable for crude oil washing, whichever occurs later. Unless such oil tanker carries crude oil which is not suitable for crude oil washing, the

oil tanker shall operate the system in accordance with the requirements of that regulation.

Existing crude oil tankers of 40,000 tons deadweight and above

(7) Subject to the provisions of paragraphs (8) and (9) of this regulation, every existing crude oil tanker of 40,000 tons deadweight and above shall be provided with segregated ballast tanks and shall comply with the requirements of paragraphs (2) and (3) of this regulation from the date of entry into force of the present Convention.

(8) Existing crude oil tankers referred to in paragraph (7) of this regulation may, in lieu of being provided with segregated ballast tanks, operate with a cargo tank cleaning procedure using crude oil washing in accordance with regulation 13B of this Annex unless the crude oil tanker is intended to carry crude oil which is not suitable for crude oil washing.

SEE INTERPRETATION 4.4

(9) Existing crude oil tankers referred to in paragraphs (7) or (8) of this regulation may, in lieu of being provided with segregated ballast tanks or operating with a cargo tank cleaning procedure using crude oil washing, operate with dedicated clean ballast tanks in accordance with the provisions of regulation 13A of this Annex for the following period:

   (a) for crude oil tankers of 70,000 tons deadweight and above, until two years after the date of entry into force of the present Convention; and

   (b) for crude oil tankers of 40,000 tons deadweight and above but below 70,000 tons deadweight, until four years after the date of entry into force of the present Convention.

SEE INTERPRETATION 4.5

Existing product carriers of 40,000 tons deadweight and above

(10) From the date of entry into force of the present Convention, every existing product carrier of 40,000 tons deadweight and above shall be provided with segregated ballast tanks and shall comply with the requirements of paragraphs (2) and (3) of this regulation, or, alternatively, operate with dedicated clean ballast tanks in accordance with the provisions of regulation 13A of this Annex.

SEE INTERPRETATION 4.5

An oil tanker qualified as a segregated ballast oil tanker

(11) Any oil tanker which is not required to be provided with segregated ballast tanks in accordance with paragraphs (1), (7) or (10) of this regulation may, however, be qualified as a segregated ballast tanker, provided that it complies with the requirements of paragraphs (2) and (3), or paragraph (5) as appropriate, of this regulation.

Regulation 13A
Requirements for oil tankers with dedicated clean ballast tanks

SEE INTERPRETATION 4.6

(1) An oil tanker operating with dedicated clean ballast tanks in accordance with the provisions of regulation 13(9) or (10) of this Annex shall have adequate tank capacity, dedicated solely to the carriage of clean ballast as defined in regulation 1(16) of this Annex, to meet the requirements of regulation 13(2) and (3) of this Annex.

(2) The arrangements and operational procedures for dedicated clean ballast tanks shall comply with the requirements established by the Administration. Such requirements shall contain at least all the provisions of the Specifications for Oil Tankers with Dedicated Clean Ballast Tanks adopted by the International Conference on Tanker Safety and Pollution Prevention, 1978, in resolution 14 and as may be revised by the Organization.*

(3) An oil tanker operating with dedicated clean ballast tanks shall be equipped with an oil content meter, approved by the Administration on the basis of specifications recommended by the Organization,[†] to enable supervision of the oil content in ballast water being discharged. The oil content meter shall be installed no later than at the first scheduled shipyard visit of the tanker following the entry into force of the present Convention. Until such time as the oil content meter is installed, it shall immediately before discharge of ballast be established

---

* Refer to the Revised specifications for oil tankers with dedicated clean ballast tanks adopted by the Organization by resolution A.495(XII); see IMO sales publication IMO-619E.

† For oil content meters installed on oil tankers built prior to 2 October 1986, refer to the Recommendation on international performance and test specifications for oily-water separating equipment and oil content meters adopted by the Organization by resolution A.393(X). For oil content meters as part of discharge monitoring and control systems installed on oil tankers built on or after 2 October 1986, refer to the Revised guidelines and specifications for oil discharge monitoring and control systems for oil tankers adopted by the Organization by resolution A.586(14); see IMO sales publications IMO-608E and IMO-646E, respectively.

by examination of the ballast water from dedicated tanks that no contamination with oil has taken place.

SEE INTERPRETATIONS 4.7 AND 4.8

(4)   Every oil tanker operating with dedicated clean ballast tanks shall be provided with a Dedicated Clean Ballast Tank Operation Manual* detailing the system and specifying operational procedures. Such a Manual shall be to the satisfaction of the Administration and shall contain all the information set out in the Specifications referred to in paragraph (2) of this regulation. If an alteration affecting the dedicated clean ballast tank system is made, the Operation Manual shall be revised accordingly.

Regulation 13B
Requirements for crude oil washing

SEE INTERPRETATIONS 4.6 AND 4.9

(1)   Every crude oil washing system required to be provided in accordance with regulation 13(6) and (8) of this Annex shall comply with the requirements of this regulation.

(2)   The crude oil washing installation and associated equipment and arrangements shall comply with the requirements established by the Administration. Such requirements shall contain at least all the provisions of the Specifications for the Design, Operation and Control of Crude Oil Washing Systems adopted by the International Conference on Tanker Safety and Pollution Prevention, 1978, in resolution 15 and as may be revised by the Organization.[†]

(3)   An inert gas system shall be provided in every cargo tank and slop tank in accordance with the appropriate regulations of chapter II-2 of the International Convention for the Safety of Life at Sea, 1974, as modified and added to by the Protocol of 1978 relating to the International Convention for the Safety of Life at Sea, 1974 and as may be further amended.

(4)   With respect to the ballasting of cargo tanks, sufficient cargo tanks shall be crude oil washed prior to each ballast voyage in order that, taking into account the tanker's trading pattern and expected weather

---

[*] See resolution A.495(XII) for the standard format of the Manual; see IMO sales publication IMO-619E.

[†] Refer to the Revised specifications for the design, operation and control of crude oil washing systems adopted by the Organization by resolution A.446(XI) and amended by the Organization by resolutions A.497(XII) and A.897(21); see IMO sales publication IMO-617E.

conditions, ballast water is put only into cargo tanks which have been crude oil washed.

(5) Every oil tanker operating with crude oil washing systems shall be provided with an Operations and Equipment Manual* detailing the system and equipment and specifying operational procedures. Such a Manual shall be to the satisfaction of the Administration and shall contain all the information set out in the Specifications referred to in paragraph (2) of this regulation. If an alteration affecting the crude oil washing system is made, the Operations and Equipment Manual shall be revised accordingly.

Regulation 13C
Existing tankers engaged in specific trades

SEE INTERPRETATION 4.6

(1) Subject to the provisions of paragraph (2) of this regulation, regulation 13(7) to (10) of this Annex shall not apply to an existing oil tanker solely engaged in specific trades between:

(a) ports or terminals within a State Party to the present Convention; or

(b) ports or terminals of States Parties to the present Convention, where:

(i) the voyage is entirely within a special area as defined in regulation 10(1) of this Annex; or

(ii) the voyage is entirely within other limits designated by the Organization.

(2) The provisions of paragraph (1) of this regulation shall only apply when the ports or terminals where cargo is loaded on such voyages are provided with reception facilities adequate for the reception and treatment of all the ballast and tank washing water from oil tankers using them and all the following conditions are complied with:

(a) subject to the exceptions provided for in regulation 11 of this Annex, all ballast water, including clean ballast water, and tank washing residues are retained on board and transferred to the reception facilities and the appropriate entry in the Oil Record Book referred to in regulation 20 of this Annex is endorsed by the competent port State authority;

---

* Refer to the Standard format of the Crude Oil Washing Operation and Equipment Manual adopted by the Marine Environment Protection Committee of the Organization by resolution MEPC.3(XII) and amended by MEPC.81(43); see IMO sales publication IMO-617E.

(b)   agreement has been reached between the Administration and the Governments of the port States referred to in subparagraph (1)(a) or (b) of this regulation concerning the use of an existing oil tanker for a specific trade;

(c)   the adequacy of the reception facilities in accordance with the relevant provisions of this Annex at the ports or terminals referred to above, for the purpose of this regulation, is approved by the Governments of the States Parties to the present Convention within which such ports or terminals are situated; and

(d)   the International Oil Pollution Prevention Certificate is endorsed to the effect that the oil tanker is solely engaged in such specific trade.

Regulation 13D
Existing oil tankers having special ballast arrangements

SEE INTERPRETATION 4.6

(1)   Where an existing oil tanker is so constructed or operates in such a manner that it complies at all times with the draught and trim requirements set out in regulation 13(2) of this Annex without recourse to the use of ballast water, it shall be deemed to comply with the segregated ballast tank requirements referred to in regulation 13(7) of this Annex, provided that all of the following conditions are complied with:

(a)   operational procedures and ballast arrangements are approved by the Administration;

(b)   agreement is reached between the Administration and the Governments of the port States Parties to the present Convention concerned when the draught and trim requirements are achieved through an operational procedure; and

(c)   the International Oil Pollution Prevention Certificate is endorsed to the effect that the oil tanker is operating with special ballast arrangements.

(2)   In no case shall ballast water be carried in oil tanks except on those rare voyages when weather conditions are so severe that, in the opinion of the master, it is necessary to carry additional ballast water in cargo tanks for the safety of the ship. Such additional ballast water shall be processed and discharged in compliance with regulation 9 of this Annex and in accordance with the requirements of regulation 15 of this Annex, and entry shall be made in the Oil Record Book referred to in regulation 20 of this Annex.

(3)   An Administration which has endorsed a Certificate in accordance with subparagraph (1)(c) of this regulation shall communicate to the Organization the particulars thereof for circulation to the Parties to the present Convention.

Regulation 13E
Protective location of segregated ballast spaces

SEE INTERPRETATIONS 2.1, 4.6, 4.10 AND 4.11

(1)   In every new crude oil tanker of 20,000 tons deadweight and above and every new product carrier of 30,000 tons deadweight and above, the segregated ballast tanks required to provide the capacity to comply with the requirements of regulation 13 of this Annex which are located within the cargo tank length, shall be arranged in accordance with the requirements of paragraphs (2), (3) and (4) of this regulation to provide a measure of protection against oil outflow in the event of grounding or collision.

(2)   Segregated ballast tanks and spaces other than oil tanks within the cargo tank length (L_t) shall be so arranged as to comply with the following requirement:

$$\Sigma PA_c + \Sigma PA_s \geqslant J[L_t(B + 2D)]$$

where: PA_c = the side shell area in square metres for each segregated ballast tank or space other than an oil tank based on projected moulded dimensions,

PA_s = the bottom shell area in square metres for each such tank or space based on projected moulded dimensions,

L_t = length in metres between the forward and after extremities of the cargo tanks,

B = maximum breadth of the ship in metres as defined in regulation 1(21) of this Annex,

D = moulded depth in metres measured vertically from the top of the keel to the top of the freeboard deck beam at side amidships. In ships having rounded gunwales, the moulded depth shall be measured to the point of intersection of the moulded lines of the deck and side shell plating, the lines extending as though the gunwale were of angular design,

J = 0.45 for oil tankers of 20,000 tons deadweight, 0.30 for oil tankers of 200,000 tons deadweight and above, subject to the provisions of paragraph (3) of this regulation.

For intermediate values of deadweight the value of J shall be determined by linear interpolation.

Whenever symbols given in this paragraph appear in this regulation, they have the meaning as defined in this paragraph.

(3) For tankers of 200,000 tons deadweight and above the value of J may be reduced as follows:

$$J_{reduced} = \left[J - \left(a - \frac{O_c + O_s}{4O_A} \right) \right] \text{ or } 0.2 \text{ whichever is greater}$$

where: a = 0.25 for oil tankers of 200,000 tons deadweight,

a = 0.40 for oil tankers of 300,000 tons deadweight,

a = 0.50 for oil tankers of 420,000 tons deadweight and above.

For intermediate values of deadweight the value of a shall be determined by linear interpolation.

O_c = as defined in regulation 23(1)(a) of this Annex,

O_s = as defined in regulation 23(1)(b) of this Annex,

O_A = the allowable oil outflow as required by regulation 24(2) of this Annex.

(4) In the determination of PA_c and PA_s for segregated ballast tanks and spaces other than oil tanks the following shall apply:

(a) the minimum width of each wing tank or space either of which extends for the full depth of the ship's side or from the deck to the top of the double bottom shall be not less than 2 metres. The width shall be measured inboard from the ship's side at right angles to the centreline. Where a lesser width is provided the wing tank or space shall not be taken into account when calculating the protecting area PA_c; and

(b) the minimum vertical depth of each double bottom tank or space shall be $B/15$ or 2 metres, whichever is the lesser. Where a lesser depth is provided the bottom tank or space shall not be taken into account when calculating the protecting area PA_s.

The minimum width and depth of wing tanks and double bottom tanks shall be measured clear of the bilge area and, in the case of minimum width, shall be measured clear of any rounded gunwale area.

Annex I

Regulation 13F
Prevention of oil pollution in the event of collision or stranding

SEE INTERPRETATION 4.6

(1) This regulation shall apply to oil tankers of 600 tons deadweight and above:

    (a) for which the building contract is placed on or after 6 July 1993, or

    (b) in the absence of a building contract, the keels of which are laid or which are at a similar stage of construction on or after 6 January 1994, or

    (c) the delivery of which is on or after 6 July 1996, or

    (d) which have undergone a major conversion:

        (i) for which the contract is placed after 6 July 1993; or

        (ii) in the absence of a contract, the construction work of which is begun after 6 January 1994; or

        (iii) which is completed after 6 July 1996.

SEE INTERPRETATION 1.2

(2) Every oil tanker of 5,000 tons deadweight and above shall:

    (a) in lieu of regulation 13E, as applicable, comply with the requirements of paragraph (3) unless it is subject to the provisions of paragraphs (4) and (5); and

    (b) comply, if applicable, with the requirements of paragraph (6).

(3) The entire cargo tank length shall be protected by ballast tanks or spaces other than cargo and fuel oil tanks as follows:

    (a) *Wing tanks or spaces*

    Wing tanks or spaces shall extend either for the full depth of the ship's side or from the top of the double bottom to the uppermost deck, disregarding a rounded gunwale where fitted. They shall be arranged such that the cargo tanks are located inboard of the moulded line of the side shell plating, nowhere less than the distance w which, as shown in figure 1, is measured at any cross-section at right angles to the side shell, as specified below:

$$w = 0.5 + \frac{DW}{20,000} \text{ (m) or}$$
$$w = 2.0 \text{ m, whichever is the lesser.}$$

The minimum value of $w = 1.0$ m.

(b) *Double bottom tanks or spaces*

At any cross-section the depth of each double bottom tank or space shall be such that the distance h between the bottom of the cargo tanks and the moulded line of the bottom shell plating measured at right angles to the bottom shell plating as shown in figure 1 is not less than specified below:

$h = B/15$ (m) or

$h = 2.0$ m, whichever is the lesser.

The minimum value of $h = 1.0$ m.

(c) *Turn of the bilge area or at locations without a clearly defined turn of the bilge*

When the distances h and w are different, the distance w shall have preference at levels exceeding $1.5h$ above the baseline as shown in figure 1.

(d) *The aggregate capacity of ballast tanks*

On crude oil tankers of 20,000 tons deadweight and above and product carriers of 30,000 tons deadweight and above, the aggregate capacity of wing tanks, double bottom tanks, forepeak tanks and afterpeak tanks shall not be less than the capacity of segregated ballast tanks necessary to meet the requirements of regulation 13. Wing tanks or spaces and double bottom tanks used to meet the requirements of regulation 13 shall be located as uniformly as practicable along the cargo tank length. Additional segregated ballast capacity provided for reducing longitudinal hull girder bending stress, trim, etc., may be located anywhere within the ship.

SEE INTERPRETATION 4.12

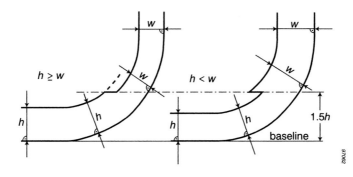

Figure 1 – Cargo tank boundary lines for the purpose of paragraph (3)

(e)   *Suction wells in cargo tanks*

Suction wells in cargo tanks may protrude into the double bottom below the boundary line defined by the distance *h* provided that such wells are as small as practicable and the distance between the well bottom and bottom shell plating is not less than 0.5*h*.

(f)   *Ballast and cargo piping*

Ballast piping and other piping such as sounding and vent piping to ballast tanks shall not pass through cargo tanks. Cargo piping and similar piping to cargo tanks shall not pass through ballast tanks. Exemptions to this requirement may be granted for short lengths of piping, provided that they are completely welded or equivalent.

(4)   (a)   Double bottom tanks or spaces as required by paragraph (3)(b) may be dispensed with, provided that the design of the tanker is such that the cargo and vapour pressure exerted on the bottom shell plating forming a single boundary between the cargo and the sea does not exceed the external hydrostatic water pressure, as expressed by the following formula:

$$f \times h_c \times \rho_c \times g + 100\Delta p \leqslant d_n \times \rho_s \times g$$

where:

h_c = height of cargo in contact with the bottom shell plating in metres

ρ_c = maximum cargo density in t/m^3

d_n = minimum operating draught under any expected loading condition in metres

ρ_s = density of seawater in t/m^3

Δp = maximum set pressure of pressure/vacuum valve provided for the cargo tank in bars

f = safety factor = 1.1

g = standard acceleration of gravity (9.81 m/s^2).

(b)   Any horizontal partition necessary to fulfil the above requirements shall be located at a height of not less than *B*/6 or 6 m, whichever is the lesser, but not more than 0.6*D*, above the baseline where *D* is the moulded depth amidships.

78

(c) The location of wing tanks or spaces shall be as defined in paragraph (3)(a) except that, below a level 1.5*h* above the baseline where *h* is as defined in paragraph (3)(b), the cargo tank boundary line may be vertical down to the bottom plating, as shown in figure 2.

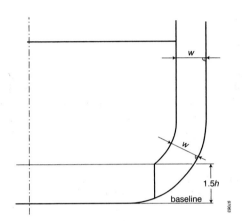

Figure 2 – Cargo tank boundary lines for the purpose of paragraph (4)

(5) Other methods of design and construction of oil tankers may also be accepted as alternatives to the requirements prescribed in paragraph (3), provided that such methods ensure at least the same level of protection against oil pollution in the event of collision or stranding and are approved in principle by the Marine Environment Protection Committee based on guidelines developed by the Organization.*

(6) For oil tankers of 20,000 tons deadweight and above the damage assumptions prescribed in regulation 25(2)(b) shall be supplemented by the following assumed bottom raking damage:

(a) longitudinal extent:

(i) ships of 75,000 tons deadweight and above: 0.6*L* measured from the forward perpendicular;

(ii) ships of less than 75,000 tons deadweight: 0.4*L* measured from the forward perpendicular;

(b) transverse extent: *B*/3 anywhere in the bottom;

(c) vertical extent: breach of the outer hull.

---

* Refer to the Interim guidelines for the approval of alternative methods of design and construction of oil tankers under regulation 13F(5) of Annex I of MARPOL 73/78 adopted by the Marine Environment Protection Committee of the Organization by resolution MEPC.66(37); see appendix 7 to Unified Interpretations of Annex I.

(7) Oil tankers of less than 5,000 tons deadweight shall:

(a) at least be fitted with double bottom tanks or spaces having such a depth that the distance h specified in paragraph (3)(b) complies with the following:

$$h = B/15 \text{ (m)}$$

with a minimum value of $h = 0.76$ m;

in the turn of the bilge area and at locations without a clearly defined turn of the bilge, the cargo tank boundary line shall run parallel to the line of the midship flat bottom as shown in figure 3; and

(b) be provided with cargo tanks so arranged that the capacity of each cargo tank does not exceed 700 m³ unless wing tanks or spaces are arranged in accordance with paragraph (3)(a) complying with the following:

$$w = 0.4 + \frac{2.4\text{DW}}{20,000} \text{ (m)}$$

with a minimum value of $w = 0.76$ m.

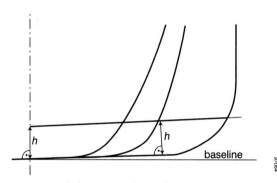

Figure 3 – Cargo tank boundary lines for the purpose of paragraph (7)

(8) Oil shall not be carried in any space extending forward of a collision bulkhead located in accordance with regulation II-1/11 of the International Convention for the Safety of Life at Sea, 1974, as amended. An oil tanker that is not required to have a collision bulkhead in accordance with that regulation shall not carry oil in any space extending forward of the transverse plane perpendicular to the centreline that is located as if it were a collision bulkhead located in accordance with that regulation.

(9) In approving the design and construction of oil tankers to be built in accordance with the provisions of this regulation, Administrations

shall have due regard to the general safety aspects including the need for the maintenance and inspections of wing and double bottom tanks or spaces.

Regulation 13G
Prevention of oil pollution in the event of collision or stranding –
Measures for existing tankers

SEE INTERPRETATION 4.6

(1) This regulation shall:

  (a) apply to

  (i) oil tankers of 20,000 tons deadweight and above carrying crude oil, fuel oil, heavy diesel oil or lubricating oil as cargo; and

  (ii) oil tankers of 30,000 tons deadweight and above other than those referred to in subparagraph (i),

  which are contracted, the keels of which are laid, or which are delivered before the dates specified in regulation 13F(1) of this Annex; and

  (b) not apply to oil tankers complying with regulation 13F of this Annex, which are contracted, the keels of which are laid, or are delivered before the dates specified in regulation 13F(1) of this Annex; and

  (c) not apply to oil tankers covered by subparagraph (a) above which comply with regulation 13F(3)(a) and (b) or 13F(4) or 13F(5) of this Annex, except that the requirement for minimum distances between the cargo tank boundaries and the ship side and bottom plating need not be met in all respects. In that event, the side protection distances shall not be less than those specified in the International Bulk Chemical Code for type 2 cargo tank location and the bottom protection distances shall comply with regulation 13E(4)(b) of this Annex.

(2) The requirements of this regulation shall take effect as from 6 July 1995, except that the requirements of paragraph (1)(a) applicable to oil tankers of 20,000 tons deadweight and above but less than 30,000 tons deadweight carrying fuel oil, heavy diesel oil or lubricating oil as cargo shall take effect as from 1 January 2003.

(2*bis*) For the purpose of paragraphs (1) and (2) of this regulation:

  (a) *Heavy diesel oil* means marine diesel oil, other than those distillates of which more than 50% by volume distils at a temperature not exceeding 340°C when tested by the method acceptable to the Organization.[*]

(b)   *Fuel oil* means heavy distillates or residues from crude oil or blends of such materials intended for use as a fuel for the production of heat or power of a quality equivalent to the specification acceptable to the Organization.[†]

(3)   (a)   An oil tanker to which this regulation applies shall be subject to an enhanced programme of inspections during periodical, intermediate and annual surveys, the scope and frequency of which shall at least comply with the guidelines developed by the Organization.[‡]

(b)   An oil tanker over five years of age to which this regulation applies shall have on board, available to the competent authority of any Government of a State Party to the present Convention, a complete file of the survey reports, including the results of all scantling measurement required, as well as the statement of structural work carried out.

(c)   This file shall be accompanied by a condition evaluation report, containing conclusions on the structural condition of the ship and its residual scantlings, endorsed to indicate that it has been accepted by or on behalf of the flag Administration. This file and condition evaluation report shall be prepared in a standard format as contained in the guidelines developed by the Organization.

(4)   An oil tanker not meeting the requirements of a new oil tanker as defined in regulation 1(26) of this Annex shall comply with the requirements of regulation 13F of this Annex not later than 25 years after its date of delivery, unless wing tanks or double bottom spaces, not used for the carriage of oil and meeting the width and height requirements of regulation 13E(4), cover at least 30% of L_t for the full depth of the ship on each side or at least 30% of the projected bottom shell area within the length L_t, where L_t is as defined in regulation 13E(2), in which case compliance with regulation 13F is required not later than 30 years after its date of delivery.

SEE INTERPRETATION 4.13

---

[*] Refer to the American Society for Testing and Materials' Standard Test Method (Designation D86).

[†] Refer to the American Society for Testing and Materials' Specification for Number Four Fuel Oil (Designation D396) or heavier.

[‡] Refer to the Guidelines on the enhanced programme of inspections during surveys of bulk carriers and oil tankers adopted by the Organization by resolution A.744(18), as amended by MSC.49(66), by resolution 2 of the 1997 Conference of Contracting Governments to SOLAS and by MSC.105(73); see IMO sales publication IMO-265E.

(5)   An oil tanker meeting the requirements of a new oil tanker as defined in regulation 1(26) of this Annex shall comply with the requirements of regulation 13F of this Annex not later than 30 years after its date of delivery.

(6)   Any new ballast and load conditions resulting from the application of paragraph (4) of this regulation shall be subject to approval of the Administration which shall have regard, in particular, to longitudinal and local strength, intact stability and, if applicable, damage stability.

(7)   Other structural or operational arrangements such as hydrostatically balanced loading may be accepted as alternatives to the requirements prescribed in paragraph (4), provided that such alternatives ensure at least the same level of protection against oil pollution in the event of collision or stranding and are approved by the Administration based on guidelines developed by the Organization.*

Regulation 14
Segregation of oil and water ballast and carriage of oil in forepeak tanks

(1)   Except as provided in paragraph (2) of this regulation, in new ships of 4,000 tons gross tonnage and above other than oil tankers, and in new oil tankers of 150 tons gross tonnage and above, no ballast water shall be carried in any oil fuel tank.

(2)   Where abnormal conditions or the need to carry large quantities of oil fuel render it necessary to carry ballast water which is not a clean ballast in any oil fuel tank, such ballast water shall be discharged to reception facilities or into the sea in compliance with regulation 9 using the equipment specified in regulation 16(2) of this Annex, and an entry shall be made in the Oil Record Book to this effect.

SEE INTERPRETATION 5.1

(3)   All other ships shall comply with the requirements of paragraph (1) of this regulation as far as is reasonable and practicable.

SEE INTERPRETATION 5.2

(4)   In a ship of 400 tons gross tonnage and above, for which the building contract is placed after 1 January 1982 or, in the absence of a building

---

* Refer to the Guidelines for approval of alternative structural or operational arrangements as called for in regulation 13G(7) of Annex I of MARPOL 73/78 adopted by the Marine Environment Protection Committee of the Organization by resolution MEPC.64(36); see appendix 8 to Unified Interpretations of Annex I and also appendix 9.

contract, the keel of which is laid or which is at a similar stage of construction after 1 July 1982, oil shall not be carried in a forepeak tank or a tank forward of the collision bulkhead.

(5) All ships other than those subject to paragraph (4) of this regulation shall comply with the provisions of that paragraph, as far as is reasonable and practicable.

Regulation 15
Retention of oil on board

(1) Subject to the provisions of paragraphs (5) and (6) of this regulation, oil tankers of 150 tons gross tonnage and above shall be provided with arrangements in accordance with the requirements of paragraphs (2) and (3) of this regulation, provided that in the case of existing tankers the requirements for oil discharge monitoring and control systems and slop tank arrangements shall apply three years after the date of entry into force of the present Convention.

(2) (a) Adequate means shall be provided for cleaning the cargo tanks and transferring the dirty ballast residue and tank washings from the cargo tanks into a slop tank approved by the Administration. In existing oil tankers, any cargo tank may be designated as a slop tank.

(b) In this system arrangements shall be provided to transfer the oily waste into a slop tank or combination of slop tanks in such a way that any effluent discharged into the sea will be such as to comply with the provisions of regulation 9 of this Annex.

(c) The arrangements of the slop tank or combination of slop tanks shall have a capacity necessary to retain the slop generated by tank washings, oil residues and dirty ballast residues. The total capacity of the slop tank or tanks shall not be less than 3% of the oil carrying capacity of the ship, except that the Administration may accept:

(i) 2% for such oil tankers where the tank washing arrangements are such that once the slop tank or tanks are charged with washing water, this water is sufficient for tank washing and, where applicable, for providing the driving fluid for eductors, without the introduction of additional water into the system;

(ii) 2% where segregated ballast tanks or dedicated clean ballast tanks are provided in accordance with regulation 13 of this Annex, or where a cargo tank cleaning system using crude oil washing is fitted in accordance with regulation 13B of this Annex. This capacity may be further reduced to 1.5% for such oil tankers where the tank washing arrangements are such that once the slop tank or tanks are charged with washing water, this water is sufficient for tank washing and, where applicable, for providing the driving fluid for eductors, without the introduction of additional water into the system;

(iii)   1% for combination carriers where oil cargo is only carried in tanks with smooth walls. This capacity may be further reduced to 0.8% where the tank washing arrangements are such that, once the slop tank or tanks are charged with washing water, this water is sufficient for tank washing and, where applicable, for providing the driving fluid for eductors, without the introduction of additional water into the system.

SEE INTERPRETATION 6.2

New oil tankers of 70,000 tons deadweight and above shall be provided with at least two slop tanks.

(d)   Slop tanks shall be designed, particularly in respect of the position of inlets, outlets, baffles or weirs where fitted, so as to avoid excessive turbulence and entrainment of oil or emulsion with the water.

SEE INTERPRETATION 6.1

---

(3)   (a)   An oil discharge monitoring and control system approved by the Administration shall be fitted. In considering the design of the oil content meter to be incorporated in the system, the Administration shall have regard to the specification recommended by the Organization.* The system shall be fitted with a recording device to provide a continuous record of the discharge in litres per nautical mile and total quantity discharged, or the oil content and rate of discharge. This record shall be identifiable as to time and date and shall be kept for at least three years. The oil discharge monitoring and control system shall come into operation when there is any discharge of effluent into the sea and shall be such as will ensure that any discharge of oily mixture is automatically stopped when the instantaneous rate of discharge of oil exceeds that permitted by regulation 9(1)(a) of this Annex. Any failure of this monitoring and control system shall stop the discharge and be noted in the Oil Record Book. A manually operated

---

* For oil content meters installed on tankers built prior to 2 October 1986, refer to the Recommendation on international performance and test specifications for oily-water separating equipment and oil content meters adopted by the Organization by resolution A.393(X). For oil content meters as part of discharge monitoring and control systems installed on tankers built on or after 2 October 1986, refer to the Revised guidelines and specifications for oil discharge monitoring and control systems for oil tankers, adopted by the Organization by resolution A.586(14); see IMO sales publications IMO-608E and IMO-646E, respectively.

alternative method shall be provided and may be used in the event of such failure, but the defective unit shall be made operable as soon as possible. The port State authority may allow the tanker with a defective unit to undertake one ballast voyage before proceeding to a repair port. The oil discharge monitoring and control system shall be designed and installed in compliance with the guidelines and specifications for oil discharge monitoring and control systems for oil tankers developed by the Organization.[*] Administrations may accept such specific arrangements as detailed in the Guidelines and Specifications.

(b) Effective oil/water interface detectors[†] approved by the Administration shall be provided for a rapid and accurate determination of the oil/water interface in slop tanks and shall be available for use in other tanks where the separation of oil and water is effected and from which it is intended to discharge effluent direct to the sea.

SEE INTERPRETATIONS 6.1 AND 6.3

(c) Instructions as to the operation of the system shall be in accordance with an operational manual approved by the Administration. They shall cover manual as well as automatic operations and shall be intended to ensure that at no time shall oil be discharged except in compliance with the conditions specified in regulation 9 of this Annex.[‡]

(4) The requirements of paragraphs (1), (2) and (3) of this regulation shall not apply to oil tankers of less than 150 tons gross tonnage, for which the control of discharge of oil under regulation 9 of this Annex shall be effected by the retention of oil on board with subsequent discharge of all contaminated washings to reception facilities. The total quantity of oil and water used for washing and returned to a storage tank shall be recorded in the Oil Record Book. This total quantity shall be discharged to reception facilities unless adequate arrangements are made to ensure that any effluent which is allowed to be discharged

---

[*] For oil content meters installed on tankers built prior to 2 October 1986, refer to the Recommendation on international performance and test specifications for oily-water separating equipment and oil content meters adopted by the Organization by resolution A.393(X). For oil content meters as part of discharge monitoring and control systems installed on tankers built on or after 2 October 1986, refer to the Revised guidelines and specifications for oil discharge monitoring and control systems for oil tankers, adopted by the Organization by resolution A.586(14); see IMO sales publications IMO-608E and IMO-646E, respectively.

[†] Refer to the Specifications for oil/water interface detectors adopted by the Marine Environment Protection Committee of the Organization by resolution MEPC.5(XIII); see IMO sales publication IMO-646E.

[‡] Refer to *Clean Seas Guide for Oil Tankers*, published by the International Chamber of Shipping and the Oil Companies International Marine Forum.

into the sea is effectively monitored to ensure that the provisions of regulation 9 of this Annex are complied with.

(5) (a) The Administration may waive the requirements of paragraphs (1), (2) and (3) of this regulation for any oil tanker which engages exclusively on voyages both of 72 hours or less in duration and within 50 miles from the nearest land, provided that the oil tanker is engaged exclusively in trades between ports or terminals within a State Party to the present Convention. Any such waiver shall be subject to the requirement that the oil tanker shall retain on board all oily mixtures for subsequent discharge to reception facilities and to the determination by the Administration that facilities available to receive such oily mixtures are adequate.

     (b) The Administration may waive the requirements of paragraph (3) of this regulation for oil tankers other than those referred to in subparagraph (a) of this paragraph in cases where:

        (i) the tanker is an existing oil tanker of 40,000 tons deadweight or above, as referred to in regulation 13C(1) of this Annex, engaged in specific trades, and the conditions specified in regulation 13C(2) are complied with; or

        (ii) the tanker is engaged exclusively in one or more of the following categories of voyages:

           (1) voyages within special areas; or

           (2) voyages within 50 miles from the nearest land outside special areas where the tanker is engaged in:

               (aa) trades between ports or terminals of a State Party to the present Convention; or

               (bb) restricted voyages as determined by the Administration, and of 72 hours or less in duration;

           provided that all of the following conditions are complied with:

           (3) all oily mixtures are retained on board for subsequent discharge to reception facilities;

           (4) for voyages specified in subparagraph (b)(ii)(2) of this paragraph, the Administration has determined that adequate reception facilities are available to receive such oily mixtures in those oil loading ports or terminals the tanker calls at;

(5)   the International Oil Pollution Prevention Certificate, when required, is endorsed to the effect that the ship is exclusively engaged in one or more of the categories of voyages specified in subparagraphs (b)(ii)(1) and (b)(ii)(2)(bb) of this paragraph; and

(6)   the quantity, time, and port of discharge are recorded in the Oil Record Book.

SEE INTERPRETATION 6.4

(6)   Where in the view of the Organization equipment required by regulation 9(1)(a)(vi) of this Annex and specified in subparagraph (3)(a) of this regulation is not obtainable for the monitoring of discharge of light refined products (white oils), the Administration may waive compliance with such requirement, provided that discharge shall be permitted only in compliance with procedures established by the Organization which shall satisfy the conditions of regulation 9(1)(a) of this Annex except the obligation to have an oil discharge monitoring and control system in operation. The Organization shall review the availability of equipment at intervals not exceeding twelve months.

(7)   The requirements of paragraphs (1), (2) and (3) of this regulation shall not apply to oil tankers carrying asphalt or other products subject to the provisions of this Annex, which through their physical properties inhibit effective product/water separation and monitoring, for which the control of discharge under regulation 9 of this Annex shall be effected by the retention of residues on board with discharge of all contaminated washings to reception facilities.

SEE INTERPRETATION 6.5

Regulation 16
Oil discharge monitoring and control system and oil filtering equipment

(1)   Any ship of 400 tons gross tonnage and above but less than 10,000 tons gross tonnage shall be fitted with oil filtering equipment complying with paragraph (4) this regulation. Any such ship which carries large quantities of oil fuel shall comply with paragraph (2) of this regulation or paragraph (1) of regulation 14.

SEE INTERPRETATIONS 7.1 AND 7.2

(2)   Any ship of 10,000 tons gross tonnage and above shall be provided with oil filtering equipment, and with arrangements for an alarm and for automatically stopping any discharge of oily mixture when the oil content in the effluent exceeds 15 parts per million.

SEE INTERPRETATION 7.2

(3)   (a)   The Administration may waive the requirements of paragraphs (1) and (2) of this regulation for any ship engaged exclusively on voyages within special areas provided that all of the following conditions are complied with:

   (i)   the ship is fitted with a holding tank having a volume adequate, to the satisfaction of the Administration, for the total retention on board of the oily bilge water;

   (ii)   all oily bilge water is retained on board for subsequent discharge to reception facilities;

   (iii)   the Administration has determined that adequate reception facilities are available to receive such oily bilge water in a sufficient number of ports or terminals the ship calls at;

   (iv)   the International Oil Pollution Prevention Certificate, when required, is endorsed to the effect that the ship is exclusively engaged on the voyages within special areas; and

   (v)   the quantity, time, and port of the discharge are recorded in the Oil Record Book.

SEE INTERPRETATIONS 6.4 AND 7.3

   (b)   The Administration shall ensure that ships of less than 400 tons gross tonnage are equipped, as far as practicable, to retain on board oil or oily mixtures or discharge them in accordance with the requirements of regulation 9(1)(b) of this Annex.

(4)   Oil filtering equipment referred to in paragraph (1) of this regulation shall be of a design approved by the Administration and shall be such as will ensure that any oily mixture discharged into the sea after passing through the system has an oil content not exceeding 15 parts per million. In considering the design of such equipment, the Administration shall have regard to the specification recommended by the Organization.*

---

* Refer to the Guidelines and specifications for pollution prevention equipment for machinery space bilges of ships adopted by the Marine Environment Protection Committee of the Organization by resolution MEPC.60(33); see IMO sales publication IMO-646E.

(5)   Oil filtering equipment referred to in paragraph (2) of this regulation shall be of a design approved by the Administration and shall be such as will ensure that any oily mixture discharged into the sea after passing through the system or systems has an oil content not exceeding 15 parts per million. It shall be provided with alarm arrangements to indicate when this level cannot be maintained. The system shall also be provided with arrangements such as will ensure that any discharge of oily mixtures is automatically stopped when the oil content of the effluent exceeds 15 parts per million. In considering the design of such equipment and arrangements, the Administration shall have regard to the specification recommended by the Organization.*

(6)   For ships delivered before 6 July 1993 the requirements of this regulation shall apply by 6 July 1998, provided that these ships can operate with oily-water separating equipment (100 ppm equipment).

SEE INTERPRETATION 7.4

Regulation 17
Tanks for oil residues (sludge)

(1)   Every ship of 400 tons gross tonnage and above shall be provided with a tank or tanks of adequate capacity, having regard to the type of machinery and length of voyage, to receive the oil residues (sludge) which cannot be dealt with otherwise in accordance with the requirements of this Annex, such as those resulting from the purification of fuel and lubricating oils and oil leakages in the machinery spaces.

SEE INTERPRETATION 8.1

(2)   In new ships, such tanks shall be designed and constructed so as to facilitate their cleaning and the discharge of residues to reception facilities. Existing ships shall comply with this requirement as far as is reasonable and practicable.

SEE INTERPRETATION 8.2

(3)   Piping to and from sludge tanks shall have no direct connection overboard, other than the standard discharge connection referred to in regulation 19.

SEE INTERPRETATION 8.3

---

* Refer to the Guidelines and specifications for pollution prevention equipment for machinery space bilges of ships adopted by the Marine Environment Protection Committee of the Organization by resolution MEPC.60(33); see IMO sales publication IMO-646E.

Regulation 18
Pumping, piping and discharge arrangements
of oil tankers

(1) In every oil tanker, a discharge manifold for connection to reception facilities for the discharge of dirty ballast water or oil contaminated water shall be located on the open deck on both sides of the ship.

(2) In every oil tanker, pipelines for the discharge to the sea of ballast water or oil contaminated water from cargo tank areas which may be permitted under regulation 9 or regulation 10 of this Annex shall be led to the open deck or to the ship's side above the waterline in the deepest ballast condition. Different piping arrangements to permit operation in the manner permitted in subparagraphs (6)(a) to (e) of this regulation may be accepted.

SEE INTERPRETATION 9.1
---

(3) In new oil tankers means shall be provided for stopping the discharge into the sea of ballast water or oil contaminated water from cargo tank areas, other than those discharges below the waterline permitted under paragraph (6) of this regulation, from a position on the upper deck or above located so that the manifold in use referred to in paragraph (1) of this regulation and the discharge to the sea from the pipelines referred to in paragraph (2) of this regulation may be visually observed. Means for stopping the discharge need not be provided at the observation position if a positive communication system such as a telephone or radio system is provided between the observation position and the discharge control position.

(4) Every new oil tanker required to be provided with segregated ballast tanks or fitted with a crude oil washing system shall comply with the following requirements:

  (a) it shall be equipped with oil piping so designed and installed that oil retention in the lines is minimized; and

  (b) means shall be provided to drain all cargo pumps and all oil lines at the completion of cargo discharge, where necessary by connection to a stripping device. The line and pump drainings shall be capable of being discharged both ashore and to a cargo tank or a slop tank. For discharge ashore a special small diameter line shall be provided and shall be connected outboard of the ship's manifold valves.

SEE INTERPRETATIONS 9.2 AND 9.3

91

(5) Every existing crude oil tanker required to be provided with segregated ballast tanks, or to be fitted with a crude oil washing system, or to operate with dedicated clean ballast tanks, shall comply with the provisions of paragraph (4)(b) of this regulation.

(6) On every oil tanker the discharge of ballast water or oil contaminated water from cargo tank areas shall take place above the waterline, except as follows:

(a) Segregated ballast and clean ballast may be discharged below the waterline:

(i) in ports or at offshore terminals, or

(ii) at sea by gravity,

provided that the surface of the ballast water has been examined immediately before the discharge to ensure that no contamination with oil has taken place.

(b) Existing oil tankers which, without modification, are not capable of discharging segregated ballast above the waterline may discharge segregated ballast below the waterline at sea, provided that the surface of the ballast water has been examined immediately before the discharge to ensure that no contamination with oil has taken place.

(c) Existing oil tankers operating with dedicated clean ballast tanks, which without modification are not capable of discharging ballast water from dedicated clean ballast tanks above the waterline, may discharge this ballast below the waterline provided that the discharge of the ballast water is supervised in accordance with regulation 13A(3) of this Annex.

(d) On every oil tanker at sea, dirty ballast water or oil contaminated water from tanks in the cargo area, other than slop tanks, may be discharged by gravity below the waterline, provided that sufficient time has elapsed in order to allow oil/water separation to have taken place and the ballast water has been examined immediately before the discharge with an oil/water interface detector referred to in regulation 15(3)(b) of this Annex, in order to ensure that the height of the interface is such that the discharge does not involve any increased risk of harm to the marine environment.

(e) On existing oil tankers at sea, dirty ballast water or oil contaminated water from cargo tank areas may be discharged below the waterline, subsequent to or in lieu of the discharge by the method referred to in subparagraph (d) of this paragraph, provided that:

(i) a part of the flow of such water is led through permanent piping to a readily accessible location on the upper deck or above where it may be visually observed during the discharge operation; and

(ii) such part flow arrangements comply with the requirements established by the Administration, which shall contain at least all the provisions of the Specifications for the Design, Installation and Operation of a Part Flow System for Control of Overboard Discharges adopted by the Organization.*

SEE INTERPRETATION 9.4

Regulation 19
Standard discharge connection

To enable pipes of reception facilities to be connected with the ship's discharge pipeline for residues from machinery bilges, both lines shall be fitted with a standard discharge connection in accordance with the following table:

Standard dimensions of flanges for discharge connections

Description	Dimension
Outside diameter	215 mm
Inner diameter	According to pipe outside diameter
Bolt circle diameter	183 mm
Slots in flange	6 holes 22 mm in diameter equidistantly placed on a bolt circle of the above diameter, slotted to the flange periphery. The slot width to be 22 mm
Flange thickness	20 mm
Bolts and nuts: quantity, diameter	6, each of 20 mm in diameter and of suitable length
The flange is designed to accept pipes up to a maximum internal diameter of 125 mm and shall be of steel or other equivalent material having a flat face. This flange, together with a gasket of oil-proof material, shall be suitable for a service pressure of 6 kg/cm^2.	

---

* See appendix 5 to Unified Interpretations for Annex I.

Regulation 20
Oil Record Book

(1) Every oil tanker of 150 tons gross tonnage and above and every ship of 400 tons gross tonnage and above other than an oil tanker shall be provided with an Oil Record Book Part I (Machinery Space Operations). Every oil tanker of 150 tons gross tonnage and above shall also be provided with an Oil Record Book Part II (Cargo/Ballast Operations). The Oil Record Book(s), whether as a part of the ship's official log-book or otherwise, shall be in the form(s) specified in appendix III to this Annex.

(2) The Oil Record Book shall be completed on each occasion, on a tank-to-tank basis if appropriate, whenever any of the following operations take place in the ship:

(a) for machinery space operations (all ships):

   (i) ballasting or cleaning of oil fuel tanks;

   (ii) discharge of dirty ballast or cleaning water from tanks referred to under (i) of the subparagraph;

   (iii) disposal of oily residues (sludge);

   (iv) discharge overboard or disposal otherwise of bilge water which has accumulated in machinery spaces;

(b) for cargo/ballast operations (oil tankers):

   (i) loading of oil cargo;

   (ii) internal transfer of oil cargo during voyage;

   (iii) unloading of oil cargo;

   (iv) ballasting of cargo tanks and dedicated clean ballast tanks;

   (v) cleaning of cargo tanks including crude oil washing;

   (vi) discharge of ballast except from segregated ballast tanks;

   (vii) discharge of water from slop tanks;

   (viii) closing of all applicable valves or similar devices after slop tank discharge operations;

   (ix) closing of valves necessary for isolation of dedicated clean ballast tanks from cargo and stripping lines after slop tank discharge operations;

   (x) disposal of residues.

(3) In the event of such discharge of oil or oily mixture as is referred to in regulation 11 of this Annex or in the event of accidental or other exceptional discharge of oil not excepted by that regulation, a statement shall be made in the Oil Record Book of the circumstances of, and the reasons for, the discharge.

(4) Each operation described in paragraph (2) of this regulation shall be fully recorded without delay in the Oil Record Book so that all entries in the book appropriate to that operation are completed. Each completed operation shall be signed by the officer or officers in charge of the operations concerned and each completed page shall be signed by the master of ship. The entries in the Oil Record Book shall be in an official language of the State whose flag the ship is entitled to fly, and, for ships holding an International Oil Pollution Prevention Certificate, in English or French. The entries in an official national language of the State whose flag the ship is entitled to fly shall prevail in case of a dispute or discrepancy.

(5) The Oil Record Book shall be kept in such a place as to be readily available for inspection at all reasonable times and, except in the case of unmanned ships under tow, shall be kept on board the ship. It shall be preserved for a period of three years after the last entry has been made.

(6) The competent authority of the Government of a Party to the Convention may inspect the Oil Record Book on board any ship to which this Annex applies while the ship is in its port or offshore terminals and may make a copy of any entry in that book and may require the master of the ship to certify that the copy is a true copy of such entry. Any copy so made which has been certified by the master of the ship as a true copy of an entry in the ship's Oil Record Book shall be made admissible in any judicial proceedings as evidence of the facts stated in the entry. The inspection of an Oil Record Book and the taking of a certified copy by the competent authority under this paragraph shall be performed as expeditiously as possible without causing the ship to be unduly delayed.

(7) For oil tankers of less than 150 tons gross tonnage operating in accordance with regulation 15(4) of this Annex an appropriate Oil Record Book should be developed by the Administration.

Regulation 21
Special requirements for drilling rigs and other platforms

SEE INTERPRETATION 10

Fixed and floating drilling rigs when engaged in the exploration, exploitation and associated offshore processing of sea-bed mineral resources and other platforms shall comply with the requirements of this Annex applicable to ships of 400 tons gross tonnage and above other than oil tankers, except that:

Annex I

(a) they shall be equipped as far as practicable with the installations required in regulations 16 and 17 of this Annex;

(b) they shall keep a record of all operations involving oil or oily mixture discharges, in a form approved by the Administration; and

(c) subject to the provisions of regulation 11 of this Annex, the discharge into the sea of oil or oily mixture shall be prohibited except when the oil content of the discharge without dilution does not exceed 15 parts per million.

Chapter III – *Requirements for minimizing oil pollution from oil tankers due to side and bottom damages*

Regulation 22
Damage assumptions

(1) For the purpose of calculating hypothetical oil outflow from oil tankers, three dimensions of the extent of damage of a parallelepiped on the side and bottom of the ship are assumed as follows. In the case of bottom damages two conditions are set forth to be applied individually to the stated portions of the oil tanker.

(a) Side damage

  (i) Longitudinal extent (l_c): $\frac{1}{3}L^{\frac{2}{3}}$ or 14.5 metres, whichever is less

  (ii) Transverse extent (t_c) (inboard from the ship's side at right angles to the centreline at the level corresponding to the assigned summer freeboard): $\frac{B}{5}$ or 11.5 metres, whichever is less

  (iii) Vertical extent (v_c): From the baseline upwards without limit

(b) Bottom damage

		For 0.3L from the forward perpendicular of the ship	*Any other part of the ship*
(i)	Longitudinal extent (l_s):	$\frac{L}{10}$	$\frac{L}{10}$ or 5 metres, whichever is less
(ii)	Transverse extent (t_s):	$\frac{B}{6}$ or 10 metres, whichever is less, but not less than 5 metres	5 metres
(iii)	Vertical extent from the baseline (v_s):	$\frac{B}{15}$ or 6 metres, whichever is less	

SEE INTERPRETATION 11.1

(2) Wherever the symbols given in this regulation appear in this chapter, they have the meaning as defined in this regulation.

Regulation 23
Hypothetical outflow of oil

SEE INTERPRETATION 11.2

(1) The hypothetical outflow of oil in the case of side damage (O_c) and bottom damage (O_s) shall be calculated by the following formulae with respect to compartments breached by damage to all conceivable locations along the length of the ship to the extent as defined in regulation 22 of this Annex.

(a) For side damages:

$$O_c = \Sigma W_i + \Sigma K_i C_i \qquad \text{(I)}$$

(b) For bottom damages:

$$O_s = \tfrac{1}{3}(\Sigma Z_i W_i + \Sigma Z_i C_i) \qquad \text{(II)}$$

where: W_i = volume of a wing tank in cubic metres assumed to be breached by the damage as specified in regulation 22 of this Annex; W_i for a segregated ballast tank may be taken equal to zero.

C_i = volume of a centre tank in cubic metres assumed to be breached by the damage as specified in regulation 22 of this Annex; C_i for a segregated ballast tank may be taken equal to zero.

K_i = $1 - \dfrac{b_i}{t_c}$; when b_i is equal to or greater than t_c, K_i shall be taken equal to zero.

Z_i = $1 - \dfrac{h_i}{v_s}$; when h_i is equal to or greater than v_s, Z_i shall be taken equal to zero.

b_i = width of wing tank in metres under consideration measured inboard from the ship's side at right angles to the centreline at the level corresponding to the assigned summer freeboard.

h_i = minimum depth of the double bottom in metres under consideration; where no double bottom is fitted h_i shall be taken equal to zero.

Whenever symbols given in this paragraph appear in this chapter, they have the meaning as defined in this regulation.

SEE INTERPRETATION 11.3

(2) If a void space or segregated ballast tank of a length less than l_c as defined in regulation 22 of this Annex is located between wing oil tanks, O_c in formula (I) may be calculated on the basis of volume W_i being the actual volume of one such tank (where they are of equal capacity) or the smaller of the two tanks (if they differ in capacity) adjacent to such space, multiplied by S_i as defined below and taking for all other wing tanks involved in such a collision the value of the actual full volume.

$$S_i = 1 - \frac{l_i}{l_c}$$

where l_i = length in metres of void space or segregated ballast tank under consideration.

(3) (a) Credit shall only be given in respect of double bottom tanks which are either empty or carrying clean water when cargo is carried in the tanks above.

   (b) Where the double bottom does not extend for the full length and width of the tank involved, the double bottom is considered non-existent and the volume of the tanks above the area of the bottom damage shall be included in formula (II) even if the tank is not considered breached because of the installation of such a partial double bottom.

   (c) Suction wells may be neglected in the determination of the value h_i provided such wells are not excessive in area and extend below the tank for a minimum distance and in no case more than half the height of the double bottom. If the depth of such a well exceeds half the height of the double bottom, h_i shall be taken equal to the double bottom height minus the well height.

   Piping serving such wells if installed within the double bottom shall be fitted with valves or other closing arrangements located at the point of connection to the tank served to prevent oil outflow in the event of damage to the piping. Such piping shall be installed as high from the bottom shell as possible. These valves shall be kept closed at sea at any time when the tank contains oil cargo, except that they may be opened only for cargo transfer needed for the purpose of trimming of the ship.

(4) In the case where bottom damage simultaneously involves four centre tanks, the value of O_s may be calculated according to the formula:

$$O_s = \tfrac{1}{4}(\Sigma Z_i W_i + \Sigma Z_i C_i) \qquad \text{(III)}$$

99

(5) An Administration may credit as reducing oil outflow in case of bottom damage, an installed cargo transfer system having an emergency high suction in each cargo oil tank, capable of transferring from a breached tank or tanks to segregated ballast tanks or to available cargo tankage if it can be assured that such tanks will have sufficient ullage. Credit for such a system would be governed by ability to transfer in two hours of operation oil equal to one half of the largest of the breached tanks involved and by availability of equivalent receiving capacity in ballast or cargo tanks. The credit shall be confined to permitting calculation of O_s according to formula (III). The pipes for such suctions shall be installed at least at a height not less than the vertical extent of the bottom damage v_s. The Administration shall supply the Organization with the information concerning the arrangements accepted by it, for circulation to other Parties to the Convention.

Regulation 24
Limitation of size and arrangement of cargo tanks

SEE INTERPRETATION 1.2

(1) Every new oil tanker shall comply with the provisions of this regulation. Every existing oil tanker shall be required, within two years after the date of entry into force of the present Convention, to comply with the provisions of this regulation if such a tanker falls into either of the following categories:

(a) a tanker, the delivery of which is after 1 January 1977; or

(b) a tanker to which both the following conditions apply:

(i) delivery is not later than 1 January 1977; and

(ii) the building contract is placed after 1 January 1974, or in cases where no building contract has previously been placed, the keel is laid or the tanker is at a similar stage of construction after 30 June 1974.

(2) Cargo tanks of oil tankers shall be of such size and arrangements that the hypothetical outflow O_c or O_s calculated in accordance with the provisions of regulation 23 of this Annex anywhere in the length of the ship does not exceed 30,000 m³ or 400 $\sqrt[3]{DW}$, whichever is the greater, but subject to a maximum of 40,000 m³.

(3) The volume of any one wing cargo oil tank of an oil tanker shall not exceed 75% of the limits of the hypothetical oil outflow referred to in paragraph (2) of this regulation. The volume of any one centre cargo oil tank shall not exceed 50,000 m³. However, in segregated ballast oil tankers as defined in regulation 13 of this Annex, the permitted

volume of a wing cargo oil tank situated between two segregated ballast tanks, each exceeding l_c in length, may be increased to the maximum limit of hypothetical oil outflow provided that the width of the wing tanks exceeds t_c.

(4) The length of each cargo tank shall not exceed 10 m or one of the following values, whichever is the greater:

   (a) where no longitudinal bulkhead is provided inside the cargo tanks:

$$(0.5\frac{b_i}{B} + 0.1)L$$

   but not to exceed 0.2L

   (b) where a centreline longitudinal bulkhead is provided inside the cargo tanks:

$$(0.25\frac{b_i}{B} + 0.15)L$$

   (c) where two or more longitudinal bulkheads are provided inside the cargo tanks:

   (i) for wing cargo tanks: 0.2L

   (ii) for centre cargo tanks:

   (1) if $\frac{b_i}{B}$ is equal to or greater than one fifth: 0.2L

   (2) if $\frac{b_i}{B}$ is less than one fifth:

   — where no centreline longitudinal bulkhead is provided:

$$(0.5\frac{b_i}{B} + 0.1)L$$

   — where a centreline longitudinal bulkhead is provided:

$$(0.25\frac{b_i}{B} + 0.15)L$$

   (d) b_i is the minimum distance from the ship's side to the outer longitudinal bulkhead of the tank in question measured inboard at right angles to the centreline at the level corresponding to the assigned summer freeboard.

(5) In order not to exceed the volume limits established by paragraphs (2), (3) and (4) of this regulation and irrespective of the accepted type of cargo transfer system installed, when such system interconnects two or more cargo tanks, valves or other similar closing devices shall be provided for separating the tanks from each other. These valves or devices shall be closed when the tanker is at sea.

(6)   Lines of piping which run through cargo tanks in a position less than t_c from the ship's side or less than v_c from the ship's bottom shall be fitted with valves or similar closing devices at the point at which they open into any cargo tank. These valves shall be kept closed at sea at any time when the tanks contain cargo oil, except that they may be opened only for cargo transfer needed for the purpose of trimming of the ship.

Regulation 25
Subdivision and stability

(1)   Every new oil tanker shall comply with the subdivision and damage stability criteria as specified in paragraph (3) of this regulation, after the assumed side or bottom damage as specified in paragraph (2) of this regulation, for any operating draught reflecting actual partial or full load conditions consistent with trim and strength of the ship as well as specific gravities of the cargo. Such damage shall be applied to all conceivable locations along the length of the ship as follows:

(a)   in tankers of more than 225 m in length, anywhere in the ship's length;

(b)   in tankers of more than 150 m, but not exceeding 225 m in length, anywhere in the ship's length except involving either after or forward bulkhead bounding the machinery space located aft. The machinery space shall be treated as a single floodable compartment; and

(c)   in tankers not exceeding 150 m in length, anywhere in the ship's length between adjacent transverse bulkheads with the exception of the machinery space. For tankers of 100 m or less in length where all requirements of paragraph (3) of this regulation cannot be fulfilled without materially impairing the operational qualities of the ship, Administrations may allow relaxations from these requirements.

Ballast conditions where the tanker is not carrying oil in cargo tanks, excluding any oil residues, shall not be considered.

SEE INTERPRETATION 11.4

(2)   The following provisions regarding the extent and the character of the assumed damage shall apply:

(a)   Side damage

(i)   Longitudinal extent: $\frac{1}{3}L^{\frac{2}{3}}$ or 14.5 metres, whichever is less

(ii)	Transverse extent (inboard from the ship's side at right angles to the centreline at the level of the summer load line):	$\frac{B}{5}$ or 11.5 metres, whichever is less
(iii)	Vertical extent:	From the moulded line of the bottom shell plating at centreline, upwards without limit

(b) Bottom damage

		For 0.3L from the forward perpendicular of the ship	*Any other part of the ship*
(i)	Longitudinal extent:	$\frac{1}{3}L^{\frac{2}{3}}$ or 14.5 metres, whichever is less	$\frac{1}{3}L^{\frac{2}{3}}$ or 5 metres, whichever is less
(ii)	Transverse extent:	$\frac{B}{6}$ or 10 metres, whichever is less	$\frac{B}{6}$ or 5 metres, whichever is less
(iii)	Vertical extent:	$\frac{B}{15}$ or 6 metres, whichever is less, measured from the moulded line of the bottom shell plating at centreline	$\frac{B}{15}$ or 6 metres, whichever is less, measured from the moulded line of the bottom shell plating at centreline

(c) If any damage of a lesser extent than the maximum extent of damage specified in subparagraphs (a) and (b) of this paragraph would result in a more severe condition, such damage shall be considered.

(d) Where the damage involving transverse bulkheads is envisaged as specified in subparagraphs (1)(a) and (b) of this regulation, transverse watertight bulkheads shall be spaced at least at a distance equal to the longitudinal extent of assumed damage specified in subparagraph (a) of this paragraph in order to be considered effective. Where transverse bulkheads are spaced at a lesser distance, one or more of these bulkheads within such extent of damage shall be assumed as non-existent for the purpose of determining flooded compartments.

(e) Where the damage between adjacent transverse watertight bulkheads is envisaged as specified in subparagraph (1)(c) of this regulation, no main transverse bulkhead or a transverse bulkhead bounding side tanks or double bottom tanks shall be assumed damaged, unless:

   (i) the spacing of the adjacent bulkheads is less than the longitudinal extent of assumed damage specified in subparagraph (a) of this paragraph; or

   (ii) there is a step or recess in a transverse bulkhead of more than 3.05 m in length, located within the extent of penetration of assumed damage. The step formed by the after peak bulkhead and after peak tank top shall not be regarded as a step for the purpose of this regulation.

(f) If pipes, ducts or tunnels are situated within the assumed extent of damage, arrangements shall be made so that progressive flooding cannot thereby extend to compartments other than those assumed to be floodable for each case of damage.

SEE INTERPRETATION 11.5

(3) Oil tankers shall be regarded as complying with the damage stability criteria if the following requirements are met:

(a) The final waterline, taking into account sinkage, heel and trim, shall be below the lower edge of any opening through which progressive flooding may take place. Such openings shall include air-pipes and those which are closed by means of weathertight doors or hatch covers and may exclude those openings closed by means of watertight manhole covers and flush scuttles, small watertight cargo tank hatch covers which maintain the high integrity of the deck, remotely operated watertight sliding doors, and sidescuttles of the non-opening type.

(b) In the final stage of flooding, the angle of heel due to unsymmetrical flooding shall not exceed 25°, provided that this angle may be increased up to 30° if no deck edge immersion occurs.

(c) The stability in the final stage of flooding shall be investigated and may be regarded as sufficient if the righting lever curve has at least a range of 20° beyond the position of equilibrium in association with a maximum residual righting lever of at least 0.1 m within the 20° range; the area under the curve within this range shall not be less than 0.0175 metre radian. Unprotected openings shall not be immersed within this range unless the space concerned is assumed to be flooded. Within this range, the immersion of any of the openings listed in subparagraph (a) of this paragraph and other openings capable of being closed weathertight may be permitted.

(d)   The Administration shall be satisfied that the stability is sufficient during intermediate stages of flooding.

(e)   Equalization arrangements requiring mechanical aids such as valves or cross-levelling pipes, if fitted, shall not be considered for the purpose of reducing an angle of heel or attaining the minimum range of residual stability to meet the requirements of subparagraphs (a), (b) and (c) of this paragraph and sufficient residual stability shall be maintained during all stages where equalization is used. Spaces which are linked by ducts of a large cross-sectional area may be considered to be common.

(4)   The requirements of paragraph (1) of this regulation shall be confirmed by calculations which take into consideration the design characteristics of the ship, the arrangements, configuration and contents of the damaged compartments; and the distribution, specific gravities and the free surface effect of liquids. The calculations shall be based on the following:

(a)   Account shall be taken of any empty or partially filled tank, the specific gravity of cargoes carried, as well as any outflow of liquids from damaged compartments.

(b)   The permeabilities assumed for spaces flooded as a result of damage shall be as follows:

Spaces	*Permeabilities*
Appropriated to stores	0.60
Occupied by accommodation	0.95
Occupied by machinery	0.85
Voids	0.95
Intended for consumable liquids	0 to 0.95*
Intended for other liquids	0 to 0.95*

(c)   The buoyancy of any superstructure directly above the side damage shall be disregarded. The unflooded parts of super-structures beyond the extent of damage, however, may be taken into consideration provided that they are separated from the damaged space by watertight bulkheads and the requirements of

---

* The permeability of partially filled compartments shall be consistent with the amount of liquid carried in the compartment. Whenever damage penetrates a tank containing liquids, it shall be assumed that the contents are completely lost from that compartment and replaced by salt water up to the level of the final plane of equilibrium.

subparagraph (3)(a) of this regulation in respect of these intact spaces are complied with. Hinged watertight doors may be acceptable in watertight bulkheads in the superstructure.

(d)   The free surface effect shall be calculated at an angle of heel of 5° for each individual compartment. The Administration may require or allow the free surface corrections to be calculated at an angle of heel greater than 5° for partially filled tanks.

(e)   In calculating the effect of free surfaces of consumable liquids it shall be assumed that, for each type of liquid at least one transverse pair or a single centreline tank has a free surface and the tank or combination of tanks to be taken into account shall be those where the effect of free surfaces is the greatest.

(5)   The master of every new oil tanker and the person in charge of a new non-self-propelled oil tanker to which this Annex applies shall be supplied in an approved form with:

(a)   information relative to loading and distribution of cargo necessary to ensure compliance with the provisions of this regulation; and

(b)   data on the ability of the ship to comply with damage stability criteria as determined by this regulation, including the effect of relaxations that may have been allowed under subparagraph (1)(c) of this regulation.

Regulation 25A
Intact stability

(1)   This regulation shall apply to oil tankers of 5,000 tons deadweight and above:

(a)   for which the building contract is placed on or after 1 February 1999; or

(b)   in the absence of a building contract, the keels of which are laid or which are at a similar stage of construction on or after 1 August 1999; or

(c)   the delivery of which is on or after 1 February 2002; or

(d)   which have undergone a major conversion:

(i)   for which the contract is placed after 1 February 1999; or

(ii)   in the absence of a contract, the construction work of which is begun after 1 August 1999; or

(iii)   which is completed after 1 February 2002.

(2)   Every oil tanker shall comply with the intact stability criteria specified in subparagraphs (a) and (b) of this paragraph, as appropriate, for any

operating draught under the worst possible conditions of cargo and ballast loading, consistent with good operational practice, including intermediate stages of liquid transfer operations. Under all conditions the ballast tanks shall be assumed slack.

SEE INTERPRETATION 11A.1

(a)   In port, the initial metacentric height GM_o, corrected for free surface measured at $0°$ heel, shall be not less than 0.15 m;

(b)   At sea, the following criteria shall be applicable:

   (i)   the area under the righting lever curve (GZ curve) shall be not less than 0.055 m·rad up to $\theta = 30°$ angle of heel and not less than 0.09 m·rad up to $\theta = 40°$ or other angle of flooding θ_f * if this angle is less than $40°$. Additionally, the area under the righting lever curve (GZ curve) between the angles of heel of $30°$ and $40°$ or between $30°$ and θ_f, if this angle is less than $40°$, shall be not less than 0.03 m·rad;

   (ii)   the righting lever GZ shall be at least 0.20 m at an angle of heel equal to or greater than $30°$;

   (iii)   the maximum righting arm shall occur at an angle of heel preferably exceeding $30°$ but not less than $25°$; and

   (iv)   the initial metacentric height GM_o, corrected for free surface measured at $0°$ heel, shall be not less than 0.15 m.

(3)   The requirements of paragraph (2) shall be met through design measures. For combination carriers simple supplementary operational procedures may be allowed.

(4)   Simple supplementary operational procedures for liquid transfer operations referred to in paragraph (3) shall mean written procedures made available to the master which:

   (i)   are approved by the Administration;

   (ii)   indicate those cargo and ballast tanks which may, under any specific condition of liquid transfer and possible range of cargo densities, be slack and still allow the stability criteria to be met. The slack tanks may vary during the liquid transfer operations and be of any combination provided they satisfy the criteria;

   (iii)   will be readily understandable to the officer-in-charge of liquid transfer operations;

---

* θ_f is the angle of heel at which the openings in the hull, superstructures or deck-houses, which cannot be closed weathertight, immerse. In applying this criterion, small openings through which progressive flooding cannot take place need not be considered as open.

(iv)   provide for planned sequences of cargo/ballast transfer operations;

(v)   allow comparisons of attained and required stability using stability performance criteria in graphical or tabular form;

(vi)   require no extensive mathematical calculations by the officer-in-charge;

(vii)   provide for corrective actions to be taken by the officer-in-charge in case of departure from recommended values and in case of emergency situations; and

(viii) are prominently displayed in the approved trim and stability booklet and at the cargo/ballast transfer control station and in any computer software by which stability calculations are performed.

Chapter IV – Prevention of pollution arising from an oil pollution incident

Regulation 26
Shipboard oil pollution emergency plan

(1)   Every oil tanker of 150 tons gross tonnage and above and every ship other than an oil tanker of 400 tons gross tonnage and above shall carry on board a shipboard oil pollution emergency plan approved by the Administration. In the case of ships built before 4 April 1993 this requirement shall apply 24 months after that date.

SEE INTERPRETATIONS 12.1 AND 12.2

---

(2)   Such a plan shall be in accordance with guidelines* developed by the Organization and written in the working language of the master and officers. The plan shall consist at least of:

(a)   the procedure to be followed by the master or other persons having charge of the ship to report an oil pollution incident, as required in article 8 and Protocol I of the present Convention, based on the guidelines developed by the Organization;†

(b)   the list of authorities or persons to be contacted in the event of an oil pollution incident;

(c)   a detailed description of the action to be taken immediately by persons on board to reduce or control the discharge of oil following the incident; and

(d)   the procedures and point of contact on the ship for co-ordinating shipboard action with national and local authorities in combating the pollution.

---

* Refer to the Guidelines for the development of shipboard oil pollution emergency plans adopted by the Marine Environment Protection Committee of the Organization by resolution MEPC.54(32) and amended by MEPC.86(44) or the Guidelines for the development of shipboard marine pollution emergency plans for oil and/or noxious liquid substances, adopted by resolution MEPC.85(44); see IMO sales publication IMO-586E.

† Refer to the General principles for ship reporting systems and ship reporting requirements, including guidelines for reporting incidents involving dangerous goods, harmful substances and/or marine pollutants adopted by the Organization by resolution A.851(20); see IMO sales publication IMO-516E.

(3) In the case of ships to which regulation 16 of Annex II of the Convention also apply, such a plan may be combined with the shipboard marine pollution emergency plan for noxious liquid substances required under regulation 16 of Annex II of the Convention. In this case, the title of such a plan shall be "Shipboard marine pollution emergency plan".

Appendices to Annex I

Appendix I
List of oils*

Asphalt solutions
Blending stocks
Roofers flux
Straight run residue

Oils
Clarified
Crude oil
Mixtures containing crude oil
Diesel oil
Fuel oil no. 4
Fuel oil no. 5
Fuel oil no. 6
Residual fuel oil
Road oil
Transformer oil
Aromatic oil (excluding vegetable oil)
Lubricating oils and blending stocks
Mineral oil
Motor oil
Penetrating oil
Spindle oil
Turbine oil

Distillates
Straight run
Flashed feed stocks

Gas oil
Cracked

Gasoline blending stocks
Alkylates – fuel
Reformates
Polymer – fuel

Gasolines
Casinghead (natural)
Automotive
Aviation
Straight run
Fuel oil no. 1 (kerosene)
Fuel oil no. 1-D
Fuel oil no. 2
Fuel oil no. 2-D

Jet fuels
JP-1 (kerosene)
JP-3
JP-4
JP-5 (kerosene, heavy)
Turbo fuel
Kerosene
Mineral spirit

Naphtha
Solvent
Petroleum
Heartcut distillate oil

---

* This list of oils shall not necessarily be considered as comprehensive.

Appendix II

Form of IOPP Certificate and Supplements

INTERNATIONAL OIL POLLUTION PREVENTION CERTIFICATE

(*Note:* This Certificate shall be supplemented by a Record
of Construction and Equipment)

Issued under the provisions of the International Convention for the Prevention
of Pollution from Ships, 1973, as modified by the Protocol of 1978 relating
thereto, and as amended by resolution MEPC.39(29), (hereinafter referred to
as "the Convention") under the authority of the Government of:

. . . .

(full designation of the country)

by. . . . .

*(full designation of the competent person or organization
authorized under the provisions of the Convention)*

Particulars of ship*

Name of ship . . . .

Distinctive number or letters . . . .

Port of registry . . . .

Gross tonnage . . . .

Deadweight of ship (metric tons)† . . . .

IMO Number‡ . . . .

---

* Alternatively, the particulars of the ship may be placed horizontally in boxes.

† For oil tankers.

‡ In accordance with resolution A.600(15), IMO Ship Identification Number Scheme, this
information may be included voluntarily.

Type of ship:*

> Oil tanker

> Ship other than an oil tanker with cargo tanks coming under regulation 2(2) of Annex I of the Convention

> Ship other than any of the above

THIS IS TO CERTIFY:

1. That the ship has been surveyed in accordance with regulation 4 of Annex I of the Convention.

2. That the survey shows that the structure, equipment, systems, fittings, arrangement and material of the ship and the condition thereof are in all respects satisfactory and that the ship complies with the applicable requirements of Annex I of the Convention.

This certificate is valid until . . . . †
subject to surveys in accordance with regulation 4 of Annex I of the Convention.

> Issued at. . . . .
> *(Place of issue of certificate)*

. . . .
(Date of issue)   *(Signature of authorized official issuing the certificate)*

(Seal or stamp of the authority, as appropriate)

---

* Delete as appropriate.

† Insert the date of expiry as specified by the Administration in accordance with regulation 8(1) of Annex I of the Convention. The day and the month of this date correspond to the anniversary date as defined in regulation 1(31) of Annex I of the Convention, unless amended in accordance with regulation 8(8) of Annex I of the Convention.

ENDORSEMENT FOR ANNUAL AND INTERMEDIATE SURVEYS

THIS IS TO CERTIFY that at a survey required by regulation 4 of Annex I of the Convention the ship was found to comply with the relevant provisions of the Convention:

Annual survey: Signed . . . .
                                       (Signature of authorized official)

                            Place . . . .

                            Date . . . .

               (Seal or stamp of the authority, as appropriate)

Annual/Intermediate* survey: Signed . . . .
                                       (Signature of authorized official)

                            Place . . . .

                            Date . . . .

               (Seal or stamp of the authority, as appropriate)

Annual/Intermediate* survey: Signed . . . .
                                       (Signature of authorized official)

                            Place . . . .

                            Date . . . .

               (Seal or stamp of the authority, as appropriate)

Annual survey: Signed . . . .
                                         (Signature of authorized official)

                            Place . . . .

                            Date . . . .

               (Seal or stamp of the authority, as appropriate)

---

* Delete as appropriate.

115

ANNUAL/INTERMEDIATE SURVEY IN ACCORDANCE
WITH REGULATION 8(8)(c)

THIS IS TO CERTIFY that, at an annual/intermediate* survey in accordance with regulation 8(8)(c) of Annex I of the Convention, the ship was found to comply with the relevant provisions of the Convention:

Signed . . . .
(Signature of authorized official)

Place . . . .

Date . . . .

(Seal or stamp of the authority, as appropriate)

ENDORSEMENT TO EXTEND THE CERTIFICATE IF VALID
FOR LESS THAN 5 YEARS WHERE REGULATION 8(3) APPLIES

The ship complies with the relevant provisions of the Convention, and this Certificate shall, in accordance with regulation 8(3) of Annex I of the Convention, be accepted as valid until . . . .

Signed . . . .
(Signature of authorized official)

Place . . . .

Date . . . .

(Seal or stamp of the authority, as appropriate)

ENDORSEMENT WHERE THE RENEWAL SURVEY HAS BEEN
COMPLETED AND REGULATION 8(4) APPLIES

The ship complies with the relevant provisions of the Convention, and this Certificate shall, in accordance with regulation 8(4) of Annex I of the Convention, be accepted as valid until . . . .

Signed . . . .
(Signature of authorized official)

Place . . . .

Date . . . .

(Seal or stamp of the authority, as appropriate)

---

* Delete as appropriate.

ENDORSEMENT TO EXTEND THE VALIDITY OF THE CERTIFICATE UNTIL REACHING THE PORT OF SURVEY OR FOR A PERIOD OF GRACE WHERE REGULATION 8(5) OR 8(6) APPLIES

This Certificate shall, in accordance with regulation 8(5) or 8(6)* of Annex I of the Convention, be accepted as valid until . . . .

Signed . . . .
(Signature of authorized official)

Place . . . .

Date . . . .

(Seal or stamp of the authority, as appropriate)

ENDORSEMENT FOR ADVANCEMENT OF ANNIVERSARY DATE WHERE REGULATION 8(8) APPLIES

In accordance with regulation 8(8) of Annex I of the Convention, the new anniversary date is . . . .

Signed . . . .
(Signature of authorized official)

Place . . . .

Date . . . .

(Seal or stamp of the authority, as appropriate)

In accordance with regulation 8(8) of Annex I of the Convention, the new anniversary date is . . . .

Signed . . . .
(Signature of authorized official)

Place . . . .

Date . . . .

(Seal or stamp of the authority, as appropriate)

---

* Delete as appropriate.

Appendix

FORM A
(Revised 1999)

Supplement to the International Oil Pollution Prevention Certificate (IOPP Certificate)

RECORD OF CONSTRUCTION AND EQUIPMENT FOR SHIPS OTHER THAN OIL TANKERS

in respect of the provisions of Annex I of the International Convention for the Prevention of Pollution from Ships, 1973, as modified by the Protocol of 1978 relating thereto (hereinafter referred to as "the Convention").

Notes:

1. This form is to be used for the third type of ships as categorized in the IOPP Certificate, i.e. "ships other than any of the above". For oil tankers and ships other than oil tankers with cargo tanks coming under regulation 2(2) of Annex I of the Convention, Form B shall be used.

2. This Record shall be permanently attached to the IOPP Certificate. The IOPP Certificate shall be available on board the ship at all times.

3. If the language of the original Record is neither English nor French, the text shall include a translation into one of these languages.

4. Entries in boxes shall be made by inserting either a cross (×) for the answers "yes" and "applicable" or a dash (–) for the answers "no" and "not applicable" as appropriate.

5. Regulations mentioned in this Record refer to regulations of Annex I of the Convention and resolutions refer to those adopted by the International Maritime Organization.

---

1 Particulars of ship

1.1 Name of ship . . . .

1.2 Distinctive number or letters . . . .

1.3 Port of registry. . . . .

1.4 Gross tonnage. . . . .

1.5   Date of build:

1.5.1   Date of building contract . . . .

1.5.2   Date on which keel was laid or ship was at a similar stage
of construction . . . .

1.5.3   Date of delivery . . . .

1.6   Major conversion (if applicable):

1.6.1   Date of conversion contract . . . .

1.6.2   Date on which conversion was commenced . . . .

1.6.3   Date of completion of conversion. . . . .

1.7   Status of ship:

1.7.1   New ship in accordance with regulation 1(6)   ☐

1.7.2   Existing ship in accordance with regulation 1(7)   ☐

1.7.3   The ship has been accepted by the Administration as an
"existing ship" under regulation 1(7) due to unforeseen
delay in delivery   ☐

2   Equipment for the control of oil discharge from machinery space bilges and oil fuel tanks
(regulations 10 and 16)

2.1   Carriage of ballast water in oil fuel tanks:

2.1.1   The ship may under normal conditions carry ballast water
in oil fuel tanks   ☐

2.2   Type of oil filtering equipment fitted:

2.2.1   Oil filtering (15 ppm) equipment (regulation 16(4))   ☐

2.2.2   Oil filtering (15 ppm) equipment with alarm and
automatic stopping device (regulation 16(5))   ☐

2.3   The ship is allowed to operate with the existing
equipment until 6 July 1998 (regulation 16(6)) and
is fitted with:

2.3.1   Oil filtering (15 ppm) equipment without alarm   ☐

2.3.2   Oil filtering (15 ppm) equipment with alarm
and manual stopping device   ☐

2.4   Approval standards:*

2.4.1   The separating/filtering equipment:

    .1   has been approved in accordance with
       resolution A.393(X);   ☐

    .2   has been approved in accordance with
       resolution MEPC.60(33);   ☐

    .3   has been approved in accordance with
       resolution A.233(VII);   ☐

    .4   has been approved in accordance with
       national standards not based upon
       resolution A.393(X) or A.233(VII);   ☐

    .5   has not been approved.   ☐

2.4.2   The process unit has been approved in accordance
      with resolution A.444(XI).   ☐

2.4.3   The oil content meter:

    .1   has been approved in accordance with
       resolution A.393(X);   ☐

    .2   has been approved in accordance with
       resolution MEPC.60(33).   ☐

2.5   Maximum throughput of the system is .... m^3/h.

2.6   Waiver of regulation 16:

2.6.1   The requirements of regulation 16(1) and 16(2)
      are waived in respect of the ship in accordance with
      regulation 16(3)(a). The ship is engaged exclusively
      on voyages within special area(s): .... ☐

---

* Refer to the Recommendation on international performance and test specifications of oily-water separating equipment and oil content meters adopted by the Organization on 14 November 1977 by resolution A.393(X), which superseded resolution A.233(VII); see IMO sales publication IMO-608E. Further reference is made to the Guidelines and specifications for pollution prevention equipment for machinery space bilges adopted by the Marine Environment Protection Committee of the Organization by resolution MEPC.60(33), which, effective on 6 July 1993, superseded resolutions A.393(X) and A.444(XI); see IMO sales publication IMO-646E.

2.6.2 The ship is fitted with holding tank(s) for the total retention on board of all oily bilge water as follows: ☐

Tank identification	Tank location		Volume (m^3)
	Frames (from)–(to)	Lateral position	
		Total volume .... (m^3)	

3 Means for retention and disposal of oil residues (sludge) (regulation 17) and bilge water holding tank(s)*

3.1 The ship is provided with oil residue (sludge) tanks as follows:

Tank identification	Tank location		Volume (m^3)
	Frames (from)–(to)	Lateral position	
		Total volume .... (m^3)	

3.2 Means for the disposal of residues in addition to the provisions of sludge tanks:

3.2.1 Incinerator for oil residues, capacity .... l/h ☐

3.2.2 Auxiliary boiler suitable for burning oil residues ☐

3.2.3 Tank for mixing oil residues with fuel oil, capacity .... m^3 ☐

3.2.4 Other acceptable means:... ☐

---

* Bilge water holding tank(s) are not required by the Convention, entries in the table under paragraph 3.3 are voluntary.

121

3.3 The ship is fitted with holding tank(s) for the retention on board of oily bilge water as follows:

Tank identification	Tank location		Volume (m^3)
	Frames (from)–(to)	Lateral position	
			Total volume .... (m^3)

4 Standard discharge connection
(regulation 19)

4.1 The ship is provided with a pipeline for the discharge of residues from machinery bilges to reception facilities, fitted with a standard discharge connection in accordance with regulation 19 ☐

5 Shipboard oil pollution emergency plan
(regulation 26)

5.1 The ship is provided with a shipboard oil pollution emergency plan in compliance with regulation 26 ☐

6 Exemption

6.1 Exemptions have been granted by the Administration from the requirements of chapter II of Annex I of the Convention in accordance with regulation 2(4)(a) on those items listed under paragraph(s) ....
.... of this Record ☐

7 Equivalents (regulation 3)

7.1 Equivalents have been approved by the Administration for certain requirements of Annex I on those items listed under paragraph(s) .... of this Record ☐

THIS IS TO CERTIFY that this Record is correct in all respects.

Issued at. . . . .
(Place of issue of the Record)

. . . .
(Signature of duly authorized officer
issuing the Record)

(Seal or stamp of the issuing authority, as appropriate)

FORM B
(Revised 1999)

Supplement to International Oil Pollution Prevention Certificate (IOPP Certificate)

RECORD OF CONSTRUCTION AND EQUIPMENT FOR OIL TANKERS

in respect of the provisions of Annex I of the International Convention for the Prevention of Pollution from Ships, 1973, as modified by the Protocol of 1978 relating thereto (hereinafter referred to as "the Convention").

Notes:

1. This form is to be used for the first two types of ships as categorized in the IOPP Certificate, i.e. "oil tankers" and "ships other than oil tankers with cargo tanks coming under regulation 2(2) of Annex I of the Convention". For the third type of ships as categorized in the IOPP Certificate, Form A shall be used.

2. This Record shall be permanently attached to the IOPP Certificate. The IOPP Certificate shall be available on board the ship at all times.

3. If the language of the original Record is neither English nor French, the text shall include a translation into one of these languages.

4. Entries in boxes shall be made by inserting either a cross (×) for the answers "yes" and "applicable" or a dash (–) for the answers "no" and "not applicable" as appropriate.

5. Unless otherwise stated, regulations mentioned in this Record refer to regulations of Annex I of the Convention and resolutions refer to those adopted by the International Maritime Organization.

---

1   Particulars of ship

1.1   Name of ship . . . .

1.2   Distinctive number or letters . . . .

1.3   Port of registry. . . . .

1.4   Gross tonnage. . . . .

1.5   Carrying capacity of ship. . . . .(m³)

1.6   Deadweight of ship . . . . (metric tons) (regulation 1(22))

1.7   Length of ship . . . . (m) (regulation 1(18))

1.8   Date of build:

1.8.1   Date of building contract . . . .

1.8.2   Date on which keel was laid or ship was at a similar stage
of construction . . . .

1.8.3   Date of delivery . . . .

1.9   Major conversion (if applicable):

1.9.1   Date of conversion contract . . . .

1.9.2   Date on which conversion was commenced . . . .

1.9.3   Date of completion of conversion. . . . .

1.10   Status of ship:

1.10.1 New ship in accordance with regulation 1(6)   ☐

1.10.2 Existing ship in accordance with regulation 1(7)   ☐

1.10.3 New oil tanker in accordance with regulation 1(26)   ☐

1.10.4 Existing oil tanker in accordance with regulation 1(27)   ☐

1.10.5 The ship has been accepted by the Administration as an
"existing ship" under regulation 1(7) due to unforeseen
delay in delivery   ☐

1.10.6 The ship has been accepted by the Administration as an
"existing oil tanker" under regulation 1(27) due to
unforeseen delay in delivery   ☐

1.10.7 The ship is not required to comply with the provisions of
regulation 24 due to unforeseen delay in delivery   ☐

1.11   Type of ship:

1.11.1 Crude oil tanker   ☐

1.11.2 Product carrier   ☐

1.11.2 *(bis)*   Product carrier not carrying fuel oil or heavy diesel
oil as referred to in regulation 13G(2*bis*),
or lubricating oil   ☐

1.11.3 Crude oil/product carrier   ☐

1.11.4 Combination carrier   ☐

1.11.5 Ship, other than an oil tanker, with cargo tanks coming
under regulation 2(2) of Annex I of the Convention   ☐

1.11.6 Oil tanker dedicated to the carriage of products referred to
in regulation 15(7)   ☐

1.11.7 The ship, being designated as a "crude oil tanker" operating with COW, is also designated as a "product carrier" operating with CBT, for which a separate IOPP Certificate has also been issued ☐

1.11.8 The ship, being designated as a "product carrier" operating with CBT, is also designated as a "crude oil tanker" operating with COW, for which a separate IOPP Certificate has also been issued ☐

1.11.9 Chemical tanker carrying oil ☐

2 Equipment for the control of oil discharge from machinery space bilges and oil fuel tanks
(regulations 10 and 16)

2.1 Carriage of ballast water in oil fuel tanks:

2.1.1 The ship may under normal conditions carry ballast water in oil fuel tanks ☐

2.2 Type of oil filtering equipment fitted:

2.2.1 Oil filtering (15 ppm) equipment (regulation 16(4)) ☐

2.2.2 Oil filtering (15 ppm) equipment with alarm and automatic stopping device (regulation 16(5)) ☐

2.3 The ship is allowed to operate with the existing equipment until 6 July 1998 (regulation 16(6)) and is fitted with:

2.3.1 Oil filtering (15 ppm) equipment without alarm ☐

2.3.2 Oil filtering (15 ppm) equipment with alarm and manual stopping device ☐

2.4 Approval standards:*

2.4.1 The separating/filtering equipment:

.1 has been approved in accordance with resolution A.393(X) ☐

---

* Refer to the Recommendation on international performance and test specifications of oily-water separating equipment and oil content meters adopted by the Organization on 14 November 1977 by resolution A.393(X), which superseded resolution A.233(VII); see IMO sales publication IMO-608E. Further reference is made to the Guidelines and specifications for pollution prevention equipment for machinery space bilges adopted by the Marine Environment Protection Committee of the Organization by resolution MEPC.60(33), which, effective on 6 July 1993, superseded resolutions A.393(X) and A.444(XI); see IMO sales publication IMO-646E.

.2   has been approved in accordance with resolution
MEPC.60(33)   ☐

.3   has been approved in accordance with resolution
A.233(VII);   ☐

.4   has been approved in accordance with national
standards not based upon resolution A.393(X) or
A.233(VII)   ☐

.5   has not been approved   ☐

2.4.2   The process unit has been approved in accordance with
resolution A.444(XI)   ☐

2.4.3   The oil content meter:

.1   has been approved in accordance with
resolution A.393(X);   ☐

.2   has been approved in accordance with
resolution MEPC.60(33).   ☐

2.5   Maximum throughput of the system is .... m^3/h.

2.6   Waiver of regulation 16:

2.6.1   The requirements of regulation 16(1) and 16(2) are
waived in respect of the ship in accordance with regulation
16(3)(a). The ship is engaged exclusively on voyages
within special area(s): ....   ☐

2.6.2   The ship is fitted with holding tank(s) for the total
retention on board of all oily bilge water as follows: ☐

Tank identification	Tank location		Volume (m^3)
	Frames (from)–(to)	Lateral position	
			Total volume .... (m^3)

2.6.3   In lieu of the holding tank(s) the ship is provided with
arrangements to transfer bilge water to the slop tank.   ☐

3 Means for retention and disposal of oil residues (sludge) (regulation 17) **and bilge water holding tank(s)***

3.1 The ship is provided with oil residue (sludge) tanks as follows:

Tank identification	Tank location		Volume (m^3)
	Frames (from)–(to)	Lateral position	
			Total volume .... (m^3)

3.2 Means for the disposal of residues in addition to the provisions of sludge tanks:

3.2.1 Incinerator for oil residues, capacity .... l/h ☐

3.2.2 Auxiliary boiler suitable for burning oil residues ☐

3.2.3 Tank for mixing oil residues with fuel oil, capacity .... m^3 ☐

3.2.4 Other acceptable means:.. ☐

3.3 The ship is fitted with holding tank(s) for the retention on board of oily bilge water as follows:

Tank identification	Tank location		Volume (m^3)
	Frames (from)–(to)	Lateral position	
			Total volume .... (m^3)

4 Standard discharge connection (regulation 19)

4.1 The ship is provided with a pipeline for the discharge of residues from machinery bilges to reception facilities, fitted with a standard discharge connection in compliance with regulation 19 ☐

---

* Bilge water holding tank(s) are not required by the Convention, entries in the table under paragraph 3.3 are voluntary.

5 Construction (regulations 13, 24 and 25)

5.1 In accordance with the requirements of regulation 13, the ship is:

5.1.1 Required to be provided with SBT, PL and COW ☐

5.1.2 Required to be provided with SBT and PL ☐

5.1.3 Required to be provided with SBT ☐

5.1.4 Required to be provided with SBT or COW ☐

5.1.5 Required to be provided with SBT or CBT ☐

5.1.6 Not required to comply with the requirements of regulation 13 ☐

5.2 Segregated ballast tanks (SBT):

5.2.1 The ship is provided with SBT in compliance with regulation 13 ☐

5.2.2 The ship is provided with SBT, in compliance with regulation 13, which are arranged in protective locations (PL) in compliance with regulation 13E ☐

5.2.3 SBT are distributed as follows:

Tank	Volume (m^3)	Tank	Volume (m^3)
		Total volume . . . . m^3	

5.3 Dedicated clean ballast tanks (CBT):

5.3.1 The ship is provided with CBT in compliance with regulation 13A, and may operate as a product carrier ☐

5.3.2 CBT are distributed as follows:

Tank	Volume (m^3)	Tank	Volume (m^3)
		Total volume . . . . m^3	

5.3.3 The ship has been supplied with a valid Dedicated Clean Ballast Tank Operation Manual, which is dated . . . . ☐

5.3.4   The ship has common piping and pumping arrangements
for ballasting the CBT and handling cargo oil   ☐

5.3.5   The ship has separate independent piping and pumping
arrangements for ballasting the CBT   ☐

5.4   Crude oil washing (COW):

5.4.1   The ship is equipped with a COW system in compliance
with regulation 13B   ☐

5.4.2   The ship is equipped with a COW system in compliance
with regulation 13B except that the effectiveness of the
system has not been confirmed in accordance with
regulation 13(6) and paragraph 4.2.10 of the Revised
COW Specifications (resolution A.446(XI)*)   ☐

5.4.3   The ship has been supplied with a valid Crude Oil Washing
Operations and Equipment Manual, which is dated . . . . .
. . . .   ☐

5.4.4   The ship is not required to be but is equipped with COW in
compliance with the safety aspects of the Revised COW
Specifications (resolution A.446(XI)*)   ☐

5.5   Exemption from regulation 13:

5.5.1   The ship is solely engaged in trade between . . . .
. . . .
in accordance with regulation 13C and is therefore
exempted from the requirements of regulation 13   ☐

5.5.2   The ship is operating with special ballast arrangements in
accordance with regulation 13D and is therefore exempted
from the requirements of regulation 13   ☐

5.6   Limitation of size and arrangements of cargo tanks (regulation 24):

5.6.1   The ship is required to be constructed according to, and
complies with, the requirements of regulation 24   ☐

5.6.2   The ship is required to be constructed according to, and
complies with, the requirements of regulation 24(4) (see
regulation 2(2))   ☐

---

* See IMO sales publication IMO-617E.

5.7   Subdivision and stability (regulation 25):

5.7.1   The ship is required to be constructed according to, and complies with, the requirements of regulation 25 ☐

5.7.2   Information and data required under regulation 25(5) have been supplied to the ship in an approved form ☐

5.7.3   The ship is required to be constructed according to, and complies with the requirements of, regulation 25A ☐

5.7.4   Information and data required under regulation 25A for combination carriers have been supplied to the ship in a written procedure approved by the Administration ☐

5.8   Double-hull construction:

5.8.1   The ship is required to be constructed according to regulation 13F and complies with the requirements of:

    .1   paragraph (3) (double-hull construction) ☐

    .2   paragraph (4) (mid-height deck tankers with double side construction) ☐

    .3   paragraph (5) (alternative method approved by the Marine Environment Protection Committee) ☐

5.8.2   The ship is required to be constructed according to and complies with the requirements of regulation 13F(7) (double bottom requirements) ☐

5.8.3   The ship is not required to comply with the requirements of regulation 13F ☐

5.8.4   The ship is subject to regulation 13G and:

    .1   is required to comply with regulation 13F not later than . . . . ☐

    .2   is so arranged that the following tanks or spaces are not used for the carriage of oil . . . . ☐

    .3   has been accepted in accordance with regulation 13G(7) and resolution MEPC.64(36) ☐

    .4   is provided with the operational manual approved on .... in accordance with resolution MEPC.64(36) ☐

5.8.5   The ship is not subject to regulation 13G ☐

6 Retention of oil on board (regulation 15)

6.1 Oil discharge monitoring and control system:

6.1.1 The ship comes under category . . . . oil tanker as
defined in resolution A.496(XII) or A.586(14)* *(delete as
appropriate)* ☐

6.1.2 The system comprises:
.1 control unit ☐
.2 computing unit ☐
.3 calculating unit ☐

6.1.3 The system is:
.1 fitted with a starting interlock ☐
.2 fitted with automatic stopping device ☐

6.1.4 The oil content meter is approved under the terms of
resolution A.393(X) or A.586(14)† *(delete as appropriate)*
suitable for:
.1 crude oil ☐
.2 black products ☐
.3 white products ☐
.4 oil-like noxious liquid substances as listed in the
attachment to the certificate ☐

6.1.5 The ship has been supplied with an operations manual for
the oil discharge monitoring and control system ☐

6.2 Slop tanks:

6.2.1 The ship is provided with . . . . dedicated slop tank(s)
with the total capacity of . . . . m^3, which is. . . . . %
of the oil carrying capacity, in accordance with:
.1 regulation 15(2)(c) ☐
.2 regulation 15(2)(c)(i) ☐
.3 regulation 15(2)(c)(ii) ☐
.4 regulation 15(2)(c)(iii) ☐

---

* Oil tankers the keels of which are laid, or which are at a similar stage of construction, on or after 2 October 1986 should be fitted with a system approved under resolution A.586(14); see IMO sales publication IMO-646E.

† For oil content meters installed on tankers built prior to 2 October 1986, refer to the Recommendation on international performance and test specifications for oily-water separating equipment and oil content meters adopted by the Organization by resolution A.393(X). For oil content meters as part of discharge monitoring and control systems installed on tankers built on or after 2 October 1986, refer to the Guidelines and specifications for oil discharge monitoring and control systems for oil tankers adopted by the Organization by resolution A.586(14); see IMO sales publications IMO-608E and IMO-646E, respectively.

6.2.2   Cargo tanks have been designated as slop tanks   ☐

6.3   Oil/water interface detectors:

6.3.1   The ship is provided with oil/water interface detectors approved under the terms of resolution MEPC.5(XIII)*   ☐

6.4   Exemptions from regulation 15:

6.4.1   The ship is exempted from the requirements of regulation 15(1), (2) and (3) in accordance with regulation 15(7)   ☐

6.4.2   The ship is exempted from the requirements of regulation 15(1), (2) and (3) in accordance with regulation 2(2)   ☐

6.5   Waiver of regulation 15:

6.5.1   The requirements of regulation 15(3) are waived in respect of the ship in accordance with regulation 15(5)(b). The ship is engaged exclusively on:

     .1   specific trade under regulation 13C: . . . .

     . . . .

     . . . .   ☐

     .2   voyages within special area(s): . . . .

     . . . .

     . . . .   ☐

     .3   voyages within 50 miles of the nearest land outside special area(s) of 72 hours or less in duration restricted to: . . . .

     . . . .

     . . . .   ☐

7   Pumping, piping and discharge arrangements
(regulation 18)

7.1   The overboard discharge outlets for segregated ballast are located:

7.1.1   Above the waterline   ☐

7.1.2   Below the waterline   ☐

7.2   The overboard discharge outlets, other than the discharge manifold, for clean ballast are located:[†]

7.2.1   Above the waterline   ☐

7.2.2   Below the waterline   ☐

---

* Refer to the Specification for oil/water interface detectors adopted by the Marine Environment Protection Committee of the Organization by resolution MEPC.5(XIII); see IMO sales publication IMO-646E.

† Only those outlets which can be monitored are to be indicated.

7.3 The overboard discharge outlets, other than the discharge manifold, for dirty ballast water or oil-contaminated water from cargo tank areas are located:*

7.3.1 Above the waterline ☐

7.3.2 Below the waterline in conjunction with the part flow arrangements in compliance with regulation 18(6)(e) ☐

7.3.3 Below the waterline ☐

7.4 Discharge of oil from cargo pumps and oil lines (regulation 18(4) and (5)):

7.4.1 Means to drain all cargo pumps and oil lines at the completion of cargo discharge:

　.1 drainings capable of being discharged to a cargo tank or slop tank ☐

　.2 for discharge ashore a special small-diameter line is provided ☐

8　Shipboard oil pollution emergency plan
(regulation 26)

8.1 The ship is provided with a shipboard oil pollution emergency plan in compliance with regulation 26 ☐

9　Equivalent arrangements for chemical tankers carrying oil

9.1 As equivalent arrangements for the carriage of oil by a chemical tanker, the ship is fitted with the following equipment in lieu of slop tanks (paragraph 6.2 above) and oil/water interface detectors (paragraph 6.3 above):

9.1.1 Oily-water separating equipment capable of producing effluent with oil content less than 100 ppm, with the capacity of . . . . m^3/h ☐

9.1.2 A holding tank with the capacity of . . . . m^3 ☐

9.1.3 A tank for collecting tank washings which is:

　.1 a dedicated tank ☐

　.2 a cargo tank designated as a collecting tank ☐

---

* Only those outlets which can be monitored are to be indicated.

9.1.4　A permanently installed transfer pump for overboard discharge of effluent containing oil through the oily-water separating equipment　☐

9.2　The oily-water separating equipment has been approved under the terms of resolution A.393(X)* and is suitable for the full range of Annex I products　☐

9.3　The ship holds a valid Certificate of Fitness for the Carriage of Dangerous Chemicals in Bulk　☐

10　Oil-like noxious liquid substances

10.1　The ship is permitted, in accordance with regulation 14 of Annex II of the Convention, to carry the oil-like noxious liquid substances specified in the list[†] attached　☐

11　Exemption

11.1　Exemptions have been granted by the Administration from the requirements of chapters II and III of Annex I of the Convention in accordance with regulation 2(4)(a) on those items listed under paragraph(s) . . . .
. . . .
. . . . of this Record　☐

12　Equivalents (regulation 3)

12.1　Equivalents have been approved by the Administration for certain requirements of Annex I on those items listed under paragraph(s) . . . .
. . . . of this Record　☐

THIS IS TO CERTIFY that this Record is correct in all respects.

Issued at. . . . .
(Place of issue of the Record)

. . . .　　　　. . . . .
(Signature of duly authorized officer issuing the Record)

(Seal or stamp of the issuing authority, as appropriate)

---

* Refer to the Guidelines and specifications for pollution prevention equipment for machinery space bilges adopted by the Marine Environment Protection Committee of the Organization by resolution MEPC.60(33), which, effective on 6 July 1993, superseded resolution A.393(X); see IMO sales publication IMO-646E.

† The list of oil-like noxious substances permitted for carriage, signed, dated and certified by a seal or a stamp of the issuing authority, shall be attached.

Appendix III
Form of Oil Record Book

OIL RECORD BOOK

PART I – Machinery space operations
(All ships)

Name of ship:

Distinctive number or letters:

Gross tonnage:

Period from: to:

Note: Oil Record Book Part I shall be provided to every oil tanker of 150 tons gross tonnage and above and every ship of 400 tons gross tonnage and above, other than oil tankers, to record relevant machinery space operations. For oil tankers, Oil Record Book Part II shall also be provided to record relevant cargo/ballast operations.

Annex I

Introduction

The following pages of this section show a comprehensive list of items of machinery space operations which are, when appropriate, to be recorded in the Oil Record Book in accordance with regulation 20 of Annex I of the International Convention for the Prevention of Pollution from Ships, 1973, as modified by the Protocol of 1978 relating thereto (MARPOL 73/78). The items have been grouped into operational sections, each of which is denoted by a letter code.

When making entries in the Oil Record Book, the date, operational code and item number shall be inserted in the appropriate columns and the required particulars shall be recorded chronologically in the blank spaces.

Each completed operation shall be signed for and dated by the officer or officers in charge. Each completed page shall be signed by the master of the ship.

The Oil Record Book contains many references to oil quantity. The limited accuracy of tank measurement devices, temperature variations and clingage will affect the accuracy of these readings. The entries in the Oil Record Book should be considered accordingly.

LIST OF ITEMS TO BE RECORDED

(A) Ballasting or cleaning of oil fuel tanks

1. Identity of tank(s) ballasted.

2. Whether cleaned since they last contained oil and, if not, type of oil previously carried.

3. Cleaning process:
    .1 position of ship and time at the start and completion of cleaning;
    .2 identify tank(s) in which one or another method has been employed (rinsing through, steaming, cleaning with chemicals; type and quantity of chemicals used);
    .3 identity of tank(s) into which cleaning water was transferred.

4. Ballasting:
    .1 position of ship and time at start and end of ballasting;
    .2 quantity of ballast if tanks are not cleaned.

(B) Discharge of dirty ballast or cleaning water from oil fuel tanks referred to under section (A)

5. Identity of tank(s).

6. Position of ship at start of discharge.

7. Position of ship on completion of discharge.

8. Ship's speed(s) during discharge.

9. Method of discharge:
    .1 through 15 ppm equipment;
    .2 to reception facilities.

10. Quantity discharged.

(C) Collection and disposal of oil residues (sludge)

11. Collection of oil residues.
    Quantities of oil residues (sludge) retained on board at the end of a voyage, but not more frequently than once a week. When ships are on short voyages, the quantity should be recorded weekly:[1]

---

[1] Only in tanks listed in item 3 of Forms A and B of the Supplement to the IOPP Certificate.

    .1   separated sludge (sludge resulting from purification of fuel and lubricating oils) and other residues, if applicable:
- identity of tank(s)............................
- capacity of tank(s)........................ m^3
- total quantity of retention .... m^3;

    .2   other residues (such as oils residues resulting from drainages, leakages, exhausted oil, etc., in the machinery spaces), if applicable due to tank arrangement in addition to .1:
- identity of tank(s)............................
- capacity of tank(s)........................ m^3
- total quantity of retention .... m^3.

12. Methods of disposal of residue.

State quantity of oil residues disposed of, the tank(s) emptied and the quantity of contents retained:

    .1   to reception facilities (identify port);[2]

    .2   transferred to another (other) tank(s) (indicate tank(s) and the total content of tank(s));

    .3   incinerated (indicate total time of operation);

    .4   other method (state which).

(D) Non-automatic discharge overboard or disposal otherwise of bilge water which has accumulated in machinery spaces

13. Quantity discharged or disposed of.

14. Time of discharge or disposal (start and stop).

15. Method of discharge or disposal:

    .1   through 15 ppm equipment (state position at start and end);

    .2   to reception facilities (identify port);[2]

    .3   transfer to slop tank or holding tank (indicate tank(s); state quantity transferred and the total quantity retained in tank(s)).

---

[2] Ships' masters should obtain from the operator of the reception facilities, which include barges and tank trucks, a receipt or certificate detailing the quantity of tank washings, dirty ballast, residues or oily mixtures transferred, together with the time and date of the transfer. This receipt or certificate, if attached to the Oil Record Book, may aid the master of the ship in proving that his ship was not involved in an alleged pollution incident. The receipt or certificate should be kept together with the Oil Record Book.

(E) Automatic discharge overboard or disposal otherwise of bilge water which has accumulated in machinery spaces

16. Time and position of ship at which the system has been put into automatic mode of operation for discharge overboard.

17. Time when the system has been put into automatic mode of operation for transfer of bilge water to holding tank (identify tank).

18. Time when the system has been put into manual operation.

19. Method of discharge overboard:

    .1 through 15 ppm equipment.

(F) Condition of oil discharge monitoring and control system

20. Time of system failure.

21. Time when system has been made operational.

22. Reasons for failure.

(G) Accidental or other exceptional discharges of oil

23. Time of occurrence.

24. Place or position of ship at time of occurrence.

25. Approximate quantity and type of oil.

26. Circumstances of discharge or escape, the reasons therefor and general remarks.

(H) Bunkering of fuel or bulk lubricating oil

27. Bunkering:

    .1 Place of bunkering.

    .2 Time of bunkering.

    .3 Type and quantity of fuel oil and identity of tank(s) (state quantity added and total content of tank(s)).

    .4 Type and quantity of lubricating oil and identity of tank(s) (state quantity added and total content of tank(s)).

(I) Additional operational procedures and general remarks

Name of ship . . . .

Distinctive number or letters . . . .

CARGO/BALLAST OPERATIONS (OIL TANKERS)*/
MACHINERY SPACE OPERATIONS (ALL SHIPS)*

Date	Code (letter)	Item (number)	Record of operations/signature of officer in charge

Signature of master . . . .

* Delete as appropriate

OIL RECORD BOOK

PART II – Cargo/ballast operations
(Oil tankers)

Name of ship:

Distinctive number or letters:

Gross tonnage:

Period from: to:

Note: Every oil tanker of 150 tons gross tonnage and above shall be provided with Oil Record Book Part II to record relevant cargo/ballast operations. Such a tanker shall also be provided with Oil Record Book Part I to record relevant machinery space operations.

Name of ship . . . .

Distinctive number or letters . . . .

PLAN VIEW OF CARGO AND SLOP TANKS
(to be completed on board)

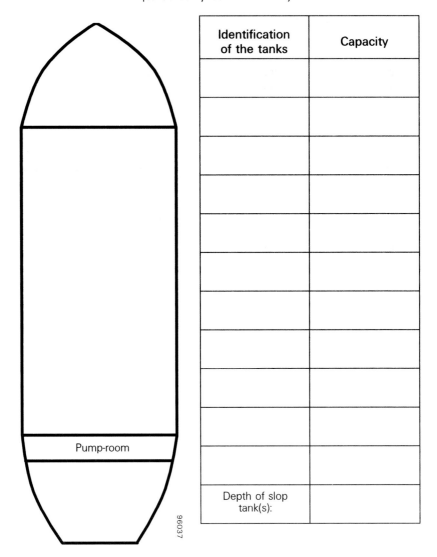

Identification of the tanks	Capacity
Depth of slop tank(s):	

Pump-room

96037

(Give the capacity of each tank
and the depth of slop tank(s))

Introduction

The following pages of this section show a comprehensive list of items of cargo and ballast operations which are, when appropriate, to be recorded in the Oil Record Book in accordance with regulation 20 of Annex I of the International Convention for the Prevention of Pollution from Ships, 1973, as modified by the Protocol of 1978 relating thereto (MARPOL 73/78). The items have been grouped into operational sections, each of which is denoted by a code letter.

When making entries in the Oil Record Book, the date, operational code and item number shall be inserted in the appropriate columns and the required particulars shall be recorded chronologically in the blank spaces.

Each completed operation shall be signed for and dated by the officer or officers in charge. Each completed page shall be countersigned by the master of the ship. In respect of the oil tankers engaged in specific trades in accordance with regulation 13C of Annex I of MARPOL 73/78, appropriate entry in the Oil Record Book shall be endorsed by the competent port State authority.*

The Oil Record Book contains many references to oil quantity. The limited accuracy of tank measurement devices, temperature variations and clingage will affect the accuracy of these readings. The entries in the Oil Record Book should be considered accordingly.

---

* This sentence should only be inserted for the Oil Record Book of a tanker engaged in a specific trade.

LIST OF ITEMS TO BE RECORDED

(A) Loading of oil cargo

1. Place of loading.

2. Type of oil loaded and identity of tank(s).

3. Total quantity of oil loaded (state quantity added and the total content of tank(s)).

(B) Internal transfer of oil cargo during voyage

4. Identity of tank(s):
   .1 from:
   .2 to: (state quantity transferred and total quantity of tank(s))

5. Was (were) the tank(s) in 4.1 emptied? (If not, state quantity retained.)

(C) Unloading of oil cargo

6. Place of unloading.

7. Identity of tank(s) unloaded.

8. Was (were) the tank(s) emptied? (If not, state quantity retained.)

(D) Crude oil washing (COW tankers only)
(To be completed for each tank being crude oil washed)

9. Port where crude oil washing was carried out or ship's position if carried out between two discharge ports.

10. Identity of tank(s) washed.[1]

11. Number of machines in use.

12. Time of start of washing.

13. Washing pattern employed.[2]

14. Washing line pressure.

15. Time washing was completed or stopped.

---

[1] When an individual tank has more machines than can be operated simultaneously, as described in the Operations and Equipment Manual, then the section being crude oil washed should be identified, e.g. No. 2 centre, forward section.

[2] In accordance with the Operations and Equipment Manual, enter whether single-stage or multi-stage method of washing is employed. If multi-stage method is used, give the vertical arc covered by the machines and the number of times that arc is covered for that particular stage of the programme.

16. State method of establishing that tank(s) was (were) dry.

17. Remarks.[3]

(E) Ballasting of cargo tanks

18. Position of ship at start and end of ballasting.

19. Ballasting process:
    .1 identity of tank(s) ballasted;
    .2 time of start and end;
    .3 quantity of ballast received. Indicate total quantity of ballast for each tank involved in the operation.

(F) Ballasting of dedicated clean ballast tanks (CBT tankers only)

20. Identity of tank(s) ballasted.

21. Position of ship when water intended for flushing, or port ballast was taken to dedicated clean ballast tank(s).

22. Position of ship when pump(s) and lines were flushed to slop tank.

23. Quantity of the oily water which, after line flushing, is transferred to the slop tank(s) or cargo tank(s) in which slop is preliminarily stored (identify tank(s)). State the total quantity.

24. Position of ship when additional ballast water was taken to dedicated clean ballast tank(s).

25. Time and position of ship when valves separating the dedicated clean ballast tanks from cargo and stripping lines were closed.

26. Quantity of clean ballast taken on board.

(G) Cleaning of cargo tanks

27. Identity of tank(s) cleaned.

28. Port or ship's position.

29. Duration of cleaning.

30. Method of cleaning.[4]

---

[3] If the programmes given in the Operations and Equipment Manual are not followed, then the reasons must be given under Remarks.

[4] Hand-hosing, machine washing and/or chemical cleaning. Where chemically cleaned, the chemical concerned and amount used should be stated.

31. Tank washings transferred to:
  .1 reception facilities (state port and quantity)[5];
  .2 slop tank(s) or cargo tank(s) designated as slop tank(s) (identify tank(s); state quantity transferred and total quantity).

(H) Discharge of dirty ballast

32. Identity of tank(s).

33. Position of ship at start of discharge into the sea.

34. Position of ship on completion of discharge into the sea.

35. Quantity discharged into the sea.

36. Ship's speed(s) during discharge.

37. Was the discharge monitoring and control system in operation during the discharge?

38. Was a regular check kept on the effluent and the surface of the water in the locality of the discharge?

39. Quantity of oily water transferred to slop tank(s) (identify slop tank(s)). State total quantity.

40. Discharged to shore reception facilities (identify port and quantity involved).[5]

(I) Discharge of water from slop tanks into the sea

41. Identity of slop tanks.

42. Time of settling from last entry of residues, or

43. Time of settling from last discharge.

44. Time and position of ship at start of discharge.

45. Ullage of total contents at start of discharge.

46. Ullage of oil/water interface at start of discharge.

47. Bulk quantity discharged and rate of discharge.

48. Final quantity discharged and rate of discharge.

49. Time and position of ship on completion of discharge.

---

[5] Ships' masters should obtain from the operator of the reception facilities, which include barges and tank trucks, a receipt or certificate detailing the quantity of tank washings, dirty ballast, residues or oily mixtures transferred, together with the time and date of the transfer. This receipt or certificate, if attached to the Oil Record Book, may aid the master of the ship in proving that his ship was not involved in an alleged pollution incident. The receipt or certificate should be kept together with the Oil Record Book.

50. Was the discharge monitoring and control system in operation during the discharge?

51. Ullage of oil/water interface on completion of discharge.

52. Ship's speed(s) during discharge.

53. Was a regular check kept on the effluent and the surface of the water in the locality of the discharge?

54. Confirm that all applicable valves in the ship's piping system have been closed on completion of discharge from the slop tanks.

(J) Disposal of residues and oily mixtures not otherwise dealt with

55. Identity of tank(s).

56. Quantity disposed of from each tank. (State the quantity retained.)

57. Method of disposal:

    .1 to reception facilities (identify port and quantity involved);[5]

    .2 mixed with cargo (state quantity);

    .3 transferred to (an)other tank(s) (identify tank(s); state quantity transferred and total quantity in tank(s));

    .4 other method (state which); state quantity disposed of.

(K) Discharge of clean ballast contained in cargo tanks

58. Position of ship at start of discharge of clean ballast.

59. Identity of tank(s) discharged.

60. Was (were) the tank(s) empty on completion?

61. Position of ship on completion if different from 58.

62. Was a regular check kept on the effluent and the surface of the water in the locality of the discharge?

(L) Discharge of ballast from dedicated clean ballast tanks (CBT tankers only)

63. Identity of tank(s) discharged.

---

[5] Ships' masters should obtain from the operator of the reception facilities, which include barges and tank trucks, a receipt or certificate detailing the quantity of tank washings, dirty ballast, residues or oily mixtures transferred, together with the time and date of the transfer. This receipt or certificate, if attached to the Oil Record Book, may aid the master of the ship in proving that his ship was not involved in an alleged pollution incident. The receipt or certificate should be kept together with the Oil Record Book.

64. Time and position of ship at start of discharge of clean ballast into the sea.

65. Time and position of ship on completion of discharge into the sea.

66. Quantity discharged:

    .1 into the sea; or

    .2 to reception facility (identify port).

67. Was there any indication of oil contamination of the ballast water before or during discharge into the sea?

68. Was the discharge monitored by an oil content meter?

69. Time and position of ship when valves separating dedicated clean ballast tanks from the cargo and stripping lines were closed on completion of deballasting.

(M) Condition of oil discharge monitoring and control system

70. Time of system failure.

71. Time when system has been made operational.

72. Reasons for failure.

(N) Accidental or other exceptional discharges of oil

73. Time of occurrence.

74. Port or ship's position at time of occurrence.

75. Approximate quantity and type of oil.

76. Circumstances of discharge or escape, the reasons therefor and general remarks.

(O) Additional operational procedures and general remarks

TANKERS ENGAGED IN SPECIFIC TRADES

(P) Loading of ballast water

77. Identity of tank(s) ballasted.

78. Position of ship when ballasted.

79. Total quantity of ballast loaded in cubic metres.

80. Remarks.

(Q) Re-allocation of ballast water within the ship

81. Reasons for re-allocation.

(R) Ballast water discharge to reception facility

82. Port(s) where ballast water was discharged.

83. Name or designation of reception facility.

84. Total quantity of ballast water discharged in cubic metres.

85. Date, signature and stamp of port authority official.

Name of ship . . . .

Distinctive number or letters . . . .

CARGO/BALLAST OPERATIONS (OIL TANKERS)*/
MACHINERY SPACE OPERATIONS (ALL SHIPS)*

Date	Code (letter)	Item (number)	Record of operations/signature of officer in charge

Signature of master . . . .

* Delete as appropriate

Annex I

Unified Interpretations of Annex I

Notes: For the purposes of the Unified Interpretations, the following abbreviations are used:

MARPOL 73/78	The 1973 MARPOL Convention as modified by the 1978 Protocol relating thereto
Regulation	Regulation in Annex I of MARPOL 73/78
IOPP Certificate	International Oil Pollution Prevention Certificate
SBT	Segregated ballast tanks
CBT	Dedicated clean ballast tanks
COW	Crude oil washing system
IGS	Inert gas systems
PL	Protective location of segregated ballast tanks
H	Date of entry into force of MARPOL 73/78 (see 2.0.1). "H + 2" means two years after the date of entry into force of MARPOL 73/78.

1 Definitions

Reg. 1(1)

1A.0 *Definition of "oil"*

1A.0.1 (Animal and vegetable oils are found to fall under the category of "noxious liquid substance", and therefore this interpretation has been deleted (see Annex II, appendix II, of MARPOL 73/78).)

Treatment for oily rags

1A.0.2 Oily rags, as defined in the Guidelines for the Implementation of Annex V of MARPOL 73/78, should be treated in accordance with Annex V and the procedures set out in the Guidelines.

Reg. 1(4)

1.0 *Definition of an "oil tanker"*

1.0.1 A gas carrier as defined in regulation 3.20 of chapter II-1 of SOLAS 74 (as amended), when carrying a cargo or part cargo of oil in bulk, should be treated as an "oil tanker" as defined in regulation 1(4).

Reg. 1(6)
1(26)

1.1 *Definition of "new ships"*

1.1.1 Regulations 1(6) and 1(26) defining "new ship" and "new oil tanker", respectively, should be construed to mean that a ship which falls into any one of the categories listed in subparagraphs (a), (b), (c), (d)(i), (d)(ii), or (d)(iii) of these paragraphs should be considered as a new ship or a new oil tanker, as appropriate.

Reg. 1(6)
1(26)
13F
24

1.2 *Unforeseen delay in delivery of ships*

1.2.1   For the purpose of defining "new" or "existing" ships under regulations 1(6), 1(26), 13F and 24, a ship for which the building contract (or keel laying) and delivery were scheduled before the dates specified in these regulations, but which has been subject to delay in delivery beyond the specified date due to unforeseen circumstances beyond the control of the builder *and* the owner, may be accepted by the Administration as an "existing ship". The treatment of such ships should be considered by the Administration *on a case by case basis*, bearing in mind the particular circumstances.

1.2.2   It is important that ships delivered after the specified dates due to unforeseen delay and allowed to be treated as existing ships by the Administration, should also be accepted as such by port States. In order to ensure this, the following practice is recommended to Administrations when considering an application for such a ship:

   .1   the Administration should thoroughly consider applications on a case by case basis, bearing in mind the particular circumstances. In doing so in the case of a ship built in a foreign country, the Administration may require a formal report from the authorities of the country in which the ship was built, stating that the delay was due to unforeseen circumstances beyond the control of the builder and the owner;

   .2   when a ship is treated as an existing ship upon such an application, the IOPP Certificate for the ship should be endorsed to indicate that the ship is accepted by the Administration as an existing ship; and

   .3   the Administration should report to the Organization on the identity of the ship and the grounds on which the ship has been accepted as an existing ship.

1.2.3   For the purpose of the application of regulation 13F, a ship for which the building contract (or keel laying) and delivery date were scheduled before the dates specified in regulation 13F(1), but which has been subject to delay in delivery, may under the same terms and conditions given in interpretations 1.2.1 and 1.2.2 for "existing ships", be accepted by the Administration as a ship to which regulation 13F does not apply.

Reg. 1(8)

1.3 *Major conversion*

1.3.1   The deadweight to be used for determining the application of provisions of Annex I is the deadweight assigned to an oil tanker at the time of the assignment of the load lines. If the load lines are reassigned for the purpose of altering the deadweight, without alteration of the structure of the ship, any substantial alteration of the deadweight consequential upon such reassignments should not be construed as "a major conversion" as defined in regulation 1(8).

However, the IOPP Certificate should indicate only one deadweight of the ship and be renewed on every reassignment of load lines.

1.3.2   If an existing crude oil tanker of 40,000 tons deadweight and above satisfying the requirements of COW changes its trade for the carriage of product oil* conversion to CBT or SBT and reissuing of the IOPP Certificate will be necessary (see paragraph 4.5 below). Such conversion should not be considered as a "major conversion" as defined in regulation 1(8).

1.3.3   When an oil tanker is used solely for the storage of oil and is subsequently put into service in the transportation of oil, such a change of function should not be construed as a "major conversion" as defined in regulation 1(8).

1.3.4 The conversion of an existing oil tanker to a combination carrier, or the shortening of a tanker by removing a transverse section of cargo tanks, should constitute a "major conversion" as defined in regulation 1(8).

1.3.5   The conversion of an existing oil tanker to a segregated ballast tanker by the addition of a transverse section of tanks should constitute a "major conversion" as defined in regulation 1(8) only when the cargo carrying capacity of the tanker is increased.

1.3.6   When a ship built as a combination carrier operates exclusively in the bulk cargo trade, the ship may be treated as a ship other than an oil tanker and Form A of the Record of Construction and Equipment should be issued to the ship. If such a ship operates in the oil trade and is equipped to comply with the requirements for an oil tanker, the ship should be certified as an oil tanker (combination carrier) and Form B of the Record of Construction and Equipment should be issued to the ship. The change of such a ship from the bulk trade to the oil trade should not be construed as a "major conversion" as defined in regulation 1(8).

Reg. 1(17)   1.4   *Definition of "segregated ballast"*

1.4.1   The segregated ballast system should be a system which is "completely separated from the cargo oil and fuel oil systems" as required by regulation 1(17). Nevertheless, provision may be made for emergency discharge of the segregated ballast by means of a connection to a cargo pump through a portable spool piece. In this case nonreturn valves should be fitted on the segregated ballast connections to prevent the passage of oil to the segregated ballast tanks. The portable spool piece should be mounted in a conspicuous position in the pump-room and a permanent notice restricting its use should be prominently displayed adjacent to it.

1.4.2   Sliding type couplings should not be used for expansion purposes where lines for cargo oil or fuel oil pass through tanks for segregated ballast, and where lines for segregated ballast pass through cargo oil or fuel oil tanks. This interpretation is applicable to

---

* "Product oil" means any oil other than crude oil as defined in regulation 1(28).

ships, the keel of which is laid, or which are at a similar stage of construction, on or after 1 July 1992.

Reg. 3

1.5 *Equivalents*

1.5.1 Acceptance by an Administration under regulation 3 of any fitting, material, appliance, or apparatus as an alternative to that required by Annex I includes type approval of pollution prevention equipment which is equivalent to that specified in resolution A.393(X).* An Administration that allows such type approval shall communicate particulars thereof, including the test results on which the approval of equivalency was based, to the Organization in accordance with regulation 3(2).

1.5.2 With regard to the term "appropriate action, if any" in regulation 3(2), any Party to the Convention that has an objection to an equivalency submitted by another Party should communicate this objection to the Organization and to the Party which allowed the equivalency within one year after the Organization circulates the equivalency to the Parties. The Party objecting to the equivalency should specify whether the objection pertains to ships entering its ports.

1A Survey and inspection

Reg. 4(1)(c)
and 4(3)(b)

1A.1 *Intermediate and annual survey for ships not required to hold an IOPP Certificate*

1A.1.1 The applicability of regulations 4(1)(c) and 4(3)(b)[†] to ships which are not required to hold an International Oil Pollution Prevention Certificate should be determined by the Administration.

2 Certificate

Reg. 5
and others

2.0 *Date of entry into force*

2.0.1 In the application of the Protocol of 1978 relating to the International Convention for the Prevention of Pollution from Ships, 1973 (1978 Protocol) the phrase "date of entry into force of the

---

* For oily-water separating equipment for machinery space bilges of ships, refer to the Guidelines and specifications for pollution prevention equipment for machinery space bilges adopted by the Marine Environment Protection Committee of the Organization by resolution MEPC.60(33), which, effective on 6 July 1993, superseded resolution A.393(X). For oil discharge monitoring and control systems installed on oil tankers built before 2 October 1986, refer to the Guidelines and specifications for oil discharge monitoring and control systems for oil tankers, and for oil discharge monitoring and control systems installed on oil tankers built on or after 2 October 1986, refer to the Revised guidelines and specifications for oil discharge monitoring and control systems, which were adopted by the Organization by resolutions A.496(XII) and A.586(14), respectively; see IMO sales publications IMO-608E and IMO-646E.

[†] Regulation 4(1)(c) was amended and the existing 4(3)(b) was removed by the HSSC amendments of MEPC.39(29), which entered into force on 3 February 2000.

present Convention" should be construed to mean the date of entry into force of the 1978 Protocol, which was 2 October 1983.

2.1 *Designation of the type of oil tankers*

2.1.1 Oil tankers must be designated on the IOPP Certificate as either "crude oil tanker", "product carrier" or "crude oil/product carrier". Furthermore, the requirements contained in regulations 13 to 13E differ for new and existing "crude oil tankers" and "product carriers", and compliance with these provisions is recorded on the IOPP Certificate. Oil trades in which different types of oil tankers are allowed to be engaged are as follows:

.1 *Crude oil/product carrier* is allowed to carry either crude oil or product oil, or both simultaneously;

.2 *Crude oil tanker* is allowed to carry crude oil but is prohibited from carrying product oil; and

.3 *Product carrier* is allowed to carry product oil but is prohibited from carrying crude oil.

2.1.2 In determining the designation of the type of oil tanker on the IOPP Certificate based on the compliance with the provisions for SBT, PL, CBT and COW, the following standards should apply.

2.1.3 *New oil tankers* of less than 20,000 tons deadweight*

2.1.3.1 These oil tankers may be designated as "crude oil/product carriers".

2.1.4 *New oil tankers* of 20,000 tons deadweight and above*

2.1.4.1 Oil tankers satisfying the requirements for SBT + PL + COW may be designated as "crude oil/product carrier".

2.1.4.2 Oil tankers satisfying the requirements for SBT + PL but not COW should be designated as "product carrier".

2.1.4.3 Oil tankers of 20,000 tons deadweight and above but less than 30,000 tons deadweight *not* fitted with SBT + PL should be designated as "product carrier".

2.1.5 *"New" oil tankers† of 70,000 tons deadweight and above*

2.1.5.1 These oil tankers satisfying the requirements for SBT may be designated as "crude oil/product carrier".

---

* As defined in regulation 1(26).

† *"New" oil tankers* in this case means oil tankers of 70,000 tons deadweight and above built after the dates specified in regulation 1(6) but before the dates specified in regulation 1(26). The term "built" in this context means building contract or keel laying or delivery as defined in paragraph (a) or (b) or (c) of regulation 1(6).

2.1.6 *Existing oil tankers** of less than 40,000 tons deadweight*

2.1.6.1   These oil tankers may be designated as "crude oil/product carrier".

2.1.7 *Existing oil tankers** of 40,000 tons deadweight and above*

2.1.7.1   Oil tankers satisfying the requirements for SBT should be designated as "crude oil/product carrier".

2.1.7.2   Oil tankers satisfying the requirements for COW only should be designated as "crude oil tanker".

2.1.7.3   Oil tankers satisfying the requirements for CBT should be designated as "crude oil/product carrier". Such designation should be valid until the expiry date of the IOPP Certificate, which should be H + 2 (see the definition of terms) for oil tankers of 70,000 tons deadweight and above and H + 4 for oil tankers of 40,000 tons deadweight and above but less than 70,000 tons deadweight.

2.1.7.4   After the above expiry date of the certificate, such an oil tanker should be designated as follows:

.1   if it continues to operate with CBT, the oil tanker should be designated as "product carrier";

.2   if it is provided with COW *only*, the oil tanker should be designated as "crude oil tanker";

.3   if it is provided with SBT, the oil tanker should be designated as "crude oil/product carrier"; and

.4   if it is provided with CBT + COW, the tanker should be designated as "crude oil/product carrier" (see paragraph 4.5 below).

Reg. 5(1)   2.2 *IOPP Certificate for existing oil tankers*

2.2.1   Under regulation 5(1) the issue of the IOPP Certificate to existing ships is not mandatory until twelve months have elapsed after the date of entry into force of MARPOL 73/78. It is, however, advisable for existing oil tankers of 40,000 tons deadweight and above to carry the IOPP Certificate or an appropriate document issued by the Administration upon entry into force of MARPOL 73/78 which can be presented to the control officers at foreign ports or terminals.

2.3 *Validity of IOPP Certificate issued before the entry into force of the Convention*

2.3.1   Where ships are surveyed and IOPP Certificates issued before the entry into force of the Convention, the period of validity of such Certificates should be calculated from the date of their issue.

---

* As defined in regulation 1(27).

156

Unified Interpretations

2.4 *IOPP Certificate for crude oil/product carriers with CBT and COW*

2.4.1   When an oil tanker with CBT and COW is surveyed for the conversion from a crude oil tanker operating with COW to a product carrier operating with CBT or vice versa (see paragraph 4.5.2.1), another IOPP Certificate should be issued for a period not exceeding the remaining period of validity of the existing Certificate, unless the survey is as comprehensive as the periodical survey* required by regulation 4(1)(b) (see also paragraph 4.5.2.2).

2.4.2   The endorsement of surveys made on the existing Certificate should be recorded on another IOPP Certificate issued as above.

Reg. 7

2.4A *New form of IOPP Certificate or its Supplement*

2.4A.1   In the case where the form of the IOPP Certificate or its Supplement is amended, the existing form of the Certificate or Supplement which is current when the amendment enters into force may remain valid until the expiry of that Certificate, provided that, at the first survey after the date of entry into force of the amendment, necessary changes are indicated in the existing Certificate or Supplement by means of suitable corrections, e.g. striking over the invalid entry and typing the new entry.

Reg. 8

2.5 *Revalidation of an IOPP Certificate*

2.5.1   Where the survey required in regulation 4 of Annex I of MARPOL 73/78 is not carried out within the period specified in that regulation, the IOPP Certificate ceases to be valid. When a survey corresponding to the requisite survey is carried out subsequently, the validity of the Certificate may be restored without altering the expiry date of the original Certificate and the Certificate endorsed to this effect. The thoroughness and stringency of such survey will depend on the period for which the prescribed survey has elapsed and the conditions of the ship.

3   Controls of discharge of oil

Reg. 9(1)

3.1 *Discharges from machinery space bilges of oil tankers*

3.1.1   The wording "from machinery space bilges excluding cargo pump-room bilges of an oil tanker unless mixed with oil cargo residue" in regulation 9(1)(b) should be interpreted as follows:

.1   Regulation 9(1)(a) applies to:

.1.1   discharges of oil or oily mixture from machinery space bilges of oil tankers where mixed with cargo oil residue or when transferred to slop tanks; and

.1.2   discharges from cargo pump-room bilges of oil tankers.

.2   Regulation 9(1)(b) applies to discharges from machinery space bilges of oil tankers other than those referred to above.

---

* The text of regulation 4(1)(b) has been amended by the HSSC amendments of resolution MEPC.39(29).

3.1.2   The above interpretation should not be construed as relaxing any existing prohibition of piping arrangements connecting the engine-room and slop tanks which may permit cargo to enter the machinery spaces. Any arrangements provided for machinery space bilge discharges into slop tanks should incorporate adequate means to prevent any backflow of liquid cargo or gases into the machinery spaces. Any such arrangements do not constitute a relaxation of the requirements of regulation 16 with respect to oil discharge monitoring and control systems and oily-water separating equipment.

3.2   *Total quantity of discharge*

3.2.1   The phrase "the total quantity of the particular cargo of which the residue formed a part" in regulation 9(1)(a)(v) relates to the total quantity of the particular cargo which was carried on the previous voyage and should not be construed as relating only to the total quantity of cargo which was contained in the cargo tanks into which water ballast was subsequently loaded.

Reg. 9(4)

3.3   *Discharges from ships of 400 tons gross tonnage and above but less than 10,000 tons gross tonnage within 12 miles from the nearest land* [Deleted]

Reg. 10(3) as amended

3.4   *Automatic stopping device required by regulation 10(3) as amended*

3.4.1   Regulation 10(3)(b)(vi) requires a stopping device which will ensure that the discharge is automatically stopped when the oil content of the effluent exceeds 15 ppm. Since, however, this is not a requirement of regulation 16, ships less than 10,000 tons gross tonnage need not be required to be equipped with such stopping device if no effluent from machinery space bilges is discharged within special areas. Conversely, the discharge of effluent within special areas from ships without an automatic stopping device is a contravention of the Convention even if the oil content of the effluent is below 15 ppm.

Reg. 12(2)

3.5   *Adequate reception facilities for substances regulated by regulation 15(7)*

3.5.1   Unloading ports receiving substances regulated by regulation 15(7) (which include *inter alia* high-density oils) should have adequate facilities dedicated for such products, allowing the entire tank-cleaning operation to be carried out in the port, and should have adequate reception facilities for the proper discharge and reception of cargo residues and solvents necessary for the cleaning operations in accordance with paragraph 6.5.2.

4   SBT, CBT, COW and PL requirements

Reg. 13(3) as amended

4.1   *Capacity of SBT*

4.1.1   For the purpose of application of regulation 13(3)(b), as amended, the following operations of oil tankers are regarded as falling within the category of exceptional cases:

    .1   when combination carriers are required to operate beneath loading or unloading gantries;

.2   when tankers are required to pass under a low bridge;

.3   when local port or canal regulations require specific draughts for safe navigation; and

.4   when loading and unloading arrangements require the tanker to be at a draught deeper than that achieved when all segregated ballast tanks are full.

Reg. 13(4)   4.2   *Application of regulation 13(4) to new oil tankers of 70,000 tons deadweight and above*

4.2.1   New oil tankers referred to in regulation 13(4) should be taken to mean oil tankers constructed or converted after the dates specified in regulation 1(26). It is not therefore mandatory for crude oil tankers of 70,000 tons deadweight and above, built after the date specified in regulation 1(6) but before the date specified in regulation 1(26), to install COW, and such oil tankers are not subject to the provisions of regulation 13(4).

Reg. 13(5)   4.3   *Segregated ballast conditions for oil tankers less than 150 metres in length*

4.3.1   In determining the minimum draught and trim of oil tankers less than 150 metres in length to be qualified as SBT oil tankers, the Administration should follow the guidance set out in appendix 1 hereto.*

4.3.2   The formulae set out in appendix 1 replace those set out in regulation 13(2), and these oil tankers should also comply with the conditions laid down in regulations 13(3) and (4) in order to be qualified as SBT oil tankers.

Reg. 13(8)   4.4   *Capacity of CBT*

4.4.1   For the purposes of determining the capacity of CBT, the following tanks may be included:

.1   segregated ballast tanks; and

.2   cofferdams and fore and after peak tanks, provided that they are exclusively used for the carriage of ballast water and are connected with permanent piping to ballast water pumps.

Reg. 13(9)   4.5   *Existing oil tankers with CBT and COW*
13(10)
4.5.1   Existing oil tankers which are fitted with CBT and COW and designated as "crude oil/product carriers" in the IOPP Certificate (see paragraph 2.1.7.4.4) should, after the expiry of the date specified in regulation 13(9), operate as follows:

.1   They should always operate with CBT when carrying crude oil or product oil or both simultaneously, and neither crude oil nor product oil should be carried in dedicated clean ballast tanks; and

---

* See appendix 1 to Unified Interpretations.

.2   When carrying crude oil and product oil simultaneously, or only crude oil, they should operate also with COW for sludge control.

4.5.2   If a crude oil tanker operating with COW is to change its designation to a product carrier operating with CBT, or vice versa, the following provisions shall apply:

.1   If the tanker has *common* piping and pump arrangements for ballast and cargo handling of the CBT, such tanker should be surveyed and a new IOPP Certificate should be issued. Such survey should ensure that cargo oil tanks to be designated as CBT have been thoroughly cleaned and ballast water which CBT will take can be treated as clean ballast as defined in regulation 1(16).

.2   If the tanker has *separate independent* piping and pump arrangements for ballasting the CBT, the Administration may issue to such a tanker two IOPP Certificates, the tanker being designated "crude oil tanker" on one of the Certificates and "product carrier" on the other. Only one of these Certificates which corresponds to the particular operation of the tanker should be valid at a time, but entries should be made on each of the Certificates in the remarks column as to the existence of the other Certificate. Such tanker need not be surveyed prior to each conversion of trade. When carrying only crude oil such tanker should be allowed to carry crude oil in those tanks which were designated as CBT when carrying product. When carrying only product no cargo should be carried in the CBT. The approved CBT and COW Manuals must include a chapter describing procedures necessary for the conversion from crude oil service to product service and vice versa.

Reg. 13   4.6   *Oil tankers used for the storage of oil*

4.6.1   When an oil tanker is used as a floating storage unit (FSU) or floating production storage and offloading facility (FPSO) which is used solely for the storage or storage and production of oil and is moored on a fixed location except in extreme environmental or emergency conditions, such a unit is not required to comply with the provisions of regulations 13 to 13G, unless as specified in whole or in part by the coastal State.

4.6.2   When an oil tanker is used as a floating facility to receive dirty ballast discharged from oil tankers, such a tanker is not required to comply with the provisions of regulations 13 to 13G.

Reg. 13A(3)   4.7   *Installation of oil content meter for CBT tankers*

4.7.1   The phrase "first scheduled shipyard visit" in regulation 13A(3) should be interpreted to mean that the oil content meter must be installed not later than at the first scheduled shipyard visit when cargo tanks are gas-freed and in any case not later than three years after the date of entry into force of MARPOL 73/78 as required by regulation 15(1).

4.7.2   It should be noted that ships built after the dates specified in regulation 1(6) but before the dates specified in regulation 1(26) are treated as new ships as far as the application of regulation 15(3) is concerned. Consequently these ships must be fitted with the required oil discharge monitoring and control systems upon entry into force of the Convention.

4.8   *CBT oil content meter*

4.8.1   The discharge of ballast from the dedicated clean ballast tanks should be continuously monitored (but not necessarily recorded) by the oil content meter required by regulation 13A(3) so that the oil content, if any, in the ballast water can be observed from time to time. This oil content meter is not required to come into operation automatically.

Reg. 13B

4.9   *COW system fitted voluntarily*

4.9.1   A COW system fitted on an oil tanker as an addition to the requirements of MARPOL 73/78 should at least comply with those provisions of the revised COW Specifications relating to safety.

Reg. 13E

4.10   *Application of PL requirements to oil tankers of 70,000 tons deadweight and above*

4.10.1   Oil tankers of 70,000 tons deadweight and above built after the dates specified in regulation 1(6) but before the dates specified in regulation 1(26) must be provided with SBT but they need not be protectively located in accordance with regulation 13E.

4.11   *Protective location of SBT*

4.11.1   The measurement of the minimum width of wing tanks and of the minimum vertical depth of double bottom tanks should be taken and value of protective areas (PA_c and PA_s) should be calculated in accordance with the Interim Recommendation for a Unified Interpretation of regulation 13E – Protective Location of Segregated Ballast Spaces – set out in appendix 2 hereto.[*]

4.11.2   Ships being built in accordance with this interpretation should be regarded as meeting the requirements of regulation 13E and would not need to be altered if different requirements were to result from a later interpretation.

4.11.3   If, in the opinion of the Administration, any oil tanker the keel of which was laid or which was at a similar stage of construction before 1 July 1980 complies with the requirements of regulation 13E without taking into account the above Interim Recommendation, the Administration may accept such tanker as complying with regulation 13E.

---

[*] See appendix 2 to Unified Interpretations.

Reg. 13F(3)(d) 4.12 *Aggregate capacity of ballast tanks*

4.12.1 Any ballast carried in localized inboard extensions, indentations or recesses of the double hull, such as bulkhead stools, should be excess ballast above the minimum requirement for segregated ballast capacity according to regulation 13 of Annex I of MARPOL 73/78.

4.12.2 In calculating the aggregate capacity under regulation 13F(3)(d), the following should be taken into account:

.1 the capacity of engine-room ballast tanks should be excluded from the aggregate capacity of ballast tanks;

.2 the capacity of ballast tank located inboard of double hull should be excluded from the aggregate capacity of ballast tanks (see figure 1).

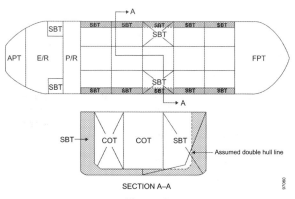

SECTION A–A

Figure 1

.3 spaces such as void spaces located in the double hull within the cargo tank length should be included in the aggregate capacity of ballast tanks (see figure 2).

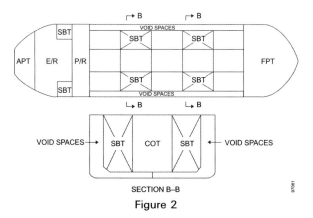

SECTION B–B

Figure 2

162

Annex I

Reg. 13F(5)   [See appendix 7 to Unified Interpretations]

Reg. 13G(4)   4.13 *Wing tanks and double bottom spaces
                         of existing oil tankers used for water ballast*

4.13.1   If the wing tanks and double bottom tanks referred to in regulation 13G(4) are used for water ballast, the ballast arrangement should at least be in compliance with the Revised Specifications for Oil Tankers with Dedicated CBT (resolution A.495(XII)).[*]

Reg. 13G(7)   [See appendices 8 and 9 to Unified Interpretations]

5   Fuel oil

Reg. 14(2)   5.1 *Large quantities of oil fuel*

5.1.1   The phrase "large quantities of oil fuel" in regulation 14(2) was formulated in drafting MARPOL 73/78 to take account of those ships which are required to stay at sea for extended periods because of the particular nature of their operation and trade. Under the circumstances considered these ships would be required to fill their empty oil fuel tanks with water ballast in order to maintain sufficient stability and safe navigation conditions.

5.1.2   Such ships may include *inter alia* certain large fishing vessels or ocean-going tugs. Certain other types of ships which for reasons of safety, such as stability, may be required to carry ballast in oil fuel tanks may also be included in this category.

Reg. 14(3)   5.2 *Application of regulation 14(3)*

5.2.1   The phrase "all other ships" in regulation 14(3) should include:

.1   new ships other than oil tankers of less than 4,000 tons gross tonnage;

.2   new oil tankers of less than 150 tons gross tonnage; and

.3   all existing ships irrespective of tonnage.

5.2.2   When the separation of oil fuel tanks and water ballast tanks is unreasonable or impracticable for ships mentioned in paragraph 5.2.1 above, ballast water may be carried in oil fuel tanks, provided that such ballast water is discharged into the sea in compliance with regulation 9(1)(b), 10(2) or 10(3) or into reception facilities in compliance with regulation 10(4).

6   Retention of oil on board

Reg. 1(4)   6.1 *Equivalent provisions for the carriage of oil by
15(2)   a chemical tanker*
15(3)(b)

6.1.1   Under regulation 1(4) of Annex I of MARPOL 73/78 any chemical tanker when carrying a cargo or part cargo of oil in bulk is

---

[*] See IMO sales publication IMO-619E.

Unified Interpretations

defined as an oil tanker and consequently must comply with the requirements of Annex I applicable to oil tankers. Such a tanker, if it is impracticable for it to be provided with slop tank arrangements in compliance with regulation 15(2) and oil/water interface detectors in accordance with regulation 15(3)(b), should comply with the equivalent provisions set out in appendix 3.[*]

Reg. 15(2)(c) as amended

6.2 *Tanks with smooth walls*

6.2.1 The term "tanks with smooth walls" should be taken to include the main cargo tanks of oil/bulk/ore carriers which may be constructed with vertical framing of a small depth. Vertically corrugated bulkheads are considered smooth walls.

Reg. 15(3)(b)

6.3 *Oil/water interface detectors*

6.3.1 In the case of existing tankers, the oil/water interface detector referred to in regulation 15(3)(b) should be provided no later than on the date of entry into force of MARPOL 73/78.

Reg. 15(5) 16(3)(a)

6.4 *Conditions for waiver*

6.4.1 The International Oil Pollution Prevention Certificate, when required, should contain sufficient information to permit the port State to determine if the ship complies with the waiver conditions regarding the phrase "restricted voyages as determined by the Administration". This may include a list of ports, the maximum duration of the voyage between ports having reception facilities, or similar conditions as established by the Administration.

Reg. 15(5)

6.4.1A The time limitation "of 72 hours or less in duration" in regulation 15(5)(b)(ii)(2) should be counted:

    (a)   from the time the tanker leaves the special area, when a voyage starts within a special area, or

    (b)   from the time the tanker leaves a port situated outside the special area to the time the tanker approaches a special area.

6.4.2 The phrase "all oily mixtures" in regulation 15(5)(a) and 15(5)(b)(ii)(3) includes all ballast water and tank washing residues from cargo oil tanks.

Reg. 15(7)

6.5 *Annex I substances which through their physical properties inhibit effective product/water separation and monitoring*

6.5.1 The Government of the receiving Party should establish appropriate measures in order to ensure that the provisions of 6.5.2 are complied with.

6.5.2 A tank which has been unloaded should, subject to the provisions of 6.5.3, be washed and all contaminated washings

---

[*] See appendix 3 to Unified Interpretations.

should be discharged to a reception facility before the ship leaves the port of unloading for another port.

6.5.3   At the request of the ship's master, the Government of the receiving Party may exempt the ship from the requirements referred to in 6.5.2, where it is satisfied that:

.1   the tank unloaded is to be reloaded with the same substance or another substance compatible with the previous one and that the tanker will not be washed or ballasted prior to loading;

.2   the tank unloaded is neither washed nor ballasted at sea if the ship is to proceed to another port unless it has been confirmed in writing that a reception facility at that port is available and adequate for the purpose of receiving the residues and solvents necessary for the cleaning operations.

6.5.4   An exemption referred to in 6.5.3 should only be granted by the Government of the receiving Party to a ship engaged in voyages to ports or terminals under the jurisdiction of other Parties to the Convention. When such an exemption has been granted it should be certified in writing by the Government of the receiving Party.

6.5.5   In the case of ships retaining their residues on board and proceeding to ports or terminals under the jurisdiction of other Parties to the Convention, the Government of the receiving Party is advised to inform the next port of call of the particulars of the ship and cargo residues, for their information and appropriate action for the detection of violations and enforcement of the Convention.

7   Oil discharge monitoring and control system and oil filtering equipment

Reg. 16(1)

7.1   *Control of discharge of ballast water from oil fuel tanks*

7.1.1   The second sentence of regulation 16(1) should be interpreted as follows:

.1   Any ship of 400 tons gross tonnage and above but less than 10,000 tons gross tonnage:

.1.1   which does not carry water ballast in oil fuel tanks should be fitted with 15 ppm oil filtering equipment for the control of discharge of machinery space bilges;

.1.2   which carries water ballast in oil fuel tanks should be fitted with the equipment required by regulation 16(2) for the control of machinery space bilges and dirty ballast water from oil fuel tanks. Ships on which it is not reasonable to fit this equipment should retain on board dirty ballast water from oil fuel tanks and discharge it to reception facilities.

7.1.2 The above equipment should be of adequate capacity to deal with the quantities of effluent to be discharged.

Reg. 16(1)
16(2)

7.2 *Oil filtering equipment*

7.2.1 Oil filtering equipment referred to in regulation 16(1) and 16(2) may include any combination of a separator, filter or coalescer and also a single unit designed to produce an effluent with oil content not exceeding 15 ppm.

Reg. 16(3)(a)

7.3 *Waivers for restricted voyages*

7.3.1 The International Oil Pollution Prevention Certificate, when required, should contain sufficient information to permit the port State to determine if the ship complies with the waiver conditions regarding the phrase "restricted voyages as determined by the Administration". This may include a list of ports, the maximum duration of the voyage between ports having reception facilities, or similar conditions as established by the Administration.

Reg. 16(6)

7.4 *Automatic stopping device for existing ships having operated with 15 ppm oil filtering equipment*

7.4.1 The requirements for the fitting of an automatic stopping device under regulation 16(2) need not be applied to existing ships until 6 July 1998, if such ships are fitted with 15 ppm oil filtering equipment.

8 Sludge tanks

Reg. 17(1)

8.1 *Capacity of sludge tanks*

8.1.1 To assist Administrations in determining the adequate capacity of sludge tanks, the following criteria may be used as guidance. These criteria should not be construed as determining the amount of oily residues which will be produced by the machinery installation in a given period of time. The capacity of sludge tanks may, however, be calculated upon any other reasonable assumptions. For a ship the keel of which is laid or which is at a similar stage of construction on or after 31 December 1990, the guidance given in items .4 and .5 below should be used in lieu of the guidance contained in items .1 and .2.

.1 For ships which do not carry ballast water in oil fuel tanks, the minimum sludge tank capacity (V_1) should be calculated by the following formula:

$$V_1 = K_1CD \ (\text{m}^3)$$

where: K_1 = 0.01 for ships where heavy fuel oil is purified for main engine use, or 0.005 for ships using diesel oil or heavy fuel oil which does not require purification before use,

C = daily fuel oil consumption (metric tons),

D = maximum period of voyage between ports where sludge can be discharged ashore (days). In the absence of precise data a figure of 30 days should be used.

.2 When such ships are fitted with homogenizers, sludge incinerators or other recognized means on board for the control of sludge, the minimum sludge tank capacity (V_1) should, in lieu of the above, be:

V_1 = 1 m^3 for ships of 400 tons gross and above but less than 4,000 tons gross tonnage, or 2 m^3 for ships of 4,000 tons gross tonnage and above.

.3 For ships which carry ballast water in fuel oil tanks, the minimum sludge tank capacity (V_2) should be calculated by the following formula:

$$V_2 = V_1 + K_2 B \ (\text{m}^3)$$

where: V_1 = sludge tank capacity specified in .1 or .2 above,

K_2 = 0.01 for heavy fuel oil bunker tanks, or 0.005 for diesel oil bunker tanks,

B = capacity of water ballast tanks which can also be used to carry oil fuel (metric tons).

.4 For ships which do not carry ballast water in fuel oil tanks, the minimum sludge tank capacity (V_1) should be calculated by the following formula:

$$V_1 = K_1 CD \ (\text{m}^3)$$

where: K_1 = 0.015 for ships where heavy fuel oil is purified for main engine use or 0.005 for ships using diesel oil or heavy fuel oil which does not require purification before use,

C = daily fuel oil consumption (m^3),

D = maximum period of voyage between ports where sludge can be discharged ashore (days). In the absence of precise data, a figure of 30 days should be used.

.5 For ships fitted with homogenizers, sludge incinerators or other recognized means on board for the control of sludge, the minimum sludge tank capacity should be:

.5.1 50% of the value calculated according to item .4 above; or

.5.2 1 m^3 for ships of 400 gross tonnage and above but less than 4,000 gross tonnage or 2 m^3 for ships of 4,000 gross tonnage and above; whichever is the greater.

8.1.2 Administrations should establish that in a ship the keel of which is laid or which is at a similar stage of construction on or after 31 December 1990, adequate tank capacity, which may include the sludge tank(s) referred to under 8.1.1 above, is available also for leakage, drain and waste oils from the machinery installations. In existing installations this should be taken into consideration as far as reasonable and practicable.

Reg. 17(2)   8.2   *Cleaning of sludge tanks and discharge of residues*

8.2.1   To assist Administrations in determining the adequacy of the design and construction of sludge tanks to facilitate their cleaning and the discharge of residues to reception facilities, the following guidance is provided, having effect on ships the keel of which is laid or which is at a similar stage of construction on or after 31 December 1990:

> .1   sufficient man-holes should be provided such that, taking into consideration the internal structure of the sludge tanks, all parts of the tank can be reached to facilitate cleaning;
>
> .2   sludge tanks in ships operating with heavy oil, that needs to be purified for use, should be fitted with adequate heating arrangements or other suitable means to facilitate the pumpability and discharge of the tank content;
>
> .3   there should be no interconnections between the sludge tank discharge piping and bilge-water piping other than possible common piping leading to the standard discharge connection referred to in regulation 19. However, arrangements may be made for draining of settled water from the sludge tanks by means of manually operated self-closing valves or equivalent arrangements; and
>
> .4   the sludge tank should be provided with a designated pump for the discharge of the tank content to reception facilities. The pump should be of a suitable type, capacity and discharge head, having regard to the characteristics of the liquid being pumped and the size and position of tank(s) and the overall discharge time.

Reg. 17(3)   8.3   *Overboard connection of sludge tanks*

8.3.1   Ships with existing installations having piping to and from sludge tanks to overboard discharge outlets, other than the standard discharge connection referred to in regulation 19, may comply with regulation 17(3) by the installation of blanks in this piping.

9   Pumping and piping arrangements

Reg. 18(2)   9.1   *Piping arrangements for discharge above the waterline*
as amended
9.1.1   Under regulation 18(2), pipelines for discharge to the sea above the waterline must be led either:

> .1   to a ship's discharge outlet located above the waterline in the deepest ballast condition; or
>
> .2   to a midship discharge manifold or, where fitted, a stern or bow loading/discharge facility above the upper deck.

9.1.2   The ship's side discharge outlet referred to in 9.1.1.1 should be so located that its lower edge will not be submerged when the ship carries the maximum quantity of ballast during its ballast voyages, having regard to the type and trade of the ship. The discharge outlet located above the waterline in the following ballast condition will be accepted as complying with this requirement:

.1   on oil tankers not provided with SBT or CBT, the ballast condition when the ship carries both normal departure ballast and normal clean ballast simultaneously;

.2   on oil tankers provided with SBT or CBT, the ballast condition when the ship carries ballast water in segregated or dedicated clean ballast tanks, together with additional ballast in cargo oil tanks in compliance with regulation 13(3).

9.1.3   The Administration may accept piping arrangements which are led to the ship's side discharge outlet located above the departure ballast waterline but not above the waterline in the deepest ballast condition, if such arrangements have been fitted before 1 January 1981.

9.1.4   Although regulation 18(2) does not preclude the use of the facility referred to in 9.1.1.2 for the discharge of ballast water, it is recognized that the use of this facility is not desirable, and it is strongly recommended that ships be provided with either the side discharge outlets referred to in 9.1.1.1 or the part flow arrangements referred to in regulation 18(6)(e).

**Reg. 18(4)(b)
as amended**

9.2   *Small diameter line*

9.2.1   For the purpose of application of regulation 18(4)(b), the cross-sectional area of the small diameter line should not exceed:

.1   10% of that of a main cargo discharge line for new oil tankers or existing oil tankers not already fitted with a small diameter line; or

.2   25% of that of a main cargo discharge line for existing oil tankers already fitted with such a line.

(See paragraph 4.4.5 of the revised COW Specifications contained in resolution A.446(XI)).[*]

Reg. 18(4)(b)

9.3   *Connection of small diameter line to the manifold valve*

9.3.1   The phrase "connected outboard of" with respect to the small diameter line for discharge ashore should be interpreted to mean a connection on the downstream side of the tanker's deck manifold valves, both port and starboard, when the cargo is being discharged.

This arrangement would permit drainage back from the tanker's cargo lines to be pumped ashore with the tanker's manifold valves

---

[*] See IMO sales publication IMO-617E.

Annex I

Unified Interpretations

closed through the same connections as for main cargo lines (see the sketch shown in appendix 4).*

Reg. 18(6)(e)(ii) as amended

9.4 *Part flow system specifications*

9.4.1   The Specifications for the Design, Installation and Operation of a Part Flow System for Control of Overboard Discharges referred to in regulation 18(6)(e)(ii) is set out in appendix 5.[†]

10 Requirements for drilling rigs and other platforms

Reg. 21 Art. 2(3)(b)(ii)

10.1 *Application of MARPOL 73/78*

10.1.1   There are four categories of discharges associated with the operation of offshore platforms when engaged in the exploration and exploitation of mineral resources, i.e.:

.1   machinery space drainage;

.2   offshore processing drainage;

.3   production water discharge; and

.4   displacement water discharge.

Only the discharge of machinery space drainage should be subject to MARPOL 73/78 (see the diagram shown in appendix 6).[‡]

10.1.2   When an oil tanker is used as a floating storage unit (FSU) or floating production storage and offloading facility (FPSO) referred to in Unified Interpretation 4.6.1 it is to be regarded as an "other platform" for the purpose of the discharge requirements of regulation 21.

11 Tank size limitation and damage stability

Reg. 22(1)(b)

11.1 *Bottom damage assumptions*

11.1.1   When applying the figures for bottom damage within the forward part of the ship as specified in regulation 22(1)(b) for the purpose of calculating both oil outflow and damage stability, $0.3L$ from the forward perpendicular should be the aftermost point of the extent of damage.

Reg. 23

11.2 *Hypothetical oil outflow for combination carriers*

11.2.1   For the purpose of calculation of the hypothetical oil outflow for combination carriers:

.1   the volume of a cargo tank should include the volume of the hatchway up to the top of the hatchway coamings, regardless of the construction of the hatch, but may not include the volume of any hatch cover; and

---

* See appendix 4 to Unified Interpretations.

† See appendix 5 to Unified Interpretations.

‡ See appendix 6 to Unified Interpretations.

Annex I

.2   for the measurement of the volume to moulded lines, no deduction should be made for the volume of internal structures.

Reg. 23(1)(b)   11.3   *Calculation of hypothetical oil outflow*

11.3.1   In a case where the width b_i is not constant along the length of a particular wing tank, the smallest b_i value in the tank should be used for the purposes of assessing the hypothetical outflows of oil O_c and O_s.

Reg. 25(1)   11.4   *Operating draught*

11.4.1   With regard to the term "any operating draught reflecting actual partial or full load conditions", the information required should enable the damage stability to be assessed under conditions the same as or similar to those under which the ship is expected to operate.

Reg. 25(2)   11.5   *Suction wells*

11.5.1   For the purpose of determining the extent of assumed damage under regulation 25(2), suction wells may be neglected, provided such wells are not excessive in area and extend below the tank for a minimum distance and in no case more than half the height of the double bottom.

Reg. 25A(2)   **11A   Intact stability**

11A.1   The vessel should be loaded with all cargo tanks filled to a level corresponding to the maximum combined total of vertical moment of volume plus free surface inertia moment at 0° heel, for each individual tank. Cargo density should correspond to the available cargo deadweight at the displacement at which transverse KM reaches a minimum value, assuming full departure consumables and 1% of the total water ballast capacity. The maximum free surface moment should be assumed in all ballast tanks. For the purpose of calculating GM_o, liquid free surface corrections should be based on the appropriate upright free surface inertia moment. The righting lever curve may be corrected on the basis of liquid transfer moments.

12   Shipboard oil pollution emergency plan

Reg. 26(1)   12.1   *Definition of new ships*

12.1.1   The phrase "ships built" referred to in the last sentence of regulation 26(1) should be taken to mean "ships delivered".

12.2   *Equivalent provision for application of requirement for oil pollution emergency plans*

12.2.1   Any fixed or floating drilling rig or other offshore installation when engaged in the exploration, exploitation or associated offshore processing of sea-bed mineral resources, which has an oil pollution emergency plan co-ordinated with, and approved in accordance with procedures established by, the coastal State, should be regarded as complying with regulation 26.

Unified Interpretations

Appendices to Unified Interpretations of Annex I

Appendix 1

Guidance to Administrations concerning draughts recommended for segregated ballast tankers below 150 m in length

Introduction

1   Three formulations are set forth as guidance to Administrations concerning minimum draught requirements for segregated ballast tankers below 150 m in length.

2   The formulations are based both on the theoretical research and surveys of actual practice on tankers of differing configuration reflecting varying degrees of concern with propeller emergence, vibration, slamming, speed loss, rolling, docking and other matters. In addition, certain information concerning assumed sea conditions is included.

3   Recognizing the nature of the underlying work, the widely varying arrangement of smaller tankers and each vessel's unique sensitivity to wind and sea conditions, no basis for recommending a single formulation is found.

Caution

4   It must be cautioned that the information presented should be used as general guidance for Administrations. With regard to the unique operating requirements of a particular vessel, the Administration should be satisfied that the tanker has sufficient ballast capacity for safe operation. In any case the stability should be examined independently.

5   *Formulation A*

    .1   mean draught (m) = $0.200 + 0.032L$

    .2   maximum trim   = $(0.024 - 6 \times 10^{-5}L)L$

6   These expressions were derived from a study of 26 tankers ranging in length from 50 to 150 m. The draughts, in some cases, were abstracted from ship's trim and stability books and represent departure ballast conditions. The ballast conditions represent sailing conditions in weather up to and including Beaufort 5.

7   *Formulation B*

.1	minimum draught at bow (m)	=	$0.700 + 0.0170L$
.2	minimum draught at stern (m)	=	$2.300 + 0.030L$
		or	
.3	minimum mean draught (m)	=	$1.550 + 0.023L$
.4	maximum trim	=	$1.600 + 0.013L$

8   These expressions resulted from investigations based on theoretical research, model and full scale tests. These formulae are based on a Sea 6 (International Sea Scale).

9   *Formulation C*

.1	minimum draught aft (m)	=	$2.0000 + 0.0275L$
.2	minimum draught forward (m)	=	$0.5000 + 0.0225L$

10   These expressions provide for certain increased draughts to aid in the prevention of propeller emergence and slamming in higher length ships.

Appendix 2

Interim recommendation for
a unified interpretation of regulation 13E

1   Regulation 13E(4) of Annex I of MARPOL 73/78 relating to the measurement of the 2 m minimum width of wing tanks and the measurement of the minimum vertical depth of double bottom tanks of 2 m or $B/15$ in respect of tanks at the ends of the ship where no identifiable bilge area exists should be interpreted as given hereunder. No difficulty exists in the measurement of the tanks in the parallel middle body of the ship where the bilge area is clearly identified. The regulation does not explain how the measurements should be taken.

2   The minimum width of wing tanks should be measured at a height of $D/5$ above the base line providing a reasonable level above which the 2 m width of collision protection should apply, under the assumption that in all cases $D/5$ is above the upper turn of bilge amidships (see figure 1). The minimum height of double bottom tanks should be measured at a vertical plane measured $D/5$ inboard from the intersection of the shell with a horizontal line $D/5$ above the base line (see figure 2).

3   The PA_c value for a wing tank which does not have a minimum width of 2 m throughout its length would be zero; no credit should be given for that part of the tank in which the minimum width is in excess of 2 m. No credit should be given in the assessment of PA_s to any double bottom tank, part of which does not meet the minimum depth requirements anywhere within its length. If, however, the projected dimensions of the bottom of the cargo tank above the double bottom fall entirely within the area of the double bottom tank or space which meets the minimum height requirement and provided the side bulkheads bounding the cargo tank above are vertical or have a slope of not more than 45° from the vertical, credit may be given to the part of the double bottom tank defined by the projection of the cargo tank bottom. For similar cases where the wing tanks above the double bottom are segregated ballast tanks or void spaces, such credit may also be given. This would not, however, preclude in the above cases credit being given to a PA_s value in the first case and to a PA_c value in the second case where the respective vertical or horizontal protection complies with the minimum distances prescribed in regulation 13E(4).

4   Projected dimensions should be used as shown in examples of figures 3 to 8. Figures 7 and 8 represent measurement of the height for the calculation of PA_c for double bottom tanks with sloping tank top. Figures 9 and 10 represent the cases where credit is given in calculation of PA_s to part or the whole of a double bottom tank.

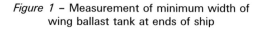

Figure 1 – Measurement of minimum width of
wing ballast tank at ends of ship

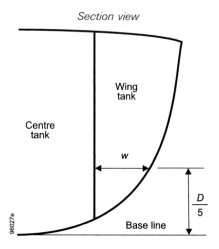

w must be at least 2 metres along the entire length of the tank for the
tank to be used in the calculation of PA_c

Figure 2 – Measurement of minimum height of
double bottom tank at ends of ship

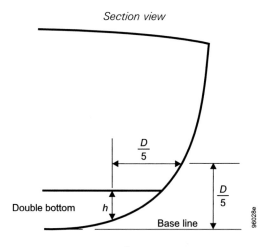

h must be at least 2 metres or $\dfrac{B}{15}$, whichever is less, along the entire
length of the tank for the tank to be used in the calculation of PA_s

Figure 3 – Calculation of PA_c and PA_s
for double bottom tank amidships

Section view

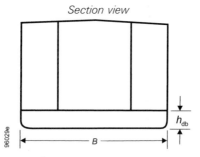

If h_{db} is at least 2 metres or $\frac{B}{15}$, whichever is less, along entire tank length,

$PA_c = h_{db} \times$ double bottom tank length $\times$ 2

$PA_s = B \times$ double bottom tank length

If h_{db} is less than 2 metres or $\frac{B}{15}$, whichever is less,

$PA_c = h_{db} \times$ double bottom tank length $\times$ 2

$PA_s = 0$

Figure 4 – Calculation of PA_c and PA_s
for double bottom tank at ends of ship

Section view

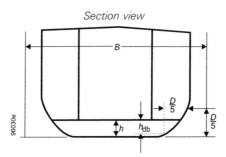

If h_{db} is at least 2 metres or $\frac{B}{15}$, whichever is less, along entire tank length,

$PA_c = h \times$ double bottom tank length $\times$ 2

$PA_s = B \times$ double bottom tank length

If h_{db} is less than 2 metres or $\frac{B}{15}$, whichever is less,

$PA_c = h \times$ double bottom tank length $\times$ 2

$PA_s = 0$

Figure 5 – Calculation of PA_c and PA_s for wing tank amidships

Plan view

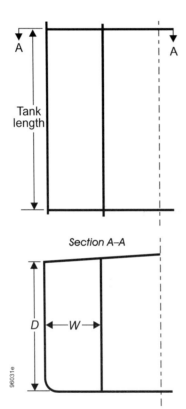

Figure 6 – Calculation of PA_c and PA_s for wing tank at end of ship

Plan view at D

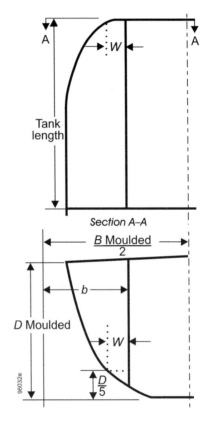

If W is 2 metres or more,

$$PA_c = D \times \text{tank length} \times 2^*$$

$$PA_s = W \times \text{tank length} \times 2^*$$

If W is less than 2 metres,

$$PA_c = 0$$

$$PA_s = W \times \text{tank length} \times 2^*$$

If W is 2 metres or more,

$$PA_c = D \times \text{tank length} \times 2^*$$

$$PA_s = b \times \text{tank length} \times 2^*$$

If W is less than 2 metres,

$$PA_c = 0$$

$$PA_s = b \times \text{tank length} \times 2^*$$

---

* To include port and starboard.

178

Figure 7 – Measurement of *h* for calculation of *PA*$_c$ for
double bottom tanks with sloping tank tops (1)

Section view

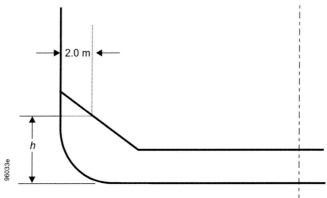

$PA_c = h \times$ double bottom tank length $\times$ 2*

Figure 8 – Measurement of *h* for calculation of *PA*$_c$ for
double bottom tanks with sloping tank tops (2)

Section view

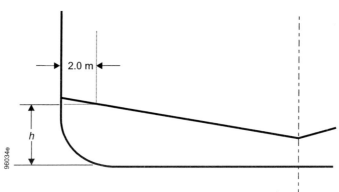

$PA_c = h \times$ double bottom tank length $\times$ 2*

---

* To include port and starboard.

Annex I

Unified Interpretations

Figure 9 – Calculation of PA_s for double bottom tank without clearly defined turn of bilge area – when wing tank is cargo tank

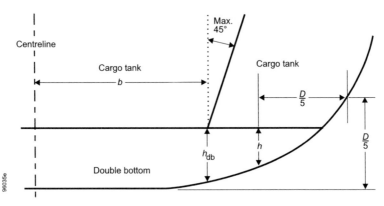

If h is less than 2 metres or $\frac{B}{15}$, whichever is less, anywhere along the tank length, but h_{db} is at least 2 metres or $\frac{B}{15}$, whichever is less, along the entire tank length within the width of $2b$, then:

$$PA_s = 2b \times \text{cargo tank length}$$

Figure 10 – Calculation of PA_s for double bottom tank without clearly defined turn of bilge area – when wing tank is segregated ballast tank or void space

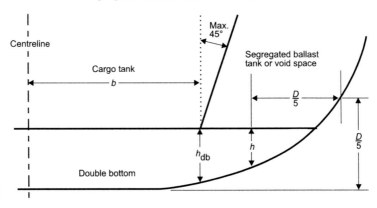

If h is less than 2 metres or $\frac{B}{15}$, whichever is less, anywhere along the tank length, but h_{db} is at least 2 metres or $\frac{B}{15}$, whichever is less, along the entire tank length within the width of $2b$, then:

$$PA_s = B \times \text{cargo tank length}$$

Appendix 3

Equivalent provisions for the carriage of oil by a chemical tanker*

1   By implication, regulation 1(4) of Annex I of MARPOL 73/78 prescribes that where a cargo subject to the provisions of Annex I of MARPOL 73/78 is carried in a cargo space of a chemical tanker, the appropriate requirements of Annex I of MARPOL 73/78 shall apply. For the purposes of application of such requirements, a chemical tanker when carrying oil, if it is impracticable to comply with the requirements of regulation 15(2) and 15(3)(b), shall comply with the following equivalent provisions in accordance with regulation 3 of Annex I.

2   A chemical tanker shall hold a valid Certificate of Fitness issued under the provision of the Code for the Construction and Equipment of Ships Carrying Dangerous Chemicals in Bulk.

3   A chemical tanker shall be fitted within the cargo tank area with the following equipment:

.1   oily-water separating equipment capable of producing effluent with oil content of less than 100 ppm, complying with the requirements of regulation 16(6) which has been demonstrated to be suitable for the full range of Annex I products and with a minimum capacity as shown in the table below:

Deadweight tons	Capacity of separating equipment (m^3/h)
Less than 2,000	5
2,000 and above but less than 5,000	7.5
5,000 and above but less than 10,000	10
10,000 and above	deadweight/1000

.2   permanently installed transfer pump for overboard discharge of effluent containing oil through the oily-water separating equipment, with a capacity not exceeding the capacity of the separating equipment;

.3   holding tank of sufficient capacity for the separated oil and with the means for discharge of such oil to reception facilities. The holding tank capacity shall be at least equal to the total quantity of residues remaining in the cargo tanks after unloading as determined by the methods prescribed in appendix A of the Standards for Procedures and Arrangements for the Discharge of Noxious Liquid Substances; and

.4   a collecting tank for collecting tank washings. Any cargo tank may be designated as a collecting tank.

---

* The 1992 amendments to regulation 16, adopted by the Marine Environment Protection Committee by resolution MEPC.51(32), have effectively made some contents of paragraphs 3 and 4 of this appendix invalid. In accordance with regulation 16(6), 100 ppm equipment was only allowed until 6 July 1998. Therefore, this appendix is subject to future amendment by the MEPC.

Annex I

Unified Interpretations

4   The equipment referred to in paragraph 3.1 shall be of the type approved under the terms of resolution A.393(X).

5   The outlet for the overboard discharge of the effluent from the oily-water separating equipment shall be located above the waterline in the deepest loaded conditions.

Appendix 4

Connection of small diameter line
to the manifold valve

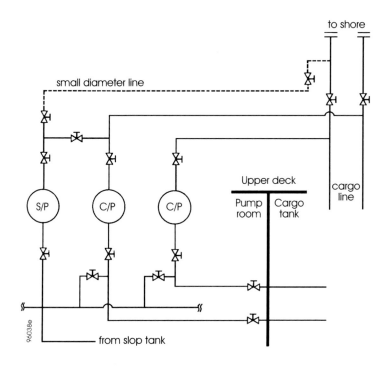

Appendix 5

Specifications for the design, installation and operation of a part flow system for control of overboard discharges

1 Purpose

1.1 The purpose of these Specifications is to provide specific design criteria and installation and operational requirements for the part flow system referred to in regulation 18(6)(e) of Annex I of the International Convention for the Prevention of Pollution from Ships, 1973, as modified by the Protocol of 1978 relating thereto (MARPOL 73/78).

2 Application

2.1 Existing oil tankers may, in accordance with regulation 18(6)(e) of Annex I of MARPOL 73/78, discharge dirty ballast water and oil-contaminated water from cargo tank areas below the waterline, provided that a part of the flow is led through permanent piping to a readily accessible location on the upper deck or above where it may be visually observed during the discharge operation and provided that the arrangements comply with the requirements established by the Administration which shall at least contain all the provisions of these Specifications.

2.2 The part flow concept is based on the principle that the observation of a representative part flow of the overboard effluent is equivalent to observing the entire effluent stream. These specifications provide the details of the design, installation and operation of a part flow system.

3 General provisions

3.1 The part flow system shall be so fitted that it can effectively provide a representative sample of the overboard effluent for visual display under all normal operating conditions.

3.2 The part flow system is in many respects similar to the sampling system for an oil discharge monitoring and control system but shall have pumping and piping arrangements separate from such a system, or combined equivalent arrangements acceptable to the Administration.

3.3 The display of the part flow shall be arranged in a sheltered and readily accessible location on the upper deck or above, approved by the Administration (e.g. the entrance to the pump-room). Regard should be given to effective communication between the location of the part flow display and the discharge control position.

3.4 Samples shall be taken from relevant sections of the overboard discharge piping and be passed to the display arrangement through a permanent piping system.

3.5 The part flow system shall include the following components:

.1 sampling probes;

.2   sample water piping system;

.3   sample feed pump(s);

.4   display arrangements;

.5   sample discharge arrangements; and, subject to the diameter of the sample piping,

.6   flushing arrangement.

3.6   The part flow system shall comply with the applicable safety requirements.

4   *System arrangement*

4.1 **Sampling points**

4.1.1 Sampling point location:

.1   Sampling points shall be so located that relevant samples can be obtained of the effluent being discharged through outlets below the waterline which are used for operational discharges.

.2   Sampling points shall as far as practicable be located in pipe sections where a turbulent flow is normally encountered.

.3   Sampling points shall as far as practicable be arranged in accessible locations in vertical sections of the discharge piping.

4.1.2 Sampling probes:

.1   Sampling probes shall be arranged to protrude into the pipe a distance of about one fourth of the pipe diameter.

.2   Sampling probes shall be arranged for easy withdrawal for cleaning.

.3   The part flow system shall have a stop valve fitted adjacent to each probe, except that where the probe is mounted in a cargo line, two stop valves shall be fitted in series, in the sample line.

.4   Sampling probes should be of corrosion-resistant and oil-resistant material, of adequate strength, properly jointed and supported.

.5   Sampling probes shall have shape that is not prone to becoming clogged by particle contaminants and should not generate high hydrodynamic pressures at the sampling probe tip. Figure 1 is an example of one suitable shape of a sampling probe.

.6   Sampling probes shall have the same nominal bore as the sample piping.

4.2 **Sample piping**

.1   The sample piping shall be arranged as straight as possible between the sampling points and the display arrangement. Sharp bends and pockets where settled oil or sediment may accumulate should be avoided.

.2   The sample piping shall be so arranged that sample water is conveyed to the display arrangement within 20 s. The flow velocity in the piping should not be less than 2 m/s.

Figure 1 – Sampling probe for a part flow display system

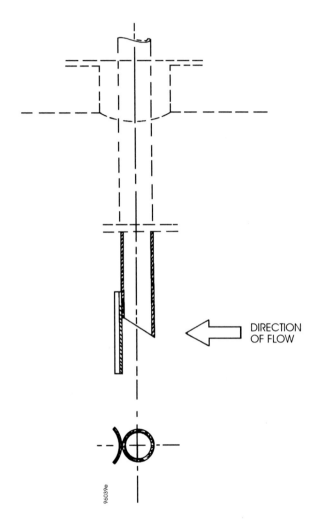

DIRECTION
OF FLOW

96039e

.3   The diameter of the piping shall not be less than 40 mm if no fixed flushing arrangement is provided and shall not be less than 25 mm if a pressurized flushing arrangement as detailed in paragraph 4.4 is installed.

.4   The sample piping should be of corrosion-resistant and oil-resistant material, of adequate strength, properly jointed and supported.

.5   Where several sampling points are installed, the piping shall be connected to a valve chest at the suction side of the sample feed pump.

4.3 Sample feed pump

.1 The sample feed pump capacity shall be suitable to allow the flow rate of the sample water to comply with 4.2.2.

4.4 Flushing arrangement

.1 If the diameter of sample piping is less than 40 mm, a fixed connection from a pressurized sea or fresh water piping system shall be installed for flushing of the sample piping system.

4.5 Display arrangement

.1 The display arrangement shall consist of a display chamber provided with a sight glass. The chamber should be of a size that will allow a free fall stream of the sample water to be clearly visible over a length of at least 200 mm. The Administration may approve equivalent arrangements.

.2 The display arrangement shall incorporate valves and piping in order to allow part of the sample flow to bypass the display chamber to obtain a laminar flow for display in the chamber.

.3 The display arrangement shall be designed to be easily opened and cleaned.

.4 The interior of the display chamber shall be white except for the background wall which shall be so coloured as to facilitate the observation of any change in the quality of the sample water.

.5 The lower part of the display chamber shall be shaped like a funnel for collection of the sample water.

.6 A test cock for taking a grab sample shall be provided in order that a sample of the water can be examined independent of that in the display chamber.

.7 The display arrangement shall be adequately lighted to facilitate visual observation of the sample water.

4.6 Sample discharge arrangement

.1 The sample water leaving the display chamber shall be routed to the sea or to a slop tank through fixed piping of adequate diameter.

5 *Operation*

5.1 When a discharge of dirty ballast water or other oil-contaminated water from the cargo tank area is taking place through an outlet below the waterline, the part flow system shall provide sample water from the relevant discharge outlet at all times.

5.2 The sample water should be observed particularly during those phases of the discharge operation when the greatest possibility of oil contamination occurs. The discharge shall be stopped whenever any traces of oil are visible in the flow and when the oil content meter reading indicates that the oil content exceeds permissible limits.

5.3   On those systems that are fitted with flushing arrangements, the sample piping should be flushed after contamination has been observed and, additionally, it is recommended that the sample piping be flushed after each period of usage.

5.4   The ship's cargo and ballast handling manuals and, where applicable, those manuals required for crude oil washing systems or dedicated clean ballast tanks operation shall clearly describe the use of the part flow system in conjunction with the ballast discharge and the slop tank decanting procedures.

Appendix 6

Offshore platform discharges

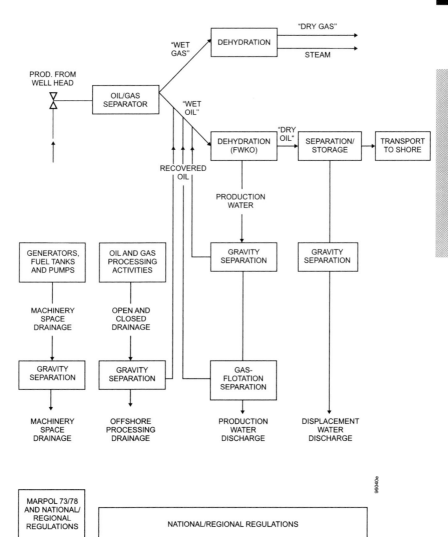

Appendix 7

Interim guidelines for the approval of alternative methods of design and construction of oil tankers under regulation 13F(5) of Annex I of MARPOL 73/78

Preamble

1   The purpose of these Interim Guidelines, hereunder referred to as "the Guidelines", is to provide an international standard for the evaluation and approval of alternative methods of design and construction of oil tankers under regulation 13F(5) of Annex I of MARPOL 73/78.

2   The basic philosophy of the Guidelines is to compare the oil outflow performance in case of collision or stranding of an alternative tanker design to that of reference double-hull designs complying with regulation 13F(3) on the basis of a calculated pollution-prevention index.

3   The oil outflow performance of double-hull tankers which comply with regulation 13F(3) may be different. The longitudinal subdivision of the cargo tanks has a major influence on the oil outflow in case of inner hull penetration. The selected reference double-hull designs exhibit a favourable oil outflow performance.

4   The calculation of oil outflow is based on the probabilistic methodology and best available tanker accident damage statistics. Reappraisal of the Guidelines may be appropriate when more information on tanker accident damage has become available and more experience with the application of these Guidelines has been gained.

5   Falling tides will have an adverse effect on oil outflow from a stranded tanker and the Guidelines take account of this. The tide values specified in section 5 represent realistic average tidal changes which have been chosen to identify the influence of tidal changes on the oil outflow in case of stranding.

1   General

1.1   Regulation 13F of Annex I of MARPOL 73/78 specifies structural requirements for new tankers of 600 tdw and above, contracted on or after 6 July 1993. Paragraph (3) of the regulation requires tankers of 5,000 tdw and above to be equipped with double hulls. Various detailed requirements and permissible exceptions are given in the regulation.

Paragraph (5) of the regulation specifies that other designs may be accepted as alternatives to double hull, provided they give at least the same level of protection against oil pollution in the event of collision or stranding and are approved in principle by the MEPC based on Guidelines developed by the Organization.

1.2   These Guidelines should be used to assess the acceptability of alternative oil tanker designs of 5,000 tdw and above with regard to the prevention of oil outflow in the event of collision or stranding as specified in paragraph (5) of regulation 13F of Annex I of MARPOL 73/78.

1.3   For any alternative design of an oil tanker not satisfying regulation 13F(3) or (4), a study of the cargo oil outflow performance should be carried out as specified in sections 4 through 6 of these Guidelines.

1.4   This study should cover the full range of ship sizes with a minimum of four different ship sizes, unless the approval is requested for only a limited range of vessel sizes. Data for four reference double-hull designs are given in section 7.

1.5   Evaluation of the cargo oil outflow performance of the proposed alternative design should be made by calculating the pollution-prevention index E as outlined in section 4 of these Guidelines.

1.6   The probabilistic methodology for the calculation of oil outflow according to these Guidelines is based on available tanker casualty statistics. With the collection of additional statistical material, the damage density distribution functions specified in 5.2 should be periodically reviewed.

1.7   In principle, and as far as applicable, the requirements of paragraphs (3)(d)–(f), (6) and (8) of regulation 13F apply also to alternative designs. The requirements of paragraph (9) of regulation 13F also apply to alternative designs. In addition, it should be demonstrated by means of a risk analysis that the new design under consideration provides an adequate safety level. Such analysis should address any specific risks associated with the alternative design, and if there are any, it should be demonstrated that safe solutions exist to cope with them.

2   Applicability

2.1   These Guidelines apply to the assessment of alternative designs of oil tankers to be constructed of steel or other equivalent material as required by regulation 42 of chapter II-2 of the 1974 SOLAS Convention as amended. Designs for tankers intended to be constructed of other materials or incorporating novel features (e.g. non-metallic materials) or designs which use impact-absorbing devices should be specially considered.

2.2   The approval procedure of these Guidelines applies to oil tankers of sizes up to 350,000 tdw. For larger sizes the approval procedure should be specially considered.

3   Approval procedure for alternative tanker designs

3.1   An Administration of a Party to MARPOL 73/78 which receives a request for approval of an alternative tanker design for the purpose of complying with regulation 13F should first evaluate the proposed design and satisfy itself that the design complies with these Guidelines and other applicable regulations of Annex I of MARPOL 73/78. That Administration should then submit the proposal and the supporting documentation, together with its own evaluation report, to the Organization for evaluation and approval of the design concept by the Marine Environment Protection Committee (MEPC) as an alternative to the requirements of regulation 13F(3). Only design concepts which have been approved in principle by the MEPC are allowed for the construction of tankers to which regulation 13F(5) applies.

3.2   The submission to the Administration and the Organization should at least include the following items:

.1   Detailed specification of the alternative design concept.

.2   Drawings showing the basic design of the tank system and, where necessary, of the entire ship.

.3   Study of the oil outflow performance as outlined in paragraphs 1.3 to 1.5.

.4   Risk analysis as outlined in paragraph 1.7.

.5   Details of the calculation procedure or computer program used for the probabilistic oil outflow analysis to satisfy the Administration that the calculation procedure used gives satisfactory results. For verification of the computer program see paragraph 6.2.

Any additional information may be required to be submitted if deemed necessary.

3.3   In addition to the approval procedure for the design concept specified in 3.1 and 3.2 above, the final shipyard design should be approved by the Flag State Administration for compliance with these Guidelines and all other applicable regulations of Annex I of MARPOL 73/78. This should include survivability considerations as referred to in 5.1.5.10.

3.4   Any approved design concept will require reconsideration if the Guidelines have been amended.

4   Oil outflow analysis

4.1   General

4.1.1   The oil pollution prevention performance of a tanker design is expressed by a non-dimensional oil pollution prevention index E which is a function of the three oil outflow parameters: "probability of zero oil outflow", "mean oil outflow" and "extreme oil outflow". The oil outflow parameters should be calculated for all conceivable damage cases within the entire envelope of damage conditions as detailed in section 5.

4.1.2   The three oil outflow parameters are defined as follows:

Probability of zero oil outflow. This parameter represents the probability that no cargo oil will escape from the tanker in case of collision or stranding. If, e.g., the parameter equals 0.6, in 60% of all collision or stranding accidents no oil outflow is expected to occur.

Mean oil outflow parameter. The mean oil outflow represents the sum of all outflow volumes multiplied by their respective probabilities. The mean oil outflow parameter is expressed as a fraction of the total cargo oil capacity at 98% tank filling.

Extreme oil outflow parameter. The extreme oil outflow is calculated – after the volumes of all outflow cases have been arranged in ascending order – as the sum of the outflow volumes between 0.9 and 1.0 cumulative probability, multiplied by their respective probabilities. The value so derived is multiplied by 10. The extreme oil outflow parameter is expressed as a fraction of the total cargo oil capacity at 98% tank filling.

4.1.3   In general, consideration of ship's survivability will not be required for the conceptual approval of an alternative design. This may, however, be required in special cases, depending on special features of the design.

4.2   Pollution-prevention index

The level of protection against oil pollution in the event of collision or stranding as compared to the reference double-hull designs should be determined by calculation of the pollution-prevention index E as follows:

$$E = k_1 \frac{P_0}{P_{OR}} + k_2 \frac{0.01 + O_{MR}}{0.01 + O_M} + k_3 \frac{0.025 + O_{ER}}{0.025 + O_E} \geq 1.0$$

where:

k_1, k_2 and k_3 are weighting factors having the values:

$$k_1 \quad = \quad 0.5$$
$$k_2 \quad = \quad 0.4$$
$$k_3 \quad = \quad 0.1$$

P_0 =   probability of zero oil outflow for the alternative design

O_M =   mean oil outflow parameter for the alternative design

O_E =   extreme oil outflow parameter for the alternative design

P_{OR}, O_{MR} and O_{ER} are the corresponding parameters for the reference double-hull design of the same cargo oil capacity as specified in section 7.

4.3   Calculation of oil outflow parameters

The oil outflow parameters P_0, O_M and O_E should be calculated as follows:

Probability of zero oil outflow, P_0:

$$P_0 = \sum_{i=1}^{n} P_i \cdot K_i$$

where:

i represents each compartment or group of compartments under consideration, running from $i = 1$ to $i = n$

P_i accounts for the probability that only the compartment or group of compartments under consideration are breached

K_i equals 0 if there is oil outflow from any of the breached cargo spaces in i. If there is no outflow, K_i equals 1.

Mean oil outflow parameter, O_M:

$$O_M = \sum_{i=1}^{n} \frac{P_i \cdot O_i}{C}$$

where:

O_i = combined oil outflow (m^3) from all cargo spaces breached in i

C = total cargo oil capacity at 98% tank filling (m^3)

Extreme oil outflow parameter, O_E:

$$O_E = 10 \left(\sum \frac{P_{ie} \cdot O_{ie}}{C} \right)$$

where the index "*ie*" represents the extreme outflow cases, which are the damage cases falling within the cumulative probability range between 0.9 and 1.0 after they have been arranged as specified in 6.1.

5 Assumptions for calculating oil outflow parameters

5.1 General

5.1.1 The assumptions specified in this section should be used when calculating the oil outflow parameters.

5.1.2 Outflow parameters should be calculated independently for collisions and strandings and then combined as follows:

0.4 of the computed value for collisions plus

0.6 of the computed value for strandings.

5.1.3 For strandings, independent calculations should be done for 0 m, 2 m and 6 m tides. The tide, however, need not be taken greater than 50% of the ship's maximum draught. Outflow parameters for the stranded conditions should be a weighted average, calculated as follows:

0.4 for 0 m tide condition

0.5 for minus 2 m tide condition

0.1 for minus 6 m tide condition.

5.1.4 The damage cases and the associated probability factor P_i for each damage case should be determined based on the damage density distribution functions as specified in paragraph 5.2.

5.1.5 The following general assumptions apply for the calculation of outflow parameters:

.1 The ship should be assumed to be loaded to the maximum assigned load line with zero trim and heel and with a cargo having a density allowing all cargo tanks to be filled to 98%.

.2 For all cases of collision damage, the entire contents of all damaged cargo oil tanks should be assumed to be spilled into the sea, unless proven otherwise.

.3 For all stranded conditions, the ship should be assumed aground on a shelf. Assumed stranded draughts prior to tidal change should be equal to the initial intact draughts. Should the ship trim or float free due to the outflow of oil, this should be accounted for in the calculations for the final shipyard design.

.4   In general, an inert gas overpressure of 0.05 bar gauge should be assumed.

.5   For the calculation of oil outflow in case of stranding, the principles of hydrostatic balance should apply, and the location of damage used for calculations of hydrostatic pressure balance and related oil outflow calculations should be the lowest point in the cargo tank.

.6   For cargo tanks bounded by the bottom shell, unless proven otherwise, oil outflow equal to 1% of the volume of the damaged tank should be assumed to account for initial exchange losses and dynamic effects due to current and waves.

.7   For breached non-cargo spaces located wholly or in part below breached cargo oil tanks, the flooded volume of these spaces at equilibrium should be assumed to contain 50% oil and 50% seawater by volume, unless proven otherwise.

.8   If deemed necessary, model tests may be required to determine the influence of tidal, current and swell effects on the oil outflow performance.

.9   For ship designs which incorporate cargo transfer systems for reducing oil outflow, calculations should be provided illustrating the effectiveness of such devices. For these calculations, damage openings consistent with the damage density distribution functions defined in 5.2 should be assumed.

.10   Where, for the final shipyard design referred to in 3.3 and in the special cases referred to in 4.1.3, damage stability calculations are required, the following should apply:

       A damage stability calculation should be performed for each damage case. The stability in the final stage of flooding should be regarded as sufficient if the requirements of regulation 25(3) of Annex I of MARPOL 73/78 are complied with.

       Should the ship fail to meet the survivability criteria as defined in regulation 25(3), 100% oil outflow from all cargo tanks should be assumed for that damage case.

5.2   Damage assumptions

5.2.1 *General, definitions*

The damage assumptions for the probabilistic oil outflow analysis are given in terms of the damage density distribution functions specified in subparagraphs 5.2.2 and 5.2.3. These functions are so scaled that the total probability for each damage parameter equals 100%, i.e. the area under each curve equals 1.0.

The location of a damage refers always to the centre of a damage. Damage location and extent to an inner horizontal bottom or vertical bulkhead should be assumed to be the same as the statistically derived damage to the outer hull.

The location and extent of damage to compartment boundaries should be assumed to be of rectangular shape, following the hull surface in the extents defined in subparagraphs 5.2.2 and 5.2.3.

The following definitions apply for the purpose of subparagraphs 5.2.2 and 5.2.3.

x = dimensionless distance from A.P. relative to the ship's length between perpendiculars

y = dimensionless longitudinal extent of damage relative to the ship's length between perpendiculars

z_t = dimensionless transverse penetration extent relative to the ship's breadth

z_v = dimensionless vertical penetration extent relative to the ship's depth

z_l = dimensionless vertical distance between the baseline and the centre of the vertical extent z_v relative to the distance between baseline and deck level (normally the ship's depth)

b = dimensionless transverse extent of bottom damage relative to the ship's breadth

b_l = dimensionless transverse location of bottom damage relative to the ship's breadth.

5.2.2 *Side damage due to collision*

Function for longitudinal location:

$f_{s1} = 1.0$ for $0 \leqslant x \leqslant 1.0$;

function for longitudinal extent:

$f_{s2} = 11.95 - 84.5y$ for $y \leqslant 0.1$

$f_{s2} = 6.65 - 31.5y$ for $0.1 < y \leqslant 0.2$

$f_{s2} = 0.35$ for $0.2 < y \leqslant 0.3$;

function for transverse penetration:

$f_{s3} = 24.96 - 399.2z_t$ for $z_t \leqslant 0.05$

$f_{s3} = 9.44 - 88.8z_t$ for $0.05 < z_t \leqslant 0.1$

$f_{s3} = 0.56$ for $0.1 < z_t \leqslant 0.3$;

function for vertical extent:

$f_{s4} = 3.83 - 11.1z_v$ for $z_v \leqslant 0.3$

$f_{s4} = 0.5$ for $z_v > 0.3$;

function for vertical location:

$f_{s5} = z_l$ for $z_l \leqslant 0.25$

$f_{s5} = 5z_l - 1.0$ for $0.25 < z_l \leqslant 0.50$

$f_{s5} = 1.50$ for $0.50 < z_l \leqslant 1.00$.

Graphs of the functions f_{s1}, f_{s2}, f_{s3}, f_{s4} and f_{s5} are shown in figures 1 and 2.

Unified Interpretations

5.2.3   *Bottom damage due to stranding*

Function for longitudinal location:

$$f_{b1} = 0.2 + 0.8x \qquad\qquad \text{for } x \leqslant 0.5$$
$$f_{b1} = 4x - 1.4 \qquad\qquad \text{for } 0.5 < x \leqslant 1.0;$$

function for longitudinal extent:

$$f_{b2} = 4.5 - 13.33y \qquad\qquad \text{for } y \leqslant 0.3$$
$$f_{b2} = 0.5 \qquad\qquad \text{for } 0.3 < y \leqslant 0.8;$$

function for vertical penetration:

$$f_{b3} = 14.5 - 134z_v \qquad\qquad \text{for } z_v \leqslant 0.1$$
$$f_{b3} = 1.1 \qquad\qquad \text{for } 0.1 < z_v \leqslant 0.3;$$

function for transverse extent:

$$f_{b4} = 4.0 - 12b \qquad\qquad \text{for } b \leqslant 0.3$$
$$f_{b4} = 0.4 \qquad\qquad \text{for } 0.3 < b \leqslant 0.9$$
$$f_{b4} = 12b - 10.4 \qquad\qquad \text{for } b > 0.9;$$

function for transverse location:

$$f_{b5} = 1.0 \qquad\qquad \text{for } 0 \leqslant b_l \leqslant 1.0.$$

Graphs of the functions f_{b1}, f_{b2}, f_{b3}, f_{b4} and f_{b5} are shown in figures 3 and 4.

6   *Probabilistic methodology for calculating oil outflow*

6.1   **Damage cases**

6.1.1   Using the damage probability distribution functions specified in paragraph 5.2, all damage cases *n* as per paragraph 4.3 should be evaluated and placed in ascending order of oil outflow. The cumulative probability for all damage cases should be computed, being the running sum of probabilities beginning at the minimum outflow damage case and proceeding to the maximum outflow damage case. The cumulative probability for all damage cases should be 1.0.

6.1.2   For each damage case the damage consequences in terms of penetrations (breaching) of cargo tank boundaries should be evaluated and the related oil outflow calculated. A cargo tank should be considered as being breached in a damage case under consideration if the applied damage envelope reaches any part of the cargo tank boundaries.

6.1.3   When determining the damage cases, it should be assumed for the purpose of these calculations that the location, extent and penetration of damages are independent of each other.

6.2   **Oil outflow calculations**

6.2.1   The probabilistic oil outflow calculations may be done as outlined by the "Example for the Application of the Interim Guidelines" given in the appendix to these

Guidelines. Other calculation procedures may be accepted, provided they show acceptable accuracy.

6.2.2   The computer program used for the oil outflow analysis should be verified against the data for oil outflow parameters for the reference double-hull designs given in section 7.

6.2.3   After the final waterline has been determined, the oil outflow from each damaged cargo tank should be computed for each damage case under the assumptions specified in 5.1.5.

7   Reference double-hull designs

Data for four reference double-hull designs of 5,000 tdw, 60,000 tdw, 150,000 tdw and 283,000 tdw are summarized in tables 7.1 and 7.2 and are illustrated in figures 5 to 8.

Table 7.1 contains the data for the oil outflow parameters P_{OR}, O_{MR} and O_{ER} to be used for the concept approval (ship survivability not considered). Table 7.2 contains the corresponding data to be used for the shipyard design approval (ship survivability considered).

Table 7.1 – Oil outflow parameters (ship survivability not considered)

Reference design number	Deadweight (t)	Oil outflow parameters (ship survivability not considered)		
		P_{OR}	O_{MR}	O_{ER}
1	5,000	0.81	0.017	0.127
2	60,000	0.81	0.014	0.104
3	150,000	0.79	0.016	0.113
4	283,000	0.77	0.013	0.085

Table 7.2 – Oil outflow parameters (ship survivability considered)

Reference design number	Deadweight (t)	Oil outflow parameters (ship survivability considered)		
		P_{OR}	O_{MR}	O_{ER}
1	5,000	0.72	0.113	0.469
2	60,000	0.81	0.021	0.173
3	150,000	0.79	0.017	0.124
4	283,000	0.77	0.015	0.098

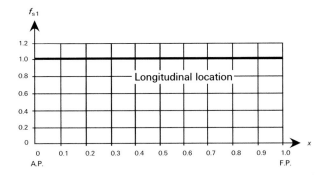

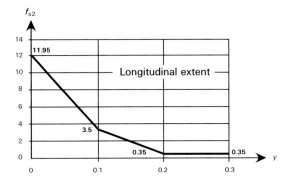

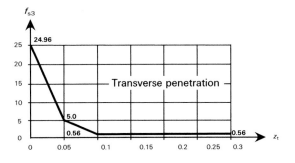

Figure 1 – Side damage due to collision:
density distribution functions f_{s1}, f_{s2}, f_{s3}

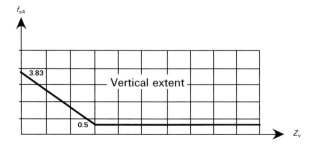

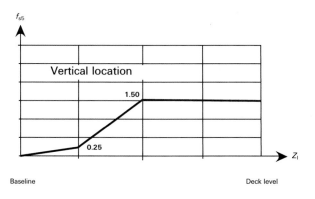

Figure 2 – Side damage due to collision:
density distribution functions f_{s4} and f_{s5}

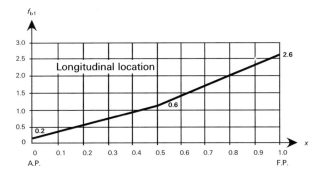

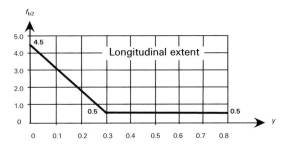

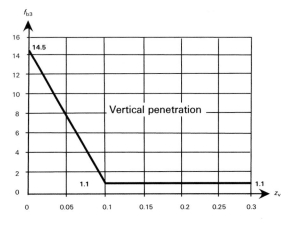

Figure 3 – Bottom damage due to stranding:
density distribution functions f_{b1}, f_{b2}, f_{b3}

201

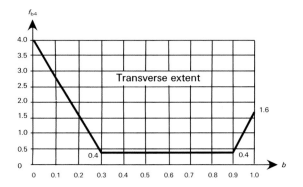

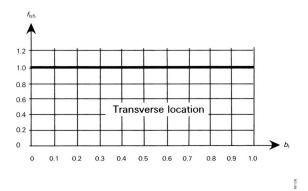

Figure 4 – Bottom damage due to stranding:
density distribution functions f_{b4} and f_{b5}

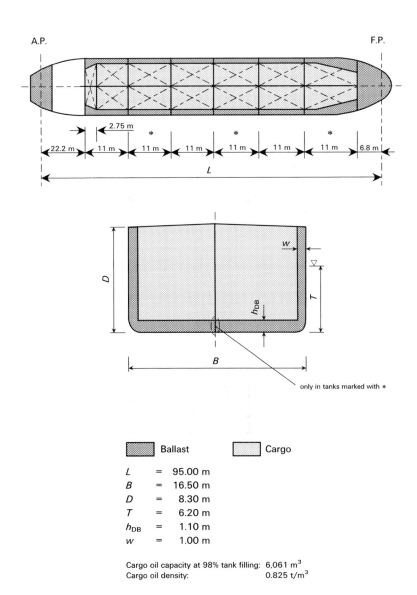

Ballast ☐ Cargo

L = 95.00 m
B = 16.50 m
D = 8.30 m
T = 6.20 m
h_{DB} = 1.10 m
w = 1.00 m

Cargo oil capacity at 98% tank filling: 6,061 m³
Cargo oil density: 0.825 t/m³

Figure 5 – Reference double-hull design no. 1
Deadweight: 5,000 tdw

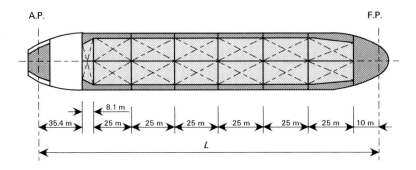

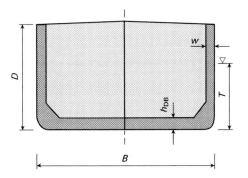

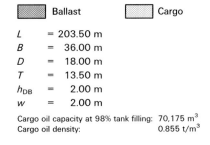

Ballast   Cargo

L	=	203.50 m
B	=	36.00 m
D	=	18.00 m
T	=	13.50 m
h_{DB}	=	2.00 m
w	=	2.00 m

Cargo oil capacity at 98% tank filling: 70,175 m³
Cargo oil density: 0.855 t/m³

95160

Figure 6 – Reference double-hull design no. 2
Deadweight: 60,000 tdw

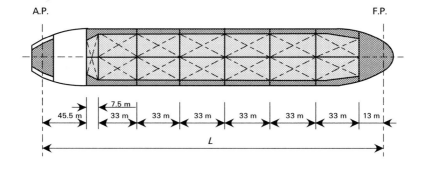

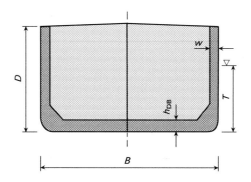

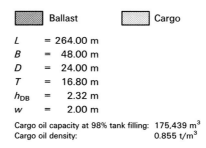

| | Ballast | | | Cargo |

L = 264.00 m
B = 48.00 m
D = 24.00 m
T = 16.80 m
h_{DB} = 2.32 m
w = 2.00 m

Cargo oil capacity at 98% tank filling: 175,439 m³
Cargo oil density: 0.855 t/m³

95161

Figure 7 – Reference double-hull design no. 3
Deadweight: 150,000 tdw

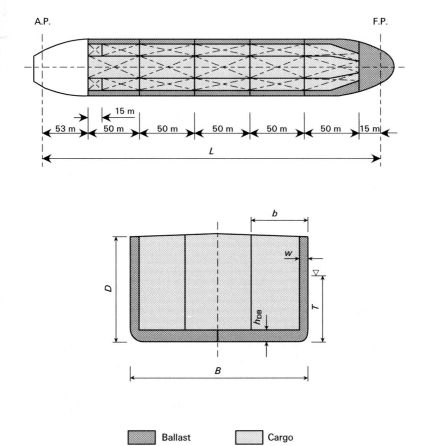

Ballast   Cargo

L	=	318.00 m
B	=	57.00 m
D	=	31.00 m
T	=	22.00 m
b	=	18.00 m
h_{DB}	=	4.20 m
w	=	2.00 m

Cargo oil capacity at 98% tank filling: 330,994 m³
Cargo oil density: 0.855 t/m³

Figure 8 – Reference double-hull design no. 4
Deadweight: 283,000 tdw

Unified Interpretations

Appendix

Example for the application of the "Interim Guidelines"

1   General

The application of the Interim Guidelines, hereunder referred to as "the Guidelines", is shown in the following worked example illustrating the calculation procedure of the oil outflow parameters for a tank barge. For presentation purposes, a simplified hull form and level of compartmentation have been assumed. The procedures described herein are readily adaptable as a computer application, which will be necessary as more complicated arrangements are evaluated. This example is evaluated in accordance with the requirements for "concept approval". Additional requirements for a shipyard design approval are noted where applicable.

An application of the Guidelines will typically follow these seven basic steps:

1)   **Vessel design**: In accordance with paragraph 3.1 of the Guidelines, the vessel is designed to meet all applicable regulations of Annex I of MARPOL 73/78.

2)   **Establishing of the full load condition**: In accordance with paragraph 5.1.5 of the Guidelines, a full load condition is developed.

3)   **Assembling of the damage cases**: By applying the damage density distribution functions provided in the Guidelines, determine each unique grouping of damaged compartments and the probability associated with that damage condition. Independent sets of damage cases are derived for side (collision) and bottom (stranding) damage.

4)   **Computation of the equilibrium condition for each damage case**: Compute the final equilibrium condition for all side and bottom damage conditions. This step is only required for the final shipyard design, in accordance with paragraph 5.1.5.10 of the Guidelines.

5)   **Computation of the oil outflow for each damage case**: Calculate the oil outflow for each damage case. Separate calculations are done for side damage, and for bottom damage at the 0.0 m, 2.0 m and 6.0 m tide conditions. For side damage, all oil is assumed to escape from damaged tanks. For bottom damage, a hydrostatic balance method is applied. For the final shipyard design, survivability is evaluated in accordance with the requirements of regulation 25(3) of Annex I of MARPOL 73/78.

6)   **Computation of the oil outflow parameters**: The cumulative probability of occurrence of each level of oil outflow is developed. This is done for the side damage and for each bottom damage tide condition. The associated oil outflow parameters are then computed. The bottom damage tidal parameters are combined in accordance with paragraph 5.1.3 and the side and bottom

damage parameters are then combined in accordance with paragraph 5.1.2 of the Guidelines.

7) **Computation of the Pollution Prevention Index E**: The new design has satisfactory characteristics if E as defined in paragraph 4.2 of the Guidelines is greater than or equal to 1.0.

2   Analysis procedure

The basic steps Nos. 1 through 6 are described in this section.

2.1   Step 1: Vessel design

The arrangement and dimensions of the example barge are as shown in figure A1 (Barge arrangement). For clarity purposes, a simple arrangement has been selected which does not meet all MARPOL 73/78 requirements. However, for actual designs submitted for approval as an alternative to double hull, the vessel must meet all applicable regulations of Annex I of MARPOL 73/78.

2.2   Step 2: Establishing of the full load condition

An intact load condition shall be developed with the vessel at its maximum assigned load line with zero trim and heel. Departure quantities of constants and consumables (fuel oil, diesel oil, fresh water, lube oil, etc.) should be assumed. Capacities of cargo oil tanks should be based on actual permeabilities for these compartments. All cargo oil tanks shall be assumed to be filled to 98% of their capacities. All cargo oil shall be taken at a homogeneous density.

For this example, it is assumed that the permeability of the cargo oil tanks is 0.99 and 0.95 for the double bottom/wing tank ballast spaces. The 100% capacity of the cargo oil tanks CO1 and CO2 is:

CO1:	9,623 m³
CO2:	28,868 m³
Total:	38,491 m³.

Cargo tank capacity at 98% filling: $C = 0.98 \times 38,491 = 37,721 \, \text{m}^3$.

For this barge, for simplicity reasons, zero weight for the constants and consumables has been assumed. At the 9.0 m assigned load line the following values for the cargo oil mass (W) and density (ρ_c) are obtained:

$$W = \text{displacement} - \text{light barge weight} = 33,949 \, \text{t}$$
$$\rho_c = 33,949 \, \text{t}/C = 0.90 \, \text{t/m}^3.$$

2.3   Step 3: Assembling of the damage cases

In this step the damage cases have to be developed. This involves applying the probability density distribution functions for side damage (figures 1 and 2) and the probability density distribution functions for bottom damage (figures 3 and 4). Each unique grouping of damaged compartments is determined together with its associated probability. The sum of the probabilities should equal 1.0 for both the side and the bottom damage evaluations.

There are different methods available for developing the compartment groupings and probabilities, each of which should converge on the same results.

In this example, the compartment groupings and the use of the probability density functions is shown by a "step-wise" evaluation method. This method involves stepping through each damage location and extent at a sufficiently fine increment. For instance, it is assumed (for the side damage) to step through the functions as follows: longitudinal location = 100 steps, longitudinal extent = 100 steps, transverse penetration = 100 steps, vertical location = 10 steps, and vertical extent = 100 steps. You will then be developing 10^9 damage incidents. The probability of each step is equal to the area under the probability density distribution curve over that increment. The probability for each damage incident is the product of the probabilities of the five functions. There are many redundant incidents which damage identical compartments. These are combined by summing their probabilities. For a typical double-hull tanker, the 10^9 damage incidents reduce down to 100 to 400 unique groupings of compartments.

2.3.1 *Side damage evaluation*

The damage density distribution functions provide independent statistics for location, length, and penetration. For side damage, the probability of a given damage longitudinal location, longitudinal extent, transverse penetration, vertical location and vertical extent is the product of the probabilities of these five damage characteristics.

To maintain the example at a manageable size, fairly coarse increments have been assumed:

Longitudinal location at 10 steps	$= L/10$	$= 0.10L$ per step
Longitudinal extent at 3 steps	$= 0.3L/3$	$= 0.10L$ per step
Transverse penetration at 6 steps	$= 0.3B/6$	$= 0.05B$ per step.

To further simplify the evaluation, each damage is assumed to extend vertically without limit. Therefore, the probabilities of vertical location and vertical extent are taken as 1.0 for each damage case. This is a reasonable assumption as the double bottom height is only 10% of the depth. Taking the area under the density distribution function for vertical location up to $0.1D$ (see figure 2, function f_{s5}) yields a value of 0.005. This means that the probability of the centre of damage location falling within the double bottom region is 1/200.

Figure A2 (Side damage definition) shows the steps for longitudinal location, longitudinal extent and transverse penetration in relation to the barge. Table A1 (Increments for step-wise side damage evaluation) gives the range for each step, the mean or average value over the step, and the probability of occurrence of that particular step. For instance, Z_1 covers the range of transverse penetration beginning at the side shell and extending inboard 5% of the breadth. The average penetration is $0.025B$ or 2.5% of the breadth. The probability of occurrence is the likelihood that the penetration will fall within the range of 0% to 5% of the breadth. The probability equals 0.749, which is the area under the density distribution function for transverse penetration (figure 1, function f_{s3}) between $0.0B$ and $0.05B$. The area under each probability density function is 1.0, and therefore the sum of the probabilities for all increments for each function is 1.0.

209

A total of ten longitudinal locations, three longitudinal extents and six transverse penetrations will be evaluated. All combinations of damages must be considered for a total of $(10) \times (3) \times (6) = 180$ separate incidents. The damaged compartments are found by overlaying each combination of location/extent/penetration onto the barge. These damage boundaries define a rectangular box. Any compartment which extends into this damage zone is considered damaged. Each of the 180 incidents results in damage to one or more compartments. Incidents with identical damaged compartments are collected into a single damage case by summing the probabilities of the individual damage incidents.

Let us begin at the aft end of the barge and proceed forward. The first damage location X_1 is centred $0.05L$ forward of the transom. The first damage extent Y_1 has an average length of $0.05L$. The average value for the first transverse penetration Z_1 is $0.025B$. The resulting damage box lies entirely within the WB1 compartment and therefore damages that compartment only. The probability of this incident is:

$$P_{111}(X_1Y_1Z_1) = (0.1000) \times (0.7725) \times (0.7490) = 0.05786$$

If we step through the transverse penetrations Z_2 through Z_6, we find that only the WB1 compartment is damaged for each of these cases. The probabilities for these cases are 0.01074, 0.00216, 0.00216, 0.00216, 0.00216, and 0.00216 respectively. The combined probability for the six cases at longitudinal damage location X_1 is:

$$P_{111-6}(X_1Y_1Z_{1-6}) = 0.05786 + 0.01074 + 0.00216 + 0.00216$$
$$+ 0.00216 + 0.00216 = 0.07725$$

Next, we move to damage extent Y_2. The damage box $X_1Y_2Z_1$ once again falls within the WB1 compartment. Likewise, transverse penetrations Z_2 through Z_6 fall within this compartment. We compute the probability for these cases and find that $P_{121-6}(X_1Y_2Z_{1-6}) = 0.01925$.

Similarly, the damage boxes defined by $X_1Y_3Z_{1-6}$ lie within the WB1 compartment and have a combined probability $P_{131-6}(X_1Y_3Z_{1-6}) = 0.00350$.

We now move to the next longitudinal location, X_2. With longitudinal extent Y_1, the damage stays within the WB1 compartment. The combined probability is $P_{211-6}(X_2Y_1Z_{1-6}) = 0.07725$.

The forward bound of the damage box $X_2Y_2Z_1$ extends forward of the transverse bulkhead located 20.0 m from the transom, damaging compartments both fore and aft of this bulkhead. Transverse penetration Z_1 extends to a point just outboard of the longitudinal bulkhead. Therefore, this combination damages both the WB1 and WB2S compartments. The probability is $P_{221}(X_2Y_2Z_1) = 0.01442$.

We find that the damage box $X_2Y_2Z_2$ extends inboard of the longitudinal bulkhead, damaging compartments WB1, WB2S and CO1. A cargo oil tank has been damaged and oil outflow will occur. Similarly, damage penetrations Z_3 through Z_6 result in breaching of the three compartments. The combined probability for these five incidents is:

$$P_{222-6}(X_2Y_2Z_{2-6}) = 0.00268 + 0.00054 + 0.00054$$
$$+ 0.00054 + 0.00054 = 0.00483$$

By stepping through the barge for all 180 incidents and combining unique damage compartment groupings, we obtain the compartment grouping and probability values shown in table A2 (Probability values for side damage). Each compartment group represents a unique set of compartments. The associated probability is the probability that each particular group of compartments will be damaged in a collision which breaches the hull. For instance, the probability of damaging the WB1 compartment is 0.17725. This means there is approximately a 17.7% likelihood that only this compartment will be damaged. Likewise, the probability of concurrently damaging the WB1 and WB2S compartments is 0.03408 or about 3.4%. Note that the cumulative probability of occurrence for all groups equals 1.0.

2.3.2 *Bottom damage evaluation*

For bottom damage, the probability of a given damage longitudinal location, longitudinal extent, vertical penetration, transverse location and transverse extent is, analogously to the side damage evaluation, the product of the probabilities of these five damage characteristics.

The following increments are assumed for the bottom damage evaluation:

Longitudinal location at 10 steps	$= L/10$	$= 0.10L$ per step
Longitudinal extent at 8 steps	$= 0.8L/8$	$= 0.10L$ per step
Vertical penetration at 6 steps	$= 0.3D/6$	$= 0.05D$ per step.

To further simplify the evaluation, all damage is assumed to extend transversely without limit. Therefore, the probabilities of transverse extent and transverse location are taken as 1.0 for each damage case.

Compartment groupings are developed using the same process as previously described for side damage.

Analogously, a total of ten longitudinal locations, eight longitudinal extents and six vertical penetrations need to be evaluated. The damage incidents to be taken into account for groundings sum up to a total of $(10) \times (8) \times (6) = 480$ separate incidents.

Figure A3 (Bottom damage definition) shows the steps for longitudinal location, longitudinal extent and vertical penetration in relation to the barge. Table A3 (Increments for step-wise bottom damage definition) gives the range for each step, the mean or average value over the step, and the probability of occurrence of that particular step.

Again, putting the aftmost compartment WB1 together in terms of damage increments, the following probabilities have to be summed up:

$$P_{111-6}(X_1 Y_1 Z_{1-6}) = (0.0240) \times (0.38333) \times (1.0) \quad = \quad 0.00920$$

$$P_{121-6}(X_1 Y_2 Z_{1-6}) = (0.0240) \times (0.2500) \times (1.0) \quad = \quad 0.00600$$

$$P_{131-6}(X_1 Y_3 Z_{1-6}) = (0.0240) \times (0.11677) \times (1.0) \quad = \quad 0.00280$$

$$P_{211-6}(X_2 Y_1 Z_{1-6}) = (0.0320) \times (0.38333) \times (1.0) \quad = \quad 0.01227.$$

Therefore the likelihood of damaging the WB1 compartment sums up to:

$$P_{WB1} = P_{11} + P_{12} + P_{13} + P_{21} = 0.03027.$$

By addressing each of the 480 incidents to the relevant compartment (or groups of compartments) the likelihood of a damage to these resulting from a grounding is obtained. This is shown in table A4 (Probability values for bottom damage).

2.4 Step 4: Computation of the equilibrium condition for each damage case

This example describes the concept analysis only. Damage stability analyses to determine the equilibrium conditions are only required for the final shipyard design, in accordance with paragraph 5.1.5.10 of the Guidelines.

2.5 Step 5: Computation of the oil outflow for each damage case

In this step the oil outflow associated with each of the compartment groupings is calculated for side and bottom damage as outlined below.

2.5.1 *Side damage evaluation*

For side damage, 100% of the oil in a damaged cargo oil tank is assumed to outflow into the sea. If we review the eleven compartment groupings for side damage, we find that oil tank damage occurs in three combinations: CO1 only, CO2 only, and concurrent damage to CO1 and CO2. The oil outflow for these tanks is as follows:

CO1 (98% full volume)	=	9,430 m^3
CO2 (98% full volume)	=	28,291 m^3
CO1 + CO2 (98% full volume)	=	37,721 m^3.

2.5.2 *Bottom damage evaluation*

For bottom damage, a pressure balance calculation must be carried out. The vessel is assumed to remain stranded on a shelf at its original intact draught. For the concept analysis, zero trim and zero heel are assumed. An inert gas overpressure of 0.05 bar gauge is assumed in accordance with paragraph 5.1.5.4 of the Guidelines. The double bottom spaces located below the cargo oil tanks "capture" some portion of the oil outflow. In accordance with paragraph 5.1.5.7 of the Guidelines, the flooded volume of such spaces should be assumed to contain 50% oil and 50% seawater by volume at equilibrium. When calculating the oil volume captured in these spaces, no assumptions are made on how the oil and seawater is distributed in these spaces.

The calculations are generally carried out for three tidal conditions: 0.0 m tide, with a 2.0 m tidal drop, and with a 6.0 m tidal drop. In accordance with paragraph 5.1.3 of the Guidelines, the tidal drop need not be taken greater than 50% of the ship's maximum draught. For this example, the appropriate tidal conditions are therefore 0.0 m, 2.0 m and 4.5 m.

The actual oil volume lost from a cargo tank is calculated for each of the three tidal conditions, assuming hydrostatic balance as follows:

$$z_c \cdot \rho_c \cdot g + 100\Delta p = z_s \cdot \rho_s \cdot g$$

where:

z_c = height of remaining oil in the damaged tank (m)

ρ_c = cargo oil density (0.9 t/m^3)

g = gravitational acceleration (9.81 m/s^2)

Δp = set pressure of cargo tank pressure/vacuum valves (0.05 bar gauge)

z_s = external seawater head above inner bottom (m)

z_s = $T - 2 = 7.00$ m

ρ_s = seawater density (1.025 t/m^3)

See also figure A4.

From the above equation one obtains for the height of remaining oil z_c for the zero-tide condition:

z_c = 7.40 m.

Thus, the height of lost oil ($h_l = 0.98 h_c - z_c$) is:

h_l = $17.64 - 7.40 = 10.24$ m.

The volume of lost oil (V_l) of cargo tank CO1 is:

V_l = $10.24 \times 36 \times 15 \times 0.99 = 5{,}474$ m^3.

In this case the total volume (V_{wo}) of oil and water in the water ballast tanks is:

V_{wo} = $2 \times [20 \times 2 + z_{wo} \times 2] \times 60 \times 0.95 = 6{,}202$ m^3

where:

z_{wo} = $0.5(z_c + z_s) = 7.20$ m.

If one assumes that 50% of V_{wo} is occupied by captured oil, one obtains for the total oil outflow ($V_{outflow}$) of cargo tank CO1:

$V_{outflow}$ = $V_l - 0.5 V_{wo} = 2{,}373$ m^3.

The oil outflow of cargo tank CO2 is:

$V_{outflow}$ = $10.24 \times 36 \times 45 \times 0.99 - 0.5 \times 6{,}202 = 13{,}322$ m^3

and the total oil outflow of cargo tanks CO1 and CO2 is:

$V_{outflow}$ = $10.24 \times 36 \times 60 \times 0.99 - 0.5 \times 6{,}202 = 18{,}796$ m^3.

Step-wise application of the damage extents and assumed increments results in fourteen compartment groupings for bottom damage. Oil tank and double bottom damage occurs in three combinations. The oil outflows for these tanks at 0.0 m, 2.0 m and 4.5 m tide are summarized in the table below:

Tank combination	Oil outflow (m^3) at		
	0.0 m tide	2.0 m tide	4.5 m tide
WB2S + WB2P + CO1	2,373	3,832	5,658
WB2S + WB2P + CO2	13,322	17,210	22,081
WB2S + WB2P + CO1 + CO2	18,796	23,898	30,292

2.6   Step 6: Computation of the oil outflow parameters

In this step the oil outflow parameters are computed in accordance with paragraph 4.3 of the Guidelines. To facilitate calculation of these parameters, place the damage groupings in a table in ascending order as a function of oil outflow. A running sum of probabilities is computed, beginning at the minimum outflow damage case and proceeding to the maximum outflow damage case. Tables A5 and A6 (Cumulative probability and oil outflow values) contain the outflow values for the side damage and bottom damage for the three tide conditions.

Probability of zero oil outflow, P_0: This parameter equals the cumulative probability for all damage cases for which there is no oil outflow. From table A5, we see that the probability of zero outflow for the side damage condition is 0.83798, and the probability of zero outflow for the bottom damage (0.0 m tide) condition is 0.84313.

Mean oil outflow parameter, O_M: This is the weighted average of all cases, and is obtained by summing the products of each damage case probability and the computed outflow for that damage case.

Extreme oil outflow parameter, O_E: This represents the weighted average of the damage cases falling within the cumulative probability range between 0.9 and 1.0. It equals the sum of the products of each damage case probability with a cumulative probability between 0.90 and 1.0 and its corresponding oil outflow, with the result multiplied by 10.

For this example, the computed outflow values are as shown in tables A5 and A6. In accordance with paragraph 5.1.3 of the Guidelines, the bottom damage outflow parameters for the 0.0 m, 2.0 m and 4.5 m tides are combined in a ratio of 0.4 : 0.5 : 0.1 respectively. In accordance with paragraph 5.1.2, the collision (side damage) and stranding (bottom damage) parameters are then combined in a ratio of 0.4 : 0.6 respectively. In table A7 (Summary of oil outflow parameters) the oil outflow parameters P_0, O_M and O_E for the example tank barge are listed.

Table A1 – Increments for step-wise side damage evaluation

Longitudinal location (step $= 0.1L$)

	range of increments			
	minimum	maximum	midpoint	probability
X_1	0.0L	0.1L	0.05L	0.1000
X_2	0.1L	0.2L	0.15L	0.1000
X_3	0.2L	0.3L	0.25L	0.1000
X_4	0.3L	0.4L	0.35L	0.1000
X_5	0.4L	0.5L	0.45L	0.1000
X_6	0.5L	0.6L	0.55L	0.1000
X_7	0.6L	0.7L	0.65L	0.1000
X_8	0.7L	0.8L	0.75L	0.1000
X_9	0.8L	0.9L	0.85L	0.1000
X_{10}	0.9L	1.0L	0.95L	0.1000

1.0000

Longitudinal extent (step = $0.1L$)

	range of extents			
	minimum	maximum	average	probability
Y_1	0.0L	0.1L	0.05L	0.7725
Y_2	0.1L	0.2L	0.15L	0.1925
Y_3	0.2L	0.3L	0.25L	0.0350

1.0000

Transverse penetration (step = $0.05B$)

	range of penetration			
	minimum	maximum	average	probability
Z_1	0.0B	0.05B	0.025B	0.7490
Z_2	0.05B	0.10B	0.075B	0.1390
Z_3	0.10B	0.15B	0.125B	0.0280
Z_4	0.15B	0.20B	0.175B	0.0280
Z_5	0.20B	0.25B	0.225B	0.0280
Z_6	0.25B	0.30B	0.275B	0.0280

1.0000

Table A2 – Probability values for side damage

No.	Unique compartment groupings	Damage extents and probabilities							Group probability
1	WB1	$X_1Y_1Z_{1-6}$ 0.07725	$X_1Y_2Z_{1-6}$ 0.01925	$X_1Y_3Z_{1-6}$ 0.00350	$X_2Y_1Z_{1-6}$ 0.07725				0.17725
2	WB1+WB2S	$X_2Y_2Z_1$ 0.01442	$X_2Y_3Z_1$ 0.00262	$X_3Y_3Z_1$ 0.00262	$X_3Y_2Z_1$ 0.01442				0.03408
3	WB1+WB2S+CO1	$X_2Y_2Z_{2-6}$ 0.00483	$X_2Y_3Z_{2-6}$ 0.00088	$X_3Y_2Z_{2-6}$ 0.00483					0.01054
4	WB2S	$X_3Y_1Z_1$ 0.05786	$X_4Y_1Z_1$ 0.05786	$X_4Y_2Z_1$ 0.01442	$X_4Y_3Z_1$ 0.00262	$X_5Y_1Z_1$ 0.05786	$X_5Y_2Z_1$ 0.01442	$X_5Y_3Z_1$ 0.00262	0.41532
		$X_6Y_1Z_1$ 0.05786	$X_6Y_2Z_1$ 0.01442	$X_6Y_3Z_1$ 0.00262	$X_7Y_1Z_1$ 0.05786	$X_7Y_2Z_1$ 0.01442	$X_7Y_3Z_1$ 0.00262	$X_8Y_1Z_1$ 0.05786	
5	WB2S+CO1	$X_3Y_1Z_{2-6}$ 0.01939							0.01939
6	WB2S+CO1+CO2	$X_4Y_1Z_{2-6}$ 0.01939	$X_4Y_2Z_{2-6}$ 0.00483	$X_4Y_3Z_{2-6}$ 0.00088	$X_5Y_3Z_{2-6}$ 0.00088				0.02598
7	WB1+WB2S+CO1+CO2	$X_3Y_3Z_{2-6}$ 0.00088							0.00088
8	WB2S+CO2	$X_5Y_1Z_{2-6}$ 0.01939	$X_5Y_2Z_{2-6}$ 0.00483	$X_6Y_1Z_{2-6}$ 0.01939	$X_6Y_2Z_{2-6}$ 0.00483	$X_6Y_3Z_{2-6}$ 0.00088	$X_7Y_1Z_{2-6}$ 0.01939	$X_7Y_2Z_{2-6}$ 0.00483	0.09381
		$X_7Y_3Z_{2-6}$ 0.00088	$X_8Y_1Z_{2-6}$ 0.01939						
9	WB2S+WB3	$X_8Y_2Z_1$ 0.01442	$X_8Y_3Z_1$ 0.00262	$X_9Y_2Z_1$ 0.01442	$X_9Y_3Z_1$ 0.00262				0.03408
10	WB2S+CO2+WB3	$X_8Y_2Z_{2-6}$ 0.00483	$X_8Y_3Z_{2-6}$ 0.00088	$X_9Y_2Z_{2-6}$ 0.00483	$X_9Y_3Z_{2-6}$ 0.00088				0.01142
11	WB3	$X_9Y_1Z_{1-6}$ 0.07725	$X_{10}Y_1Z_{1-6}$ 0.07725	$X_{10}Y_2Z_{1-6}$ 0.01925	$X_{10}Y_3Z_{1-6}$ 0.00350				0.17725
									1.00000

Table A3 – Increments for step-wise bottom damage definition

Longitudinal location (step = $0.1L$)

	range of increments			
	minimum	maximum	midpoint	probability
X_1	$0.0L$	$0.1L$	$0.05L$	0.0240
X_2	$0.1L$	$0.2L$	$0.15L$	0.0320
X_3	$0.2L$	$0.3L$	$0.25L$	0.0400
X_4	$0.3L$	$0.4L$	$0.35L$	0.0480
X_5	$0.4L$	$0.5L$	$0.45L$	0.0560
X_6	$0.5L$	$0.6L$	$0.55L$	0.0800
X_7	$0.6L$	$0.7L$	$0.65L$	0.1200
X_8	$0.7L$	$0.8L$	$0.75L$	0.1600
X_9	$0.8L$	$0.9L$	$0.85L$	0.2000
X_{10}	$0.9L$	$1.0L$	$0.95L$	0.2400
				1.0000

Longitudinal extent (step = $0.1L$)

	range of extents			
	minimum	maximum	average	probability
Y_1	$0.0L$	$0.1L$	$0.05L$	0.3833
Y_2	$0.1L$	$0.2L$	$0.15L$	0.2500
Y_3	$0.2L$	$0.3L$	$0.25L$	0.1167
Y_4	$0.3L$	$0.4L$	$0.35L$	0.0500
Y_5	$0.4L$	$0.5L$	$0.45L$	0.0500
Y_6	$0.5L$	$0.6L$	$0.55L$	0.0500
Y_7	$0.6L$	$0.7L$	$0.65L$	0.0500
Y_8	$0.7L$	$0.8L$	$0.75L$	0.0500
				1.0000

Vertical penetration (step = $0.05D$)

	range of penetration			
	minimum	maximum	average	probability
Z_1	$0.0D$	$0.05D$	$0.025D$	0.5575
Z_2	$0.05D$	$0.10D$	$0.075D$	0.2225
Z_3	$0.10D$	$0.15D$	$0.125D$	0.0550
Z_4	$0.15D$	$0.20D$	$0.175D$	0.0550
Z_5	$0.20D$	$0.25D$	$0.225D$	0.0550
Z_6	$0.25D$	$0.30D$	$0.275D$	0.0550
				1.0000

Table A4 – Probability values for bottom damage

	Unique compartment groupings	Damage extents and probabilities								Group probability
1	WB1	$X_{1-2}Y_1Z_{1-6}$ 0.02147	$X_1Y_2Z_{1-6}$ 0.006	$X_1Y_3Z_{1-6}$ 0.0028						0.03027
2	WB1+WB2S+WB2P		$X_{2-3}Y_2Z_{1-2}$ 0.01404	$X_{2-3}Y_3Z_{1-2}$ 0.00655	$X_{1-4}Y_4Z_{1-2}$ 0.00562	$X_{1-4}Y_5Z_{1-2}$ 0.00562	$X_{1-5}Y_6Z_{1-2}$ 0.00078	$X_{1-5}Y_7Z_{1-2}$ 0.0078	$X_{1-4}Y_8Z_{1-2}$ 0.00562	0.05305
3	WB2S+WB2P+WB3		$X_{8-9}Y_2Z_{1-2}$ 0.0702	$X_{8-9}Y_3Z_{1-2}$ 0.03276	$X_{7-10}Y_4Z_{1-2}$ 0.02808	$X_{7-10}Y_5Z_{1-2}$ 0.02808	$X_{6-10}Y_6Z_{1-2}$ 0.0312	$X_{6-10}Y_7Z_{1-2}$ 0.0312	$X_{7-10}Y_8Z_{1-2}$ 0.02808	0.24960
4	WB1+WB2S+WB2P+WB3								$X_{5-6}Y_8Z_{1-2}$ 0.00530	0.00530
5	WB2S+WB2P	$X_{3-8}Y_1Z_{1-2}$ 0.1507	$X_{4-7}Y_2Z_{1-2}$ 0.05928	$X_{4-7}Y_3Z_{1-2}$ 0.02766	$X_{5-6}Y_4Z_{1-2}$ 0.0053	$X_{5-6}Y_5Z_{1-2}$ 0.0053				0.24824
6	WB3	$X_{9-10}Y_1Z_{1-6}$ 0.16867	$X_{10}Y_2Z_{1-6}$ 0.06	$X_{10}Y_3Z_{1-6}$ 0.028						0.25667
7	WB1+WB2S+WB2P+CO1		$X_{2-3}Y_2Z_{3-6}$ 0.00396	$X_2Y_3Z_{3-6}$ 0.00082	$X_{1-2}Y_4Z_{3-6}$ 0.00062	$X_1Y_5Z_{3-6}$ 0.00026	$X_1Y_6Z_{3-6}$ 0.00026			0.00592
8	WB2S+WB2P+CO1	$X_3Y_1Z_{3-6}$ 0.00337								0.00337
9	WB2S+WB2P+CO2	$X_{5-8}Y_1Z_{3-6}$ 0.03508	$X_{5-7}Y_2Z_{3-6}$ 0.01408	$X_{6-7}Y_3Z_{3-6}$ 0.00513	$X_6Y_4Z_{3-6}$ 0.00088					0.05517
10	WB2S+WB2P+WB3+CO2		$X_{8-9}Y_2Z_{3-6}$ 0.0198	$X_{8-9}Y_3Z_{3-6}$ 0.00924	$X_{7-10}Y_4Z_{3-6}$ 0.00792	$X_{7-10}Y_5Z_{3-6}$ 0.00792	$X_{7-10}Y_6Z_{3-6}$ 0.00792	$X_{8-10}Y_7Z_{3-6}$ 0.0066	$X_{8-10}Y_8Z_{3-6}$ 0.0066	0.06600
11	WB1+WB2S+WB2P+CO1+CO2			$X_{3-4}Y_3Z_{3-6}$ 0.00098	$X_{3-4}Y_4Z_{3-6}$ 0.00098	$X_{2-4}Y_5Z_{3-6}$ 0.00132	$X_{2-5}Y_6Z_{3-6}$ 0.00194	$X_{1-5}Y_7Z_{3-6}$ 0.0022	$X_{1-4}Y_8Z_{3-6}$ 0.00158	0.00903
12	WB2S+WB2P+WB3+CO1+CO2						$X_6Y_6Z_{3-6}$ 0.00088	$X_{6-7}Y_7Z_{3-6}$ 0.0022	$X_7Y_8Z_{3-6}$ 0.00132	0.00440
13	WB1+WB2S+WB2P+WB3+CO1+CO2								$X_{5-6}Y_8Z_{3-6}$ 0.0015	0.00150
14	WB2S+WB2P+CO1+CO2	$X_4Y_1Z_{3-6}$ 0.00405	$X_4Y_2Z_{3-6}$ 0.00264	$X_{4-5}Y_3Z_{3-6}$ 0.00267	$X_5Y_4Z_{3-6}$ 0.00062	$X_{5-6}Y_5Z_{3-6}$ 0.0015				0.01148
										1.00000

Table A5 – Cumulative probability and oil outflow values

Side damage

Compartment groupings	Oil outflow O_i (m³)	Probability P_i	Cumulative probability [sum of P_i]	Mean oil outflow $P_i \times O_i$ (m³)	Probability P_{ie}	Extreme outflow $O_{ie} \times P_{ie} \times 10$ (m³)
WB1	0.00	0.17725	0.17725	0.00		
WB1+WB2S	0.00	0.03408	0.21133	0.00		
WB2S	0.00	0.41532	0.62665	0.00		
WB2S+WB3	0.00	0.03408	0.66073	0.00		
WB3	0.00	0.17725	0.83798	0.00		
WB1+WB2S+CO1	9430.00	0.01054	0.84852	99.39		
WB2S+CO1	9430.00	0.01939	0.86791	182.85		
WB2S+CO2	28291.00	0.09381	0.96172	2653.98	0.06172	17461.2052
WB2S+CO2+WB3	28291.00	0.01142	0.97314	323.08	0.01142	3230.8322
WB1+WB2S+CO1+CO2	37721.00	0.00088	0.97402	33.19	0.00088	331.9448
WB2S+CO1+CO2	37721.00	0.02598	1.00000	979.99	0.02598	9799.9158
				4272.48	0.10000	30823.898

Table A5 – Cumulative probability and oil outflow values *(continued)*

Bottom damage (0.0 metre tide)

Compartment groupings	Oil outflow O_i (m^3)	Probability P_i	Cumulative probability [sum of P_i]	Mean oil outflow $P_i \times O_i$ (m^3)	Probability P_{ie}	Extreme outflow $O_{ie} \times P_{ie} \times 10$ (m^3)
WB1	0.00	0.03027	0.03027	0.00		
WB1 + WB2S + WB2P	0.00	0.05304	0.08331	0.00		
WB1 + WB2S + WB2P + WB3	0.00	0.00530	0.08861	0.00		
WB2S + WB2P	0.00	0.24825	0.33686	0.00		
WB2S + WB2P + WB3	0.00	0.24960	0.58646	0.00		
WB3	0.00	0.25667	0.84313	0.00		
WB1 + WB2S + WB2P + CO1	2373.00	0.00592	0.84905	14.05		
WB2S + WB2P + CO1	2373.00	0.00337	0.85242	8.00		
WB2S + WB2P + CO2	13322.00	0.05518	0.90760	735.11	0.00760	1012.4720
WB2S + WB2P + WB3 + CO2	13322.00	0.06600	0.97360	879.25	0.06600	8792.5200
WB1 + WB2S + WB2P + CO1 + CO2	18796.00	0.00903	0.98263	169.73	0.00903	1697.2788
WB3 + WB2S + WB2P + CO1 + CO2	18796.00	0.00150	0.98413	28.19	0.00150	281.9400
WB1 + WB2S + WB2P + WB3 + CO1 + CO2	18796.00	0.00440	0.98853	82.70	0.00440	827.0240
WB2S + WB2P + CO1 + CO2	18796.00	0.01147	1.00000	215.59	0.01147	2155.9012
				2132.62	0.10000	14767.1360

Table A6 – Cumulative probability and oil outflow values

Bottom damage (2.0 metre tide)

Compartment groupings	Oil outflow O_i (m³)	Probability P_i	Cumulative probability [sum of P_i]	Mean oil outflow $P_i \times O_i$ (m³)	Probability P_{ie}	Extreme outflow $O_{ie} \times P_{ie} \times 10$ (m³)
WB1	0.00	0.03027	0.03027	0.00		
WB1+WB2S+WB2P	0.00	0.05304	0.08331	0.00		
WB1+WB2S+WB2P+WB3	0.00	0.00530	0.08861	0.00		
WB2S+WB2P	0.00	0.24825	0.33686	0.00		
WB2S+WB2P+WB3	0.00	0.24960	0.58646	0.00		
WB3	0.00	0.25667	0.84313	0.00		
WB1+WB2S+WB2P+CO1	3832.00	0.00592	0.84905	22.69		
WB2S+WB2P+CO1	3832.00	0.00337	0.85242	12.91		
WB2S+WB2P+CO2	17210.00	0.05518	0.90760	949.65	0.00760	1307.9600
WB2S+WB2P+WB3+CO2	17210.00	0.06600	0.97360	1135.86	0.06600	11358.6000
WB1+WB2S+WB2P+CO1+CO2	23898.00	0.00903	0.98263	215.80	0.00903	2157.9894
WB3+WB2S+WB2P+CO1+CO2	23898.00	0.00150	0.98413	35.85	0.00150	358.4700
WB1+WB2S+WB2P+WB3+CO1+CO2	23898.00	0.00440	0.98853	105.15	0.00440	1051.5120
WB2S+WB2P+CO1+CO2	23898.00	0.01147	1.00000	274.11	0.01147	2741.1006
				2752.02	0.10000	18975.6320

221

Table A6 – Cumulative probability and oil outflow values *(continued)*

Bottom damage (4.5 metre tide)

Compartment groupings	Oil outflow O_i (m³)	Probability P_i	Cumulative probability [sum of P_i]	Mean oil outflow $P_i \times O_i$ (m³)	Probability P_{ie}	Extreme outflow $O_{ie} \times P_{ie} \times 10$ (m³)
WB1	0.00	0.03027	0.03027	0.00		
WB1 + WB2S + WB2P	0.00	0.05304	0.08331	0.00		
WB1 + WB2S + WB2P + WB3	0.00	0.00530	0.08861	0.00		
WB2S + WB2P	0.00	0.24825	0.33686	0.00		
WB2S + WB2P + WB3	0.00	0.24960	0.58646	0.00		
WB3	0.00	0.25667	0.84313	0.00		
WB1 + WB2S + WB2P + CO1	5658.00	0.00592	0.84905	33.50		
WB2S + WB2P + CO1	5658.00	0.00337	0.85242	19.07		
WB2S + WB2P + CO2	22081.00	0.05518	0.90760	1218.43	0.00760	1678.1560
WB2S + WB2P + WB3 + CO2	22081.00	0.06600	0.97360	1457.35	0.06600	14573.4600
WB1 + WB2S + WB2P + CO1 + CO2	30292.00	0.00903	0.98263	273.54	0.00903	2735.3676
WB3 + WB2S + WB2P + CO1 + CO2	30292.00	0.00150	0.98413	45.44	0.00150	454.3800
WB1 + WB2S + WB2P + WB3 + CO1 + CO2	30292.00	0.00440	0.98853	133.28	0.00440	1332.8480
WB2S + WB2P + CO1 + CO2	30292.00	0.01147	1.00000	347.45	0.01147	3474.4924
				3528.05	0.10000	24248.7040

Table A7 – Summary of oil outflow parameters

Bottom damage	(40%) 0.0 m tide	(50%) 2.0 m tide	(10%) 4.5 m tide	Combined
Probability of zero outflow P_0	0.8431	0.8431	0.8431	0.8431
Mean outflow (m^3)	2133	2752	3528	2582
Extreme outflow (m^3)	14767	18976	24249	17820

Combined side and bottom damage	(40%) Side damage	(60%) Bottom damage	Combined
Probability of zero outflow P_0	0.8380	0.8431	0.8411
Mean outflow (m^3)	4272	2582	3258
Extreme outflow (m^3)	30824	17820	23021
Mean outflow parameter O_M			0.0864
Extreme outflow parameter O_E			0.6103

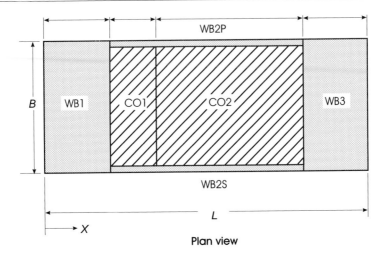

Plan view

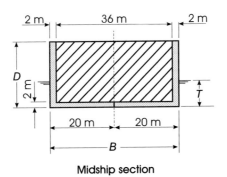

Midship section

Barge particulars

L = _____ 100 m

B = _____ 40 m

D = _____ 20 m

T = _____ 9 m

displacement = _____ 36,900 t

light barge weight = _____ 2,951 t

CO1, CO2 = _____ cargo oil tanks

WB1, WB2, WB3 = water ballast tanks

Figure A1 – Barge arrangement

224

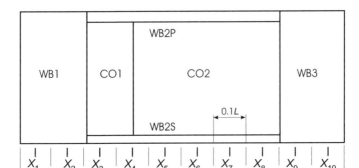

Longitudinal damage location

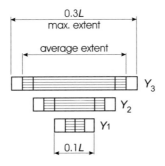

Longitudinal damage extent

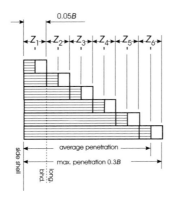

Transverse damage penetration

Figure A2 – Side damage definition

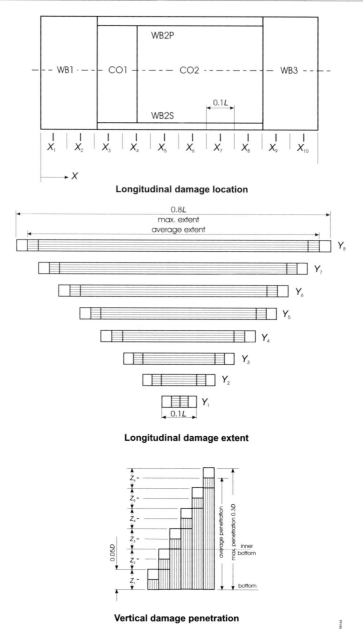

Longitudinal damage location

Longitudinal damage extent

Vertical damage penetration

Figure A3 – Bottom damage definition

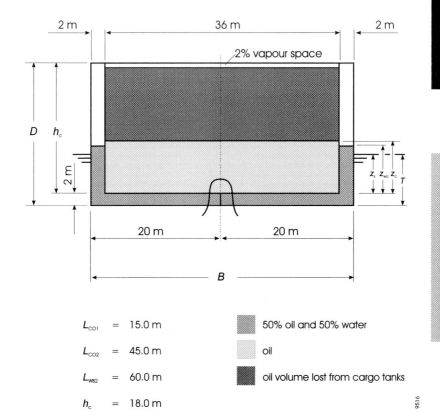

L_{CO1}	=	15.0 m
L_{CO2}	=	45.0 m
L_{WB2}	=	60.0 m
h_c	=	18.0 m

■ 50% oil and 50% water

□ oil

■ oil volume lost from cargo tanks

Figure A4 – Oil outflow scheme for bottom damage

Appendix 8

Guidelines for approval of alternative structural or operational arrangements as called for in MARPOL 73/78, Annex I, regulation 13G(7)*

Background

1   Regulation 13G(4) of Annex I of MARPOL 73/78 specifies the requirements applicable to existing crude oil tankers of 20,000 tons deadweight and above and product carriers of 30,000 tons deadweight and above to reduce the accidental outflow of oil in the event of a collision or stranding. Regulation 13G(7) permits other structural or operational arrangements to be accepted as alternatives, provided that such alternatives provide at least the same level of protection against oil pollution in the event of collision or stranding, and are approved by the Administration based on guidelines developed by the Organization.

The guidelines contained herein specify the criteria by which the acceptability of alternative arrangements should be determined. Methods approved by the MEPC at the time of development of the guidelines are detailed in the appendix.

Other alternative arrangements may be approved by the MEPC after considering their pollution-prevention and safety characteristics. A proposal for approval of a new or revised arrangement should be submitted by an Administration and contain technical and operational specifications and evaluation of any safety aspects.

Applicability

2   These guidelines apply to crude oil tankers of 20,000 tons deadweight and above and product carriers of 30,000 tons deadweight and above which are not required to comply with regulation 13F and do not satisfy the requirements of regulation 13G(1)(c).

Performance requirements

3   The required minimum protection against accidental oil outflow is governed by regulation 13G(4), which stipulates that tankers to which regulation 13G applies shall have wing tanks or double-bottom spaces, not used for the carriage of oil and meeting the width and height requirements of regulation 13E(4), covering at least 30% of L_t for the full depth of the ship on each side or at least 30% of the projected bottom shell area within the length L_t, where L_t is as defined in regulation 13E(2). Equivalent structural or operational arrangements, as permitted by regulation 13G(7), should ensure at least the same degree of protection against oil pollution in the event of collision or stranding. The equivalency should be determined by calculations in accordance with paragraphs 4 and 5 below.

---

* An interpretation (IACS Unified Interpretation MPC 7) of sections of these guidelines has been circulated as MEPC/Circ. 365. It is included as appendix 9 of these Interpretations.

Damage and outflow criteria

4   The oil outflow should be calculated for the damage cases identified in subparagraph 5.1 of these guidelines. The hypothetical outflow should be calculated for the conditions specified in subparagraphs 4.1, 4.2 and 4.3 below and in accordance with the procedures defined in subparagraphs 5.2, 5.3 and 5.4. The hypothetical outflows so calculated, divided by the volume of the cargo being carried by the ship in its original configuration, and expressed as a percentage, constitute the equivalent oil spill number (the EOS number) for the ship under each of the conditions detailed in subparagraphs 4.1, 4.2 and 4.3.

4.1   The EOS number should be calculated for the existing ship, with the ship loaded to the maximum assigned load line with zero trim and with cargo having a uniform density, allowing all cargo tanks to be loaded to 98% full. This calculation establishes the base EOS number and also the nominal cargo oil density, which should be applied in the calculations required by subparagraphs 4.2 and 4.3.

4.2   A second EOS number should be calculated for the ship arranged with non-cargo side tanks as referred to in regulation 13G(4).

4.3   A third EOS number should be calculated for the selected alternative method and should not exceed the EOS number as calculated according to subparagraph 4.2, and should furthermore not be greater than 85% of the EOS number calculated according to subparagraph 4.1.

4.4   Fuel oil tanks located within the cargo tank length should be considered as cargo oil tanks for the purpose of calculating the EOS numbers.

Methodology for calculation of the hypothetical oil outflow

5   The methodology detailed in this paragraph should be used for calculating the EOS number as required by paragraph 4.

5.1   Damage assumptions

The damage assumptions identified below should be applied to all oil tanks when calculating the EOS number.

5.1.1   *Side damage*

Longitudinal extent	l_c	=	$\frac{1}{3}L^{2/3}$ or 14.5 m whichever is less
Transverse extent	t_c	=	$B/5$ or 11.5 m whichever is less
Vertical extent	v_c	=	from the baseline upwards without limit

5.1.2   *Bottom damage*

Longitudinal extent	l_s	=	$0.2L$
Transverse extent	b_s	=	$B/6$ or 10 m whichever is less, but not less than 5 m
Vertical extent from the baseline	v_s	=	$B/15$

229

5.2 Calculation of outflow in case of side damage

Calculation of the outflow from a side damage should be done as follows:

Length between the forward and after extremities of the cargo tanks	= L_t	(m)
Length of tank number i	= l_i	(m)
Distance from hull plating to the tank boundary	= s_i	(m)
Cargo volume in tank number i	= V_i	(m³)
Length of side damage according to subparagraph 5.1.1	= l_c	(m)
Transverse extent of damage according to subparagraph 5.1.1	= t_c	(m)

Even longitudinal distribution of damage location is assumed

Probability factor for breaching tank number i due to side damage

$$q_{ci} = (1 - s_i/t_c)\frac{(l_i + l_c)}{(L_t + l_c)}$$

$(1 - s_i/t_c)$ to be ≥ 0

Total hypothetical outflow in case of a side damage

$$O_c = \sum q_{ci} \cdot V_i$$

This calculation method gives appropriate credit for any number and size of side ballast tanks. It also takes into account the effect of the cargo tank size. The risk of breaching a longitudinal bulkhead and outflow from centre tanks is also taken into account.

5.3 Calculation of outflow in case of bottom damage

Calculation of the outflow from bottom damages should be done as follows:

Length between the forward and after extremities of the cargo tanks	= L_t	(m)
Width of the cargo tank area	= B_t	(m)
Length of tank number i	= l_i	(m)
Width of tank number i	= b_i	(m)
Height of a double bottom	= h_i	(m)
Cargo volume in tank number i	= V_i	(m³)
Length of a bottom damage according to subparagraph 5.1.2	= l_s	(m)
Width of a bottom damage according to subparagraph 5.1.2	= b_s	(m)
Vertical extent of a bottom damage according to subparagraph 5.1.2	= v_s	(m)

Probability factor for breaching tank number i due to bottom damage

$$q_{si} = (1 - h_i/v_s) \frac{(l_i + l_s)(b_i + b_s)}{(L_t + l_s)(B_t + b_s)}$$

$(1 - h_i/v_s)$ to be ≥ 0

Nominal density of the cargo according to paragraph 4	$=$	ρ_c (t/m^3)
Density of the seawater (normally 1.025)	$=$	ρ_s (t/m^3)
Loaded condition draught	$=$	d (m)
Height of cargo column above the cargo tank bottom	$=$	h_c (m)
Highest normal overpressure in the inert gas system (normally 0.05 bar)	$=$	Δp (bar)
Margin for average transient loss, swell and tide effects	$=$	1.1
Standard acceleration of gravity	g $=$	9.81 m/s^2

Outflow factor due to hydrostatic overpressure in tank number i

$$q_{hi} = 1 - \frac{(\rho_s(d - h_i)g - 100\Delta p)}{1.1\rho_c \cdot h_c \cdot g}$$

q_{hi} to be ≥ 0

Outflow from tank number i

$$O_{si} = q_{si} \cdot q_{hi} \cdot V_i$$

Total hypothetical outflow in case of a bottom damage

$$O_s = \sum q_{si} \cdot q_{hi} \cdot V_i$$

In case the ship is equipped with a double bottom, the calculated outflow from tanks located above such double bottom may be assumed to be reduced by 50% of the total capacity of the affected double bottom tanks but in no case by more than 50% of the calculated outflow from each tank.

5.4 Calculation of total outflow in case of a side or bottom damage

The outflow calculated under subparagraphs 5.2 and 5.3 above should be combined to the total hypothetical outflow as follows:

$$O_{tot} = 0.4O_c + 0.6O_s$$

Outflow-reducing arrangements

6 Alternative outflow-reducing methods as permitted under regulation 13G(7) may include a single method or a combination of methods giving protection in case of collision or stranding or both. Methods that have been approved by the MEPC are identified in the appendix.

Other methods may be accepted by the Organization. Such methods should, in addition to meeting the outflow criteria given in paragraphs 4 and 5, be evaluated in each individual case for acceptability from general operational and safety points of view. In particular any such method:

> should not expose the tanker to an unacceptable stress level in intact condition and should not cause the accidental hull damage to be exacerbated;

> should not create an unacceptable additional fire or explosion hazard.

Operations Manual

7   The master should be supplied with operational instructions, approved by the Administration, in which the operational conditions required for compliance with these guidelines should be clearly described. These instructions may be contained in a separate manual or be incorporated into existing shipboard manuals. These instructions should identify approved loading conditions, including part load conditions and including any ballasting used for obtaining these conditions. It should also contain information on the use of inert gas system and relevant trim, stress and stability information.

Endorsement of the IOPP Certificate/Supplement

8   The IOPP Certificate/Supplement should be endorsed to identify the structural or operational measures approved in accordance with regulation 13(G)(7) as well as the approved operations instructions.

Appendix

Arrangements acceptable as alternatives under regulation 13G(7) of Annex I of MARPOL 73/78

This appendix contains detailed requirements on arrangements accepted by the MEPC as alternatives under the provisions of regulation 13G(7) of Annex I of MARPOL 73/78. At the time of development this appendix contains only one approved alternative method.

Requirements for application of hydrostatic balance loading in cargo tanks

Hydrostatic balance loading is based on the principle that the hydrostatic pressure at the cargo tank bottom of the cargo oil column plus the ullage space inert gas overpressure remains equal to or less than the hydrostatic pressure of the outside water column, thereby mitigating the outflow of oil in case of bottom damage.

The maximum cargo level in each tank being loaded under this criterion should therefore satisfy the following equation:

$$h_c \cdot \rho_c \cdot g + 100\Delta p \leq (d - h_i) \cdot \rho_s \cdot g$$

where:

h_c is the maximum acceptable cargo level in each tank, measured from the cargo tank bottom   (m),

ρ_c is the density of the current cargo   (t/m^3),

d is the corresponding draught of the vessel   (m),

h_i is the height of the tank bottom above the keel   (m),

Δp is the highest normal overpressure in the inert gas system, expressed in bar (normally 0.05 bar)   (bar),

ρ_s is the density of the seawater   (t/m^3),

g is the standard acceleration of gravity   ($g = 9.81$ m/s^2).

Ballast may be carried in segregated ballast tanks to increase draught to a larger value. This may be used to allow more cargo to be taken into cargo tanks within the hydrostatic equilibrium criterion and within the limits of the assigned load line.

The arrangements and procedures for operation with the hydrostatic balance method should be approved by the Administration. The approval should be based on a system specification and documentation, incorporating also:

.1   calculations made to confirm whether or not resonance can occur between the natural period of longitudinal cargo liquid motion and the natural period of pitching of the ship, and also between the natural period of transverse cargo liquid motion and the natural period of rolling of the ship under approved cargo loading conditions and in any cargo tanks. In this context 'resonance can occur' means that the natural period of longitudinal motion of cargo oil is within the range from 60% to 130% of the natural period of pitching of the ship and/or the natural period of transverse motion of cargo is within the range from 80% to 120% of the natural period of rolling of the ship. When resonance can occur between ship's motion and cargo liquid motion, the sloshing pressure caused by such resonance should be estimated, and it should be confirmed that the existing structure has sufficient strength to withstand the estimated sloshing pressure; and

.2   calculations of intact and damage stability, including the effects of free surface. Damage stability calculations are, however, only required for ships defined in regulation 1(6).

When the accidental outflow reduction requirement can be met by applying hydrostatic loading to a limited number of tanks, wing tanks should have priority, thereby ensuring some reduction also in outflow from a side damage and minimizing sloshing in part-loaded centre tanks.

When operating in a multiport loading or unloading mode using the hydrostatic balance loading method, tanks covering at least 30% of the side of the length of the cargo section should be kept empty until the last loading location or should be unloaded at the first unloading location.

Copies of certified ullage measurement reports should be kept on board, clearly identified, for at least three years.

Appendix 9

Interpretation of requirements for application of hydrostatic balance loading in cargo tanks (resolution MEPC.64(36))*

1   The Marine Environment Protection Committee, at its forty-first session (30 March to 3 April 1998), noted that a large number of tankers of 25 years of age and over would potentially use the hydrostatic balance loading operational alternative which is permitted by MARPOL regulation I/13G(7), in order to continue to trade for another five years, and recognized that there was a need to develop a unified interpretation with the purpose of avoiding any potential problems which might arise with the hydrostatic balance loading.

2   Subsequently, the Committee, at its forty-second session (2 to 6 November 1998), having considered the recommendation made by the Sub-Committee on Bulk Liquids and Gases, at its third session, regarding IACS Unified Interpretation MPC 7 "Hydrostatic Balance Loading", agreed to circulate this Unified Interpretation to Member Governments, as set out in the annex, subject to the following clarifications:

  .1 all ballast tanks should be assumed empty when calculating EOS1 and EOS2, whereas ballast water allocation may be considered when calculating EOS3; and

  .2 it is understood that ballast water may be taken on board during the voyage in order to maintain the draughts necessary for compliance and to satisfy trim, stability, strength and other requirements.

3   At its forty-third session (28 June to 2 July 1999), the Committee approved an IACS proposal to make a number of minor corrections to the original interpretations.

4   As a result, this Circular includes these corrections and replaces MEPC/ Circ.347.

5   Member Governments are invited to use the annexed Interpretation together with the above clarifications when applying the provisions of the Guidelines for approval of alternative structural or operational arrangements, as called for in regulation 13G(7) of Annex I of MARPOL 73/78 (resolution MEPC.64(36)), to tankers of 25 years of age and over referred to in regulation 13G(4) of Annex I to MARPOL 73/78.

---

* This is MEPC Circular 365 of 26 July 1999.

Annex

IACS Unified Interpretation MPC 7 – Hydrostatic Balance Loading

(May 1998)

(Annex I, Regulation 13G(7) – Guidelines for approval of alternative structural or operational method, IMO resolution MEPC.64(36))

Damage and outflow criteria (as per 4 of the IMO Guidelines)

.1   The original configuration is the configuration of the vessel, as covered by the IOPP Certificate and the current G.A. plan prior to the application of MARPOL regulation 13G(7).

In the case of a product/crude oil carrier which operates alternatively with CBT when trading as product tanker or with COW when trading as a crude-oil tanker, the assessment in accordance with MEPC.64(36) should be done for each mode separately.

Calculation of base EOS number as per 4.1 of the IMO Guidelines (EOS1)

.2   When calculating first EOS number (EOS1) as defined in 4.1 of the Guidelines, the ship is assumed to be loaded at Summer Water Line with zero trim, without consumable or ballast.

.3   For the purpose of calculating EOS1, the volume of the cargo being carried by the ship is 98% of the volume of cargo and fuel oil tanks within L_t as per the original configuration of the ship.

Refer to the annex.

.4   Nominal density of the cargo, ρ_c:

The nominal density of the cargo to be used in the calculation of EOS1, EOS2 and EOS3 is given by the following formula:

$$\rho_c = \frac{\Delta(\text{summer}) - LSW}{V_{98\% \text{ (original cargo and fuel oil tanks configuration within } L_t)}}$$

where:

$\Delta(\text{summer})$ = Displacement of the ship corresponding to the maximum assigned summer load line with zero trim

LSW = Light ship weight

$V_{98\% \text{ (original cargo and fuel oil tanks configuration within } L_t)}$ = 98% of the cargo and fuel oil tanks volume within L_t, in ship's original configuration

Footnote: Written applications for evaluation of tanker arrangements under MEPC.64(36) received on or after 8 May 1998 will be evaluated in accordance with this unified interpretation unless advised otherwise by the flag Administration.

Calculation of second EOS number for the ship arranged with non-cargo side tanks as referred to in regulation 13G(4) as per 4.2 of the IMO Guidelines (EOS2)

.5   For the purpose of calculations, in the second EOS number (EOS2), the tanker is assumed to have side protection only as referred to in paragraph 4.2 of MEPC.64(36).

.6   For the purpose of calculating EOS2, hypothetical side protection may be considered provided that the assumed positions of the longitudinal and transverse bulkheads provide at least the minimum side protection required by Reg.13G(4) and are placed in the locations which lead to the lowest EOS2.

For volumes and measurement of parameters, please refer to figure 1 below.

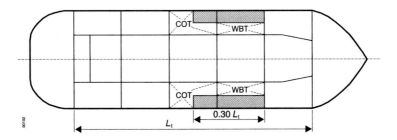

Figure 1

The remaining volumes of the tanks covered by hypothetical side protection are assumed to carry the same type of contents as before (i.e. water ballast tanks remain water ballast tanks and cargo oil tanks remain cargo oil tanks in remaining part).

The sizes of the remaining part of cargo oil and/or fuel oil tanks are as follows:

l_i: not modified.

b_i: reduced if hypothetical side protection is provided throughout the entire l_i. Not modified if hypothetical side protection is provided for partial l_i.

V_i: volume of cargo in the remaining cargo oil and/or fuel oil tanks.

.7   Draught and trim requirements of MARPOL, Annex I need not be taken into account for the purpose of calculating EOS2.

Refer to the annex.

.8   When calculating EOS2, the ship is assumed even keel at the draught of the loaded condition corresponding to the ship so arranged to comply with Reg. 13G(4) without any consumable nor ballast.

Refer to paragraph .6 and to the annex.

Calculation of third EOS number for hydrostatic balance method as per 4.3 of the IMO Guidelines (EOS3)

.9 For the purpose of calculating EOS3, the draught is that corresponding to the Hydrostatic Balance Loading (HBL) configuration. Ballast may be used to achieve an increased draught only in determining EOS3.

Filling levels in the tanks identified for HBL should be equal to the maximum level determined by the formula shown in the Appendix to the Guidelines using uniform nominal oil density and corresponding draught.

Refer to the annex.

Calculation of outflow in case of side damage as per 5.2 of the IMO Guidelines

.10 Distance from the hull boundary to the tank plating, s_i:

s_i is the minimum distance from the hull plating to the tank boundary measured at right angle to the centreline and at the level corresponding to the maximum assigned summer load line.

.11 Cargo volume in tank number i, V_i:

The maximum volume of V_i is 98% of the volume of the tank.

Calculation of outflow in case of bottom damage as per 5.3 of the IMO Guidelines

.12 Width of tank i, b_i:

- for tanks adjacent to the side shell, b_i is the width of the tank, measured inboard at $l_i/2$, at right angle to the centreline and at the level of the maximum assigned summer load line

- for a centre tank, b_i is the width of the tank bottom measured at $l_i/2$.

.13 Width of the cargo tank area, B_t:

B_t is the maximum breadth as defined by Annex I, regulation 1(21) measured within L_t.

.14 Height of double bottom, h_i:

h_i is the minimum height of the double bottom measured from the baseline.

Refer to figures 2 and 3.

.15 Height of the cargo column above the cargo tank bottom, h_c:

h_c is the height of the cargo column measured from the cargo tank bottom at the point where h_i is measured. Refer to figure 2.

Where a double bottom does not exist, then h_c is to be taken at its maximum value considering any deadrise of the ship. Refer to figure 3.

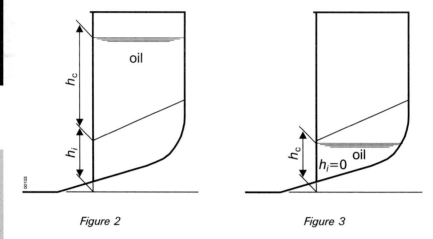

Figure 2   Figure 3

Calculation of total hypothetical outflow as per 5 of the IMO Guidelines

.16   When investigating outflow due to side damage, the side offering the more unfavourable EOS number should only be used.

Requirements for application of hydrostatic balance loading in cargo tanks as per the Appendix to the IMO Guidelines

.17   The number of tanks to be HBL is determined by the conditions underlined by EOS3 not being greater than EOS2 for the ship arranged with non-cargo side tanks and furthermore not being greater than 85% of the EOS1. Once it has been established that these conditions have been met, with zero trim, adjusted V_i, nominal density and maximum cargo level as per the Appendix formula, the configuration is considered having been validated and there is no need to recalculate EOS3 for the actual loading conditions corresponding to that configuration.

Filling levels in HBL tanks for the actual loading conditions are determined by the HBL formula of the Appendix to the Guidelines using actual density and draught at each HBL tank location.

.18   Partial filling less than HBL height may alternatively be considered. In such conditions, the cargo height in the selected cargo tanks is to be determined based on EOS3 compliance with the criteria, with zero trim, nominal density, and draught corresponding to adjusted V_i.

Where cargo levels less than maximum cargo level calculated by the HBL formula of the Appendix to the Guidelines are used for EOS3 calculation, the actual filling levels will be calculated as follows:

238

$(h_c \times K) \times \rho_c \times g + 100\Delta\rho \leq (d - h_i) \times \rho_s \times g$

where K is a correction factor $h_{(HBL)}/h_{(PF)}$ with:

$h_{(PF)}$ = maximum cargo height in partial filling condition, leading to a satisfactory EOS3 for the selected configuration (with nominal density, assumed zero trim, corresponding draught $d_{(PF)}$, adjusted V_i).

$h_{(HBL)}$ = maximum HBL cargo height for the selected configuration at draught $d_{(PF)}$.

Thus:

$$K = ((d_{(PF)} - h_i) \times \rho_s \times g - 100\Delta p)/(\rho_n \times g \times h_{(PF)}))$$

Annex

Matrix of parameters

EOS (see note 1)	Assumed trim	Draught	Density	Loaded oil volume	Consumables and ballast
EOS1	0	Maximum assigned summer water line	Nominal	V_1	none
EOS2	0	Corresponding draught	Nominal	V_2	none
EOS3	0	Corresponding draught	Nominal	V_3	See note 2

Where:

V_1 = 98% of cargo and fuel oil tanks volumes within L_t in the original configuration.

V_2 = V_1 minus 98% volume of side protection tanks corresponding to 13G(4), in way of cargo tanks.

V_3 = Oil volume of full cargo and fuel oil tanks within L_t (at 98%) and of HBL and/or partially loaded tanks.

Note 1: $EOS = \frac{O_{tot}}{V_1}$

Note 2: Ballast may be used to achieve an increased draught only in determining EOS3.

Annex II of MARPOL 73/78 (including amendments)

Regulations for the Control of Pollution by Noxious Liquid Substances in Bulk

Annex II of MARPOL 73/78
(including amendments)

Regulations for the Control of Pollution by Noxious Liquid Substances in Bulk

Regulation 1
Definitions

For the purposes of this Annex:

(1) *Chemical tanker* means a ship constructed or adapted primarily to carry a cargo of noxious liquid substances in bulk and includes an "oil tanker" as defined in Annex I of the present Convention when carrying a cargo or part cargo of noxious liquid substances in bulk.

(2) *Clean ballast* means ballast carried in a tank which, since it was last used to carry a cargo containing a substance in Category A, B, C or D, has been thoroughly cleaned and the residues resulting therefrom have been discharged and the tank emptied in accordance with the appropriate requirements of this Annex.

(3) *Segregated ballast* means ballast water introduced into a tank permanently allocated to the carriage of ballast or to the carriage of ballast or cargoes other than oil or noxious liquid substances as variously defined in the Annexes of the present Convention, and which is completely separated from the cargo and oil fuel system.

(4) *Nearest land* is as defined in regulation 1(9) of Annex I of the present Convention.

(5) *Liquid substances* are those having a vapour pressure not exceeding 2.8 kp/cm^2 at a temperature of 37.8°C.

(6) *Noxious liquid substance* means any substance referred to in appendix II to this Annex or provisionally assessed under the provisions of regulation 3(4) as falling into Category A, B, C or D.

(7)   *Special area* means a sea area where for recognized technical reasons in relation to its oceanographic and ecological condition and to the particular character of its traffic the adoption of special mandatory methods for the prevention of sea pollution by noxious liquid substances is required. Special areas shall be:

(a)   the Baltic Sea area, and

(b)   the Black Sea area, and

(c)   the Antarctic area

(8)   *Baltic Sea area* is as defined in regulation 10(1)(b) of Annex I of the present Convention.

(9)   *Black Sea area* is as defined in regulation 10(1)(c) of Annex I of the present Convention.

(9A)   *The Antarctic area* means the sea area south of latitude 60° S.

(10)   *International Bulk Chemical Code* means the International Code for the Construction and Equipment of Ships Carrying Dangerous Chemicals in Bulk[*] adopted by the Marine Environment Protection Committee of the Organization by resolution MEPC.19(22), as may be amended by the Organization, provided that such amendments are adopted and brought into force in accordance with the provisions of article 16 of the present Convention concerning amendment procedures applicable to an appendix to an Annex.

(11)   *Bulk Chemical Code* means the Code for the Construction and Equipment of Ships Carrying Dangerous Chemicals in Bulk[†] adopted by the Marine Environment Protection Committee of the Organization by resolution MEPC.20(22), as may be amended by the Organization, provided that such amendments are adopted and brought into force in accordance with the provisions of article 16 of the present Convention concerning amendment procedures applicable to an appendix to an Annex.

(12)   *Ship constructed* means a ship the keel of which is laid or which is at a similar stage of construction. A ship converted to a chemical tanker, irrespective of the date of construction, shall be treated as a chemical tanker constructed on the date on which such conversion commenced. This conversion provision shall not apply to the modification of a ship which complies with all of the following conditions:

---

[*] See IMO sales publication IMO-100E.
[†] See IMO sales publication IMO-772E.

(a)   the ship is constructed before 1 July 1986; and

(b)   the ship is certified under the Bulk Chemical Code to carry only those products identified by the Code as substances with pollution hazards only.

SEE INTERPRETATION 1.1

(13)   *Similar stage of construction* means the stage at which:

(a)   construction identifiable with a specific ship begins; and

(b)   assembly of that ship has commenced comprising at least 50 tons or 1% of the estimated mass of all structural material, whichever is less.

(14)   *Anniversary date* means the day and the month of each year which will correspond to the date of expiry of the International Pollution Prevention Certificate for the Carriage of Noxious Liquid Substances in Bulk.

Regulation 2
Application

(1)   Unless expressly provided otherwise the provisions of this Annex shall apply to all ships carrying noxious liquid substances in bulk.

(2)   Where a cargo subject to the provisions of Annex I of the present Convention is carried in a cargo space of a chemical tanker, the appropriate requirements of Annex I of the present Convention shall also apply.

(3)   Regulation 13 of this Annex shall apply only to ships carrying substances which are categorized for discharge control purposes in Category A, B or C.

(4)   For ships constructed before 1 July 1986, the provisions of regulation 5 of this Annex in respect of the requirement to discharge below the waterline and maximum concentration in the wake astern of the ship shall apply as from 1 January 1988.

(5)   The Administration may allow any fitting, material, appliance or apparatus to be fitted in a ship as an alternative to that required by this Annex if such fitting, material, appliance or apparatus is at least as effective as that required by this Annex. This authority of the Administration shall not extend to the substitution of operational methods to effect the control of discharge of noxious liquid substances

as equivalent to those design and construction features which are prescribed by regulations in this Annex.

SEE INTERPRETATION 2.1

(6)   The Administration which allows a fitting, material, appliance or apparatus as alternative to that required by this Annex, under paragraph (5) of this regulation, shall communicate to the Organization for circulation to the Parties to the Convention, particulars thereof, for their information and appropriate action, if any.

SEE INTERPRETATION 2.1

(7)   (a)   Where an amendment to this Annex and the International Bulk Chemical and Bulk Chemical Codes involves changes to the structure or equipment and fittings due to the upgrading of the requirements for the carriage of certain substances, the Administration may modify or delay for a specified period the application of such an amendment to ships constructed before the date of entry into force of that amendment, if the immediate application of such an amendment is considered unreasonable or impracticable. Such relaxation shall be determined with respect to each substance, having regard to the guidelines developed by the Organization.*

        (b)   The Administration allowing a relaxation of the application of an amendment under this paragraph shall submit to the Organization a report giving details of the ship or ships concerned, the cargoes carried, the trade in which each ship is engaged and the justification for the relaxation, for circulation to the Parties to the Convention for their information and appropriate action, if any.

Regulation 3
Categorization and listing of noxious liquid substances

(1)   For the purpose of the regulations of this Annex, noxious liquid substances shall be divided into four categories as follows:

        (a)   *Category A:* Noxious liquid substances which if discharged into the sea from tank cleaning or deballasting operations would present a major hazard to either marine resources or human

---

* Refer to the Guidelines for the Application of Amendments to the List of Substances in Annex II of MARPOL 73/78 and in the IBC Code and the BCH Code with Respect to Pollution Hazards, approved by the Marine Environment Protection Committee at its thirty-first session; see the Appendix to Unified Interpretations of Annex II.

health or cause serious harm to amenities or other legitimate uses of the sea and therefore justify the application of stringent anti-pollution measures.

(b) *Category B:* Noxious liquid substances which if discharged into the sea from tank cleaning or deballasting operations would present a hazard to either marine resources or human health or cause harm to amenities or other legitimate uses of the sea and therefore justify the application of special anti-pollution measures.

(c) *Category C:* Noxious liquid substances which if discharged into the sea from tank cleaning or deballasting operations would present a minor hazard to either marine resources or human health or cause minor harm to amenities or other legitimate uses of the sea and therefore require special operational conditions.

(d) *Category D:* Noxious liquid substances which if discharged into the sea from tank cleaning or deballasting operations would present a recognizable hazard to either marine resources or human health or cause minimal harm to amenities or other legitimate uses of the sea and therefore require some attention in operational conditions.

(2) Guidelines for use in the categorization of noxious liquid substances are given in appendix I to this Annex.

(3) Noxious liquid substances carried in bulk which are presently categorized as Category A, B, C and D and subject to the provisions of this Annex are referred to in appendix II to this Annex.

(4) Where it is proposed to carry a liquid substance in bulk which has not been categorized under paragraph (1) of this regulation or evaluated as referred to in regulation 4(1) of this Annex, the Governments of Parties to the Convention involved in the proposed operation shall establish and agree on a provisional assessment for the proposed operation on the basis of the guidelines referred to in paragraph (2) of this regulation. Until full agreement between the Governments involved has been reached, the substance shall be carried under the most severe conditions proposed. As soon as possible, but not later than 90 days after its first carriage, the Administration concerned shall notify the Organization and provide details of the substance and the provisional assessment for prompt circulation to all Parties for their information and consideration. The Government of each Party shall have a period of 90 days in which to forward its comments to the Organization, with a view to the assessment of the substance.

SEE INTERPRETATIONS 2A.1 to 2A.4

Regulation 4
Other liquid substances

(1)　The substances referred to in appendix III to this Annex have been evaluated and found to fall outside the Category A, B, C and D, as defined in regulation 3(1) of this Annex because they are at present considered to present no harm to human health, marine resources, amenities or other legitimate uses of the sea, when discharged into the sea from tank cleaning or deballasting operations.

(2)　The discharge of bilge or ballast water or other residues or mixtures containing only substances referred to in appendix III to this Annex shall not be subject to any requirement of this Annex.

(3)　The discharge into the sea of clean ballast or segregated ballast shall not be subject to any requirement of this Annex.

Regulation 5
*Discharge of noxious liquid substances**

Category A, B and C substances outside special areas and Category D substances in all areas

Subject to the provisions of paragraph (14) of this regulation and of regulation 6 of this Annex,

(1)　The discharge into the sea of substances in Category A as defined in regulation 3(1)(a) of this Annex or of those provisionally assessed as such or ballast water, tank washings, or other residues or mixtures containing such substances shall be prohibited. If tanks containing such substances or mixtures are to be washed, the resulting residues shall be discharged to a reception facility until the concentration of the substance in the effluent to such facility is at or below 0.1% by weight and until the tank is empty, with the exception of phosphorus, yellow or white, for which the residual concentration shall be 0.01% by weight. Any water subsequently added to the tank may be discharged into the sea when all the following conditions are satisfied:

(a)　the ship is proceeding *en route* at a speed of at least 7 knots in the case of self-propelled ships or at least 4 knots in the case of ships which are not self-propelled;

SEE INTERPRETATION 3.1

---

* For reference to "standards developed by the Organization" as used in this regulation, refer to the Standards for procedures and arrangements for the discharge of noxious liquid substances, adopted by resolution MEPC.18(22), as amended by resolution MEPC.62(35).

(b)   the discharge is made below the waterline, taking into account the location of the seawater intakes; and

(c)   the discharge is made at a distance of not less than 12 nautical miles from the nearest land in a depth of water of not less than 25 m.

(2)   The discharge into the sea of substances in Category B as defined in regulation 3(1)(b) of this Annex or of those provisionally assessed as such, or ballast water, tank washings, or other residues or mixtures containing such substances shall be prohibited except when all the following conditions are satisfied:

(a)   the ship is proceeding *en route* at a speed of at least 7 knots in the case of self-propelled ships or at least 4 knots in the case of ships which are not self-propelled;

SEE INTERPRETATION 3.1

(b)   the procedures and arrangements for discharge are approved by the Administration. Such procedures and arrangements shall be based upon standards developed by the Organization and shall ensure that the concentration and rate of discharge of the effluent is such that the concentration of the substance in the wake astern of the ship does not exceed 1 part per million;

(c)   the maximum quantity of cargo discharged from each tank and its associated piping system does not exceed the maximum quantity approved in accordance with the procedures referred to in subparagraph (b) of this paragraph, which shall in no case exceed the greater of 1 m^3 or 1/3,000 of the tank capacity in m^3;

(d)   the discharge is made below the waterline, taking into account the location of the seawater intakes; and

(e)   the discharge is made at a distance of not less than 12 nautical miles from the nearest land and in a depth of water of not less than 25 m.

(3)   The discharge into the sea of substances in Category C as defined in regulation 3(1)(c) of this Annex or of those provisionally assessed as such, or ballast water, tank washings, or other residues or mixtures containing such substances shall be prohibited except when all the following conditions are satisfied:

(a)   the ship is proceeding *en route* at a speed of at least 7 knots in the case of self-propelled ships or at least 4 knots in the case of ships which are not self-propelled;

(b)   the procedures and arrangements for discharge are approved by the Administration. Such procedures and arrangements shall be based upon standards developed by the Organization and shall

Annex II

ensure that the concentration and rate of discharge of the effluent is such that the concentration of the substance in the wake astern of the ship does not exceed 10 parts per million;

(c)   the maximum quantity of cargo discharged from each tank and its associated piping system does not exceed the maximum quantity approved in accordance with the procedures referred to in subparagraph (b) of this paragraph, which shall in no case exceed the greater of 3 m^3 or $1/1,000$ of the tank capacity in m^3;

(d)   the discharge is made below the waterline, taking into account the location of the seawater intakes; and

(e)   the discharge is made at a distance of not less than 12 nautical miles from the nearest land and in a depth of water of not less than 25 m.

(4)   The discharge into the sea of substances in Category D as defined in regulation 3(1)(d) of this Annex, or of those provisionally assessed as such, or ballast water, tank washings, or other residues or mixtures containing such substances shall be prohibited except when all the following conditions are satisfied:

(a)   the ship is proceeding *en route* at a speed of at least 7 knots in the case of self-propelled ships or at least 4 knots in the case of ships which are not self-propelled;

SEE INTERPRETATION 3.1

(b)   such mixtures are of a concentration not greater than one part of the substance in ten parts of water; and

(c)   the discharge is made at a distance of not less than 12 nautical miles from the nearest land.

(5)   Ventilation procedures approved by the Administration may be used to remove cargo residues from a tank. Such procedures shall be based upon standards developed by the Organization. Any water subsequently introduced into the tank shall be regarded as clean and shall not be subject to paragraph (1), (2), (3) or (4) of this regulation.

(6)   The discharge into the sea of substances which have not been categorized, provisionally assessed, or evaluated as referred to in regulation 4(1) of this Annex, or of ballast water, tank washings, or other residues or mixtures containing such substances shall be prohibited.

Category A, B and C substances within special areas

Subject to the provisions of paragraph (14) of this regulation and regulation 6 of this Annex,

(7) The discharge into the sea of substances in Category A as defined in regulation 3(1)(a) of this Annex or of those provisionally assessed as such, or ballast water, tank washings, or other residues or mixtures containing such substances shall be prohibited. If tanks containing such substances or mixtures are to be washed, the resulting residues shall be discharged to a reception facility which the States bordering the special area shall provide in accordance with regulation 7 of this Annex, until the concentration of the substance in the effluent to such facility is at or below 0.05% by weight and until the tank is empty, with the exception of phosphorus, yellow or white, for which the residual concentration shall be 0.005% by weight. Any water subsequently added to the tank may be discharged into the sea when all the following conditions are satisfied:

(a) the ship is proceeding *en route* at a speed of at least 7 knots in the case of self-propelled ships or at least 4 knots in the case of ships which are not self-propelled;

SEE INTERPRETATION 3.1

(b) the discharge is made below the waterline, taking into account the location of the seawater intakes; and

(c) the discharge is made at a distance of not less than 12 nautical miles from the nearest land and in a depth of water of not less than 25 m.

(8) The discharge into the sea of substances in Category B as defined in regulation (3)(1)(b) of this Annex or of those provisionally assessed as such, or ballast water, tank washings, or other residues or mixtures containing such substances shall be prohibited except when all the following conditions are satisfied:

(a) the tank has been prewashed in accordance with the procedure approved by the Administration and based on standards developed by the Organization and the resulting tank washings have been discharged to a reception facility;

(b) the ship is proceeding *en route* at a speed of at least 7 knots in the case of self-propelled ships or at least 4 knots in the case of ships which are not self-propelled;

(c) the procedures and arrangements for discharge and washings are approved by the Administration. Such procedures and arrangements shall be based upon standards developed by the Organization and shall ensure that the concentration and rate of discharge of the effluent is such that the concentration of the substance in the wake astern of the ship does not exceed 1 part per million;

Annex II

(d) the discharge is made below the waterline, taking into account the location of the seawater intakes; and

(e) the discharge is made at a distance of not less than 12 nautical miles from the nearest land and in a depth of water of not less than 25 m.

(9) The discharge into the sea of substances in Category C as defined in regulation 3(1)(c) of this Annex or of those provisionally assessed as such, or ballast water, tank washings, or other residues or mixtures containing such substances shall be prohibited except when all the following conditions are satisfied:

(a) the ship is proceeding *en route* at a speed of at least 7 knots in the case of self-propelled ships or at least 4 knots in the case of ships which are not self-propelled;

SEE INTERPRETATION 3.1

(b) the procedures and arrangements for discharge are approved by the Administration. Such procedures and arrangements shall be based upon standards developed by the Organization and shall ensure that the concentration and rate of discharge of the effluent is such that the concentration of the substance in the wake astern of the ship does not exceed 1 part per million;

(c) the maximum quantity of cargo discharged from each tank and its associated piping system does not exceed the maximum quantity approved in accordance with the procedures referred to in subparagraph (b) of this paragraph which shall in no case exceed the greater of 1 m^3 or 1/3,000 of the tank capacity in m^3;

(d) the discharge is made below the waterline, taking into account the location of the seawater intakes; and

(e) the discharge is made at a distance of not less than 12 nautical miles from the nearest land and in a depth of water of not less than 25 m.

(10) Ventilation procedures approved by the Administration may be used to remove cargo residues from a tank. Such procedures shall be based upon standards developed by the Organization. Any water subsequently introduced into the tank shall be regarded as clean and shall not be subject to paragraph (7), (8) or (9) of this regulation.

(11) The discharge into the sea of substances which have not been categorized, provisionally assessed or evaluated as referred to in regulation 4(1) of this Annex, or of ballast water, tank washings, or other residues or mixtures containing such substances shall be prohibited.

(12) Nothing in this regulation shall prohibit a ship from retaining on board the residues from a Category B or C cargo and discharging such residues into the sea outside a special area in accordance with paragraph (2) or (3) of this regulation, respectively.

(13) (a) The Governments of Parties to the Convention, the coastlines of which border on any given special area, shall collectively agree and establish a date by which time the requirement of regulation 7(1) of this Annex will be fulfilled and from which the requirements of paragraphs (7), (8), (9) and (10) of this regulation in respect of that area shall take effect and notify the Organization of the date so established at least six months in advance of that date. The Organization shall then promptly notify all Parties of that date.

(b) If the date of entry into force of the present Convention is earlier than the date established in accordance with subparagraph (a) of this paragraph, the requirements of paragraphs (1), (2) and (3) of this regulation shall apply during the interim period.

(14) In respect of the Antarctic area, any discharge into the sea of noxious liquid substances or mixtures containing such substances shall be prohibited.

Regulation 5A
*Pumping, piping and unloading arrangements**

(1) Every ship constructed on or after 1 July 1986 shall be provided with pumping and piping arrangements to ensure, through testing under favourable pumping conditions, that each tank designated for the carriage of a Category B substance does not retain a quantity of residue in excess of 0.1 m^3 in the tank's associated piping and in the immediate vicinity of that tank's suction point.

(2) (a) Subject to the provisions of subparagraph (b) of this paragraph, every ship constructed before 1 July 1986 shall be provided with pumping and piping arrangements to ensure, through testing under favourable pumping conditions, that each tank designated for the carriage of a Category B substance does not retain a quantity of residue in excess of 0.3 m^3 in the tank's associated piping and in the immediate vicinity of that tank's suction point.

---

* For reference to "standards developed by the Organization" as used in this regulation, refer to the Standards for procedures and arrangements for the discharge of noxious liquid substances, adopted by resolution MEPC.18(22), as amended by resolution MEPC.62(35).

(b) Until 2 October 1994 ships referred to in subparagraph (a) of this paragraph if not in compliance with the requirements of that subparagraph shall, as a minimum, be provided with pumping and piping arrangements to ensure, through testing under favourable pumping conditions and surface residue assessment, that each tank designated for the carriage of a Category B substance does not retain a quantity of residue in excess of 1 m^3 or 1/3000 of the tank capacity in m^3, whichever is greater, in that tank and the associated piping.

(3) Every ship constructed on or after 1 July 1986 shall be provided with pumping and piping arrangements to ensure, through testing under favourable pumping conditions, that each tank designated for the carriage of a Category C substance does not retain a quantity of residue in excess of 0.3 m^3 in the tank's associated piping and in the immediate vicinity of that tank's suction point.

(4) (a) Subject to the provisions of subparagraph (b) of this paragraph, every ship constructed before 1 July 1986 shall be provided with pumping and piping arrangements to ensure, through testing under favourable pumping conditions, that each tank designated for the carriage of a Category C substance does not retain a quantity of residue in excess of 0.9 m^3 in the tank's associated piping and in the immediate vicinity of that tank's suction point.

(b) Until 2 October 1994 the ships referred to in subparagraph (a) of this paragraph if not in compliance with the requirements of that subparagraph shall, as a minimum, be provided with pumping and piping arrangements to ensure, through testing under favourable pumping conditions and surface residue assessment, that each tank designated for the carriage of a Category C substance does not retain a quantity of residue in excess of 3 m^3 or 1/1000 of the tank capacity in m^3, whichever is greater, in that tank and the associated piping.

(5) Pumping conditions referred to in paragraphs (1), (2), (3) and (4) of this regulation shall be approved by the Administration and based on standards developed by the Organization. Pumping efficiency tests referred to in paragraphs (1), (2), (3) and (4) of this regulation shall use water as the test medium and shall be approved by the Administration and based on standards developed by the Organization. The residues on cargo tank surfaces, referred to in paragraphs (2)(b) and (4)(b) of this regulation shall be determined based on standards developed by the Organization.

(6) (a) Subject to the provisions of subparagraph (b) of this paragraph, the provisions of paragraphs (2) and (4) of this regulation need not apply to a ship constructed before 1 July 1986 which is

engaged in restricted voyages as determined by the Administration between:

(i)   ports or terminals within a State Party to the present Convention; or

(ii)   ports or terminals of States Parties to the present Convention.

(b)   The provisions of subparagraph (a) of this paragraph shall only apply to a ship constructed before 1 July 1986 if:

     (i)   each time a tank containing Category B or C substances or mixtures is to be washed or ballasted, the tank is washed in accordance with a prewash procedure approved by the Administration and based on standards developed by the Organization and the tank washings are discharged to a reception facility;

     (ii)   subsequent washings or ballast water are discharged to a reception facility or at sea in accordance with other provisions of this Annex;

     (iii)   the adequacy of the reception facilities at the ports or terminals referred to above, for the purpose of this paragraph, is approved by the Governments of the States Parties to the present Convention within which such ports or terminals are situated;

     (iv)   in the case of ships engaged in voyages to ports or terminals under the jurisdiction of other States Parties to the present Convention, the Administration communicates to the Organization, for circulation to the Parties to the Convention, particulars of the exemption, for their information and appropriate action, if any; and

SEE INTERPRETATION 4.1

     (v)   the certificate required under this Annex is endorsed to the effect that the ship is solely engaged in such restricted voyages.

(7)   For a ship whose constructional and operational features are such that ballasting of cargo tanks is not required and cargo tank washing is only required for repair or dry-docking, the Administration may allow exemption from the provisions of paragraphs (1), (2), (3) and (4) of this regulation, provided that all of the following conditions are complied with:

(a)   the design, construction and equipment of the ship are approved by the Administration, having regard to the service for which it is intended;

(b)   any effluent from tank washings which may be carried out before a repair or dry-docking is discharged to a reception facility, the adequacy of which is ascertained by the Administration;

(c)   the certificate required under this Annex indicates:

    (i)   that each cargo tank is certified for the carriage of only one named substance; and

    (ii)   the particulars of the exemption;

(d)   the ship carries a suitable operational manual approved by the Administration; and

(e)   in the case of ships engaged in voyages to ports or terminals under the jurisdiction of other States Parties to the present Convention, the Administration communicates to the Organization, for circulation to the Parties to the Convention, particulars of the exemption, for their information and appropriate action, if any.

SEE INTERPRETATION 4.1

Regulation 6
Exceptions

Regulation 5 of this Annex shall not apply to:

(a)   the discharge into the sea of noxious liquid substances or mixtures containing such substances necessary for the purpose of securing the safety of a ship or saving life at sea; or

(b)   the discharge into the sea of noxious liquid substances or mixtures containing such substances resulting from damage to a ship or its equipment:

    (i)   provided that all reasonable precautions have been taken after the occurrence of the damage or discovery of the discharge for the purpose of preventing or minimizing the discharge; and

    (ii)   except if the owner or the master acted either with intent to cause damage, or recklessly and with knowledge that damage would probably result; or

(c)   the discharge into the sea of noxious liquid substances or mixtures containing such substances, approved by the Administration, when being used for the purpose of combating specific pollution incidents in order to minimize the damage from pollution. Any such discharge shall be subject to the approval of any Government in whose jurisdiction it is contemplated the discharge will occur.

Regulation 7
Reception facilities and cargo unloading terminal arrangements

(1)   The Government of each Party to the Convention undertakes to ensure the provision of reception facilities according to the needs of ships using its ports, terminals or repair ports as follows:

    (a)   cargo loading and unloading ports and terminals shall have facilities adequate for reception without undue delay to ships of such residues and mixtures containing noxious liquid substances as would remain for disposal from ships carrying them as a consequence of application of this Annex; and

    (b)   ship repair ports undertaking repairs to chemical tankers shall have facilities adequate for the reception of residues and mixtures containing noxious liquid substances.

SEE INTERPRETATION 5.1

(2)   The Government of each Party shall determine the types of facilities provided for the purpose of paragraph (1) of this regulation at each cargo loading and unloading port, terminal and ship repair port in its territories and notify the Organization thereof.

(3)   The Government of each Party to the Convention shall undertake to ensure that cargo unloading terminals shall provide arrangements to facilitate stripping of cargo tanks of ships unloading noxious liquid substances at these terminals. Cargo hoses and piping systems of the terminal, containing noxious liquid substances received from ships unloading these substances at the terminal, shall not be drained back to the ship.

(4)   Each Party shall notify the Organization, for transmission to the Parties concerned, of any case where facilities required under paragraph (1) or arrangements required under paragraph (3) of this regulation are alleged to be inadequate.

Regulation 8
*Measures of control**

(1)   (a)   The Government of each Party to the Convention shall appoint or authorize surveyors for the purpose of implementing this

---

* For reference to "standards developed by the Organization" as used in this regulation, refer to the Standards for procedures and arrangements for the discharge of noxious liquid substances, adopted by resolution MEPC.18(22), as amended by resolution MEPC.62(35).

Annex II

regulation. The surveyors shall execute control in accordance with control procedures developed by the Organization.[*]

(b) The master of a ship carrying noxious liquid substances in bulk shall ensure that the provisions of regulation 5 and this regulation have been complied with and that the Cargo Record Book is completed in accordance with regulation 9 of this Annex whenever operations as referred to in that regulation take place.

(c) An exemption referred to in paragraphs (2)(b), (5)(b), (6)(c) or (7)(c) of this regulation may only be granted by the Government of the receiving Party to a ship engaged in voyages to ports or terminals under the jurisdiction of other States Parties to the present Convention. When such an exemption has been granted, the appropriate entry made in the Cargo Record Book shall be endorsed by the surveyor referred to in subparagraph (a) of this paragraph.

Category A substances in all areas

(2) With respect to Category A substances the following provisions shall apply in all areas:

(a) A tank which has been unloaded shall, subject to the provisions of subparagraph (b) of this paragraph, be washed in accordance with the requirements of paragraphs (3) or (4) of this regulation before the ship leaves the port of unloading.

(b) At the request of the ship's master, the Government of the receiving Party may exempt the ship from the requirements referred to in subparagraph (a) of this paragraph, where it is satisfied that:

　(i) the tank unloaded is to be reloaded with the same substance or another substance compatible with the previous one and that the tank will not be washed or ballasted prior to loading; or

　(ii) the tank unloaded is neither washed nor ballasted at sea and the provisions of paragraphs (3) or (4) of this regulation are complied with at another port provided that it has been confirmed in writing that a reception facility at that port is available and is adequate for such a purpose; or

　(iii) the cargo residues will be removed by a ventilation procedure approved by the Administration and based on standards developed by the Organization.

---

[*] Refer to the Procedures for port State control adopted by the Organization by resolution A.787(19) and amended by A.882(21); see IMO sales publication IMO-650E.

(3) If the tank is to be washed in accordance with subparagraph (2)(a) of this regulation, the effluent from the tank washing operation shall be discharged to a reception facility at least until the concentration of the substance in the discharge, as indicated by analyses of samples of the effluent taken by the surveyor, has fallen to the concentration specified in regulation 5(1) or 5(7), as applicable, of this Annex. When the required concentration has been achieved, remaining tank washings shall continue to be discharged to the reception facility until the tank is empty. Appropriate entries of these operations shall be made in the Cargo Record Book and endorsed by the surveyor referred to under paragraph (1)(a) of this regulation.

(4) Where the Government of the receiving party is satisfied that it is impracticable to measure the concentration of the substance in the effluent without causing undue delay to the ship, that Party may accept an alternative procedure as being equivalent to paragraph (3) of this regulation provided that:

 (a) the tank is prewashed in accordance with a procedure approved by the Administration and based on standards developed by the Organization; and

 (b) the surveyor referred to under paragraph (1)(a) certifies in the Cargo Record Book that:

  (i) the tank, its pump and piping systems have been emptied; and

  (ii) the prewash has been carried out in accordance with the prewash procedure approved by the Administration for that tank and that substance; and

  (iii) the tank washings resulting from such prewash have been discharged to a reception facility and the tank is empty.

Category B and C substances outside special areas

(5) With respect to Category B and C substances, the following provisions shall apply outside special areas:

 (a) A tank which has been unloaded shall, subject to the provisions of subparagraph (b) of this paragraph, be prewashed before the ship leaves the port of unloading, whenever:

  (i) the substance unloaded is identified in the standards developed by the Organization as resulting in a residue quantity exceeding the maximum quantity which may be discharged into the sea under regulation 5(2) or (3) of this Annex in case of Category B or C substances respectively; or

---

SEE INTERPRETATION 5A.1

Annex II

(ii)   the unloading is not carried out in accordance with the pumping conditions for the tank approved by the Administration and based on standards developed by the Organization as referred to under regulation 5A(5) of this Annex, unless alternative measures are taken to the satisfaction of the surveyor referred to in paragraph (1)(a) of this regulation to remove the cargo residues from the ship to quantities specified in regulation 5A of this Annex as applicable.

The prewash procedure used shall be approved by the Administration and based on standards developed by the Organization and the resulting tank washings shall be discharged to a reception facility at the port of unloading.

(b)   At the request of the ship's master, the Government of the receiving party may exempt the ship from the requirements of subparagraph (a) of this paragraph, where it is satisfied that:

(i)   the tank unloaded is to be reloaded with the same substance or another substance compatible with the previous one and that the tank will not be washed nor ballasted prior to loading; or

(ii)   the tank unloaded is neither washed nor ballasted at sea and the tank is prewashed in accordance with a procedure approved by the Administration and based on standards developed by the Organization and resulting tank washings are discharged to a reception facility at another port, provided that it has been confirmed in writing that a reception facility at that port is available and adequate for such a purpose; or

(iii)   the cargo residues will be removed by a ventilation procedure approved by the Administration and based on standards developed by the Organization.

Category B substances within special areas

(6)   With respect to Category B substances, the following provisions shall apply within special areas:

(a)   A tank which has been unloaded shall, subject to the provisions of subparagraphs (b) and (c), be prewashed before the ship leaves the port of unloading. The prewash procedure used shall be approved by the Administration and based on standards developed by the Organization and the resulting tank washings shall be discharged to a reception facility at the port of unloading.

(b) The requirements of subparagraph (a) of this paragraph do not apply when all the following conditions are satisfied:

   (i)   the Category B substance unloaded is identified in the standards developed by the Organization as resulting in a residue quantity not exceeding the maximum quantity which may be discharged into the sea outside special areas under regulation 5(2) of this Annex, and the residues are retained on board for subsequent discharge into the sea outside the special area in compliance with regulation 5(2) of this Annex; and

   (ii) the unloading is carried out in accordance with the pumping conditions for the tank approved by the Administration and based on standards developed by the Organization as referred to under regulation 5A(5) of this Annex, or failing to comply with the approved pumping conditions, alternative measures are taken to the satisfaction of the surveyor referred to in paragraph (1)(a) of this regulation to remove the cargo residues from the ship to quantities specified in regulation 5A of this Annex as applicable.

(c) At the request of the ship's master, the Government of the receiving party may exempt the ship from the requirements of subparagraph (a) of this paragraph, where it is satisfied that:

   (i)   the tank unloaded is to be reloaded with the same substance or another substance compatible with the previous one and that the tank will not be washed or ballasted prior to loading; or

   (ii) the tank unloaded is neither washed nor ballasted at sea and the tank is prewashed in accordance with a procedure approved by the Administration and based on standards developed by the Organization and resulting tank washings are discharged to a reception facility at another port, provided that it has been confirmed in writing that a reception facility at that port is available and adequate for such a purpose; or

   (iii) the cargo residues will be removed by a ventilation procedure approved by the Administration and based on standards developed by the Organization.

Category C substances within special areas

(7) With respect to Category C substances, the following provisions shall apply within special areas:

(a) A tank which has been unloaded shall, subject to the provisions of subparagraphs (b) and (c) of this paragraph, be prewashed before the ship leaves the port of unloading, whenever:

(i) the Category C substance unloaded is identified in the standards developed by the Organization as resulting in a residue quantity exceeding the maximum quantity which may be discharged into the sea under regulation 5(9) of this Annex; or

SEE INTERPRETATION 5A.1

(ii) the unloading is not carried out in accordance with the pumping conditions for the tank approved by the Administration and based on standards developed by the Organization as referred to under regulation 5A(5) of this Annex, unless alternative measures are taken to the satisfaction of the surveyor referred to in paragraph (1)(a) of this regulation to remove the cargo residues from the ship to quantities specified in regulation 5A of this Annex as applicable.

The prewash procedure used shall be approved by the Administration and based on standards developed by the Organization and the resulting tank washings shall be discharged to a reception facility at the port of unloading.

(b) The requirements of subparagraph (a) of this paragraph do not apply when all the following conditions are satisfied:

(i) the Category C substance unloaded is identified in the standards developed by the Organization as resulting in a residue quantity not exceeding the maximum quantity which may be discharged into the sea outside special areas under regulation 5(3) of this Annex, and the residues are retained on board for subsequent discharge into the sea outside the special area in compliance with regulation 5(3) of this Annex; and

(ii) the unloading is carried out in accordance with the pumping conditions for the tank approved by the Administration and based on standards developed by the Organization as referred to under regulation 5A(5) of this Annex, or failing to comply with the approved pumping conditions, alternative measures are taken to the satisfaction of the surveyor referred to in paragraph (1)(a) of this regulation to remove the cargo residues from the ship to quantities specified in regulation 5A of this Annex as applicable.

(c) At the request of the ship's master, the Government of the receiving party may exempt the ship from the requirements of subparagraph (a) of this paragraph, where it is satisfied that:

(i) the tank unloaded is to be reloaded with the same substance or another substance compatible with the previous one and that the tank will not be washed or ballasted prior to loading; or

(ii) the tank unloaded is neither washed nor ballasted at sea and the tank is prewashed in accordance with a procedure approved by the Administration and based on standards developed by the Organization and resulting tank washings are discharged to a reception facility at another port, provided that it has been confirmed in writing that a reception facility at that port is available and adequate for such a purpose; or

(iii) the cargo residues will be removed by a ventilation procedure approved by the Administration and based on standards developed by the Organization.

Category D substances in all areas

(8) With respect to Category D substances, a tank which has been unloaded shall either be washed and the resulting tank washings shall be discharged to a reception facility, or the remaining residues in the tank shall be diluted and discharged into the sea in accordance with regulation 5(4) of this Annex.

Discharge from a slop tank

(9) Any residues retained on board in a slop tank, including those from cargo pump-room bilges, which contain a Category A substance, or within a special area either a Category A or a Category B substance, shall be discharged to a reception facility in accordance with the provisions of regulation 5(1), (7) or (8) of this Annex, whichever is applicable.

Regulation 9
Cargo Record Book

(1) Every ship to which this Annex applies shall be provided with a Cargo Record Book, whether as part of the ship's official log-book or otherwise, in the form specified in appendix IV to this Annex.

(2)   The Cargo Record Book shall be completed, on a tank-to-tank basis, whenever any of the following operations with respect to a noxious liquid substance take place in the ship:

    (i)   loading of cargo;

    (ii)   internal transfer of cargo;

    (iii) unloading of cargo;

    (iv) cleaning of cargo tanks;

    (v)   ballasting of cargo tanks;

    (vi) discharge of ballast from cargo tanks;

    (vii) disposal of residues to reception facilities;

    (viii) discharge into the sea or removal by ventilation of residues in accordance with regulation 5 of this Annex.

(3)   In the event of any discharge of the kind referred to in article 8 of the present Convention and regulation 6 of this Annex of any noxious liquid substance or mixture containing such substance, whether intentional or accidental, an entry shall be made in the Cargo Record Book stating the circumstances of, and the reason for, the discharge.

(4)   When a surveyor appointed or authorized by the Government of the Party to the Convention to supervise any operations under this Annex has inspected a ship, then that surveyor shall make an appropriate entry in the Cargo Record Book.

(5)   Each operation referred to in paragraphs (2) and (3) of this regulation shall be fully recorded without delay in the Cargo Record Book so that all the entries in the book appropriate to that operation are completed. Each entry shall be signed by the officer or officers in charge of the operation concerned and each page shall be signed by the master of the ship. The entries in the Cargo Record Book shall be in an official language of the State whose flag the ship is entitled to fly, and, for ships holding an International Pollution Prevention Certificate for the Carriage of Noxious Liquid Substances in Bulk or a certificate referred to in regulation 12A of this Annex in English or French. The entries in an official national language of the State whose flag the ship is entitled to fly shall prevail in case of a dispute or discrepancy.

(6)   The Cargo Record Book shall be kept in such a place as to be readily available for inspection and, except in the case of unmanned ships under tow, shall be kept on board the ship. It shall be retained for a period of three years after the last entry has been made.

(7)   The competent authority of the Government of a Party may inspect the Cargo Record Book on board any ship to which this Annex applies while the ship is in its port, and may make a copy of any entry in that book and may require the master of the ship to certify that the

copy is a true copy of such entry. Any copy so made which has been certified by the master of the ship as a true copy of an entry in the ship's Cargo Record Book shall be made admissible in any judicial proceedings as evidence of the facts stated in the entry. The inspection of a Cargo Record Book and the taking of a certified copy by the competent authority under this paragraph shall be performed as expeditiously as possible without causing the ship to be unduly delayed.

Regulation 10
Surveys

(1) Ships carrying noxious liquid substances in bulk shall be subject to the surveys specified below:

(a) An initial survey before the ship is put in service or before the certificate required under regulation 11 of this Annex is issued for the first time, which shall include a complete survey of its structure, equipment, systems, fittings, arrangements and material in so far as the ship is covered by this Annex. This survey shall be such as to ensure that the structure, equipment, systems, fittings, arrangements and material fully comply with the applicable requirements of this Annex.

(b) A renewal survey at intervals specified by the Administration, but not exceeding 5 years, except where regulation 12(2), 12(5), 12(6) or 12(7) of this Annex is applicable. The renewal survey shall be such as to ensure that the structure, equipment, systems, fittings, arrangements and material fully comply with applicable requirements of this Annex.

(c) An intermediate survey within 3 months before or after the second anniversary date or within 3 months before or after the third anniversary date of the Certificate which shall take the place of one of the annual surveys specified in paragraph (1)(d) of this regulation. The intermediate survey shall be such as to ensure that the equipment and associated pump and piping systems fully comply with the applicable requirements of this Annex and are in good working order. Such intermediate surveys shall be endorsed on the Certificate issued under regulation 11 of this Annex.

SEE INTERPRETATION 6.1

(d) An annual survey within 3 months before or after each anniversary date of the Certificate including a general inspection of the structure, equipment, systems, fittings, arrangements and material referred to in paragraph (1)(a) of this regulation to ensure that they have been maintained in accordance with paragraph (3) of this regulation and that they remain satisfactory for the service for which the ship is intended. Such annual surveys shall be endorsed on the Certificate issued under regulation 11 of this Annex.

SEE INTERPRETATION 6.1

(e)   An additional survey either general or partial, according to the circumstances, shall be made after a repair resulting from investigations prescribed in paragraph (3) of this regulation, or whenever any important repairs or renewals are made. The survey shall be such as to ensure that the necessary repairs or renewals have been effectively made, that the material and workmanship of such repairs or renewals are in all respects satisfactory and that the ship complies in all respects with the requirements of this Annex.

(2) (a)   Surveys of ships as regards the enforcement of the provisions of this Annex shall be carried out by officers of the Administration. The Administration may, however, entrust the surveys either to surveyors nominated for the purpose or to organizations recognized by it.

(b)   An Administration nominating surveyors or recognizing organizations to conduct surveys as set forth in subparagraph (a) of this paragraph shall, as a minimum, empower any nominated surveyor or recognized organization to:

(i)   require repairs to a ship; and

(ii)   carry out surveys if requested by the appropriate authorities of a port State.

The Administration shall notify the Organization of the specific responsibilities and conditions of the authority delegated to the nominated surveyors or recognized organizations, for circulation to Parties to the present Convention for the information of their officers.

(c)   When a nominated surveyor or recognized organization determines that the condition of the ship or its equipment does not correspond substantially with the particulars of the Certificate, or is such that the ship is not fit to proceed to sea without presenting an unreasonable threat of harm to the marine environment, such surveyor or organization shall immediately ensure that corrective action is taken and shall in due course notify the Administration. If such corrective action is not taken the Certificate should be withdrawn and the Administration shall be notified immediately; and if the ship is in a port of another Party, the appropriate authorities of the port State shall also be notified immediately. When an officer of the Administration, a nominated surveyor or a recognized organization has notified the appropriate authorities of the port State, the Government of the port State concerned shall give such officer, surveyor or organization any necessary assistance to carry out their obligations under this regulation. When applicable, the Government of the port State concerned shall take such steps as will ensure that the ship shall not sail until it can proceed to sea or leave the port for the purpose of proceeding to the nearest appropriate repair yard available without presenting an unreasonable threat of harm to the marine environment.

(d)   In every case, the Administration concerned shall fully guarantee the completeness and efficiency of the survey and shall undertake to ensure the necessary arrangements to satisfy this obligation.

(3) (a)   The condition of the ship and its equipment shall be maintained to conform with the provisions of the present Convention to ensure that the ship in all respects will remain fit to proceed to sea without presenting an unreasonable threat of harm to the marine environment.

(b)   After any survey of the ship under paragraph (1) of this regulation has been completed, no change shall be made in the structure, equipment, systems, fittings, arrangements or material covered by the survey, without the sanction of the Administration, except the direct replacement of such equipment and fittings.

(c)   Whenever an accident occurs to a ship or a defect is discovered which substantially affects the integrity of the ship or the efficiency or completeness of its equipment covered by this Annex, the master or owner of the ship shall report at the earliest opportunity to the Administration, the recognized organization or the nominated surveyor responsible for issuing the relevant Certificate, who shall cause investigations to be initiated to determine whether a survey as required by paragraph (1) of this regulation is necessary. If the ship is in a port of another Party, the master or owner shall also report immediately to the appropriate authorities of the port State and the nominated surveyor or recognized organization shall ascertain that such report has been made.

Regulation 11
Issue or endorsement of Certificate

(1)   An International Pollution Prevention Certificate for the Carriage of Noxious Liquid Substances in Bulk shall be issued, after an initial or renewal survey in accordance with the provisions of regulation 10 of this Annex, to any ship carrying noxious liquid substances in bulk and which is engaged in voyages to ports or terminals under the jurisdiction of other Parties to the Convention.

(2)   Such Certificate shall be issued or endorsed either by the Administration or by any person or organization duly authorized by it. In every case, the Administration assumes full responsibility for the Certificate.

(3)   (a)   The Government of a Party to the Convention may, at the request of the Administration, cause a ship to be surveyed and, if satisfied that the provisions of this Annex are complied with, shall issue or authorize the issue of an International Pollution Prevention Certificate for the Carriage of Noxious Liquid Substances in Bulk to the ship and, where appropriate, endorse or authorize the endorsement of that Certificate on the ship, in accordance with this Annex.

       (b)   A copy of the Certificate and a copy of the survey report shall be transmitted as soon as possible to the requesting Administration.

       (c)   A Certificate so issued shall contain a statement to the effect that it has been issued at the request of the Administration and it shall have the same force and receive the same recognition as the Certificate issued under paragraph (1) of this regulation.

       (d)   No International Pollution Prevention Certificate for the Carriage of Noxious Liquid Substances in Bulk shall be issued to a ship which is entitled to fly the flag of a State which is not a Party.

(4)   The International Pollution Prevention Certificate for the Carriage of Noxious Liquid Substances in Bulk shall be drawn up in the form corresponding to the model given in appendix V to this Annex. If the language used is neither English nor French, the text shall include a translation into one of these languages.

(5)   Notwithstanding any other provisions of the amendments to this Annex adopted by the Marine Environment Protection Committee (MEPC) by resolution MEPC.39(29), any International Pollution Prevention Certificate for the Carriage of Noxious Liquid Substances in Bulk, which is current when these amendments enter into force, shall remain valid until it expires under the terms of this Annex prior to the amendments entering into force.

Regulation 12
Duration and validity of Certificate

(1)   An International Pollution Prevention Certificate for the Carriage of Noxious Liquid Substances in Bulk shall be issued for a period specified by the Administration which shall not exceed 5 years.

(2)   (a)   Notwithstanding the requirements of paragraph (1) of this regulation, when the renewal survey is completed within 3 months before the expiry date of the existing Certificate, the new Certificate shall be valid from the date of completion of the renewal survey to a date not exceeding 5 years from the date of expiry of the existing Certificate.

(b)   When the renewal survey is completed after the expiry date of the existing Certificate, the new Certificate shall be valid from the date of completion of the renewal survey to a date not exceeding 5 years from the date of expiry of the existing Certificate.

(c)   When the renewal survey is completed more than 3 months before the expiry date of the existing Certificate, the new Certificate shall be valid from the date of completion of the renewal survey to a date not exceeding 5 years from the date of completion of the renewal survey.

(3)   If a Certificate is issued for a period of less than 5 years, the Administration may extend the validity of the Certificate beyond the expiry date to the maximum period specified in paragraph (1) of this regulation, provided that the surveys referred to in regulation 10(1)(c) and 10(1)(d) of this Annex applicable when a Certificate is issued for a period of 5 years are carried out as appropriate.

(4)   If a renewal survey has been completed and a new Certificate cannot be issued or placed on board the ship before the expiry date of the existing Certificate, the person or organization authorized by the Administration may endorse the existing Certificate and such a Certificate shall be accepted as valid for a further period which shall not exceed 5 months from the expiry date.

(5)   If a ship at the time when a Certificate expires is not in a port in which it is to be surveyed, the Administration may extend the period of validity of the Certificate but this extension shall be granted only for the purpose of allowing the ship to complete its voyage to the port in which it is to be surveyed, and then only in cases where it appears proper and reasonable to do so. No Certificate shall be extended for a period longer than 3 months, and a ship to which an extension is granted shall not, on its arrival in the port in which it is to be surveyed, be entitled by virtue of such extension to leave that port without

having a new Certificate. When the renewal survey is completed, the new Certificate shall be valid to a date not exceeding 5 years from the date of expiry of the existing Certificate before the extension was granted.

(6)   A Certificate issued to a ship engaged on short voyages which has not been extended under the foregoing provisions of this regulation may be extended by the Administration for a period of grace of up to one month from the date of expiry stated on it. When the renewal survey is completed, the new Certificate shall be valid to a date not exceeding 5 years from the date of expiry of the existing Certificate before the extension was granted.

(7)   In special circumstances, as determined by the Administration, a new Certificate need not be dated from the date of expiry of the existing Certificate as required by paragraph (2)(b), (5) or (6) of this regulation. In these special circumstances, the new Certificate shall be valid to a date not exceeding 5 years from the date of completion of the renewal survey.

(8)   If an annual or intermediate survey is completed before the period specified in regulation 10 of this Annex, then:

(a)   the anniversary date shown on the Certificate shall be amended by endorsement to a date which shall not be more than 3 months later than the date on which the survey was completed;

(b)   the subsequent annual or intermediate survey required by regulation 10 of this Annex shall be completed at the intervals prescribed by that regulation using the new anniversary date;

(c)   the expiry date may remain unchanged provided one or more annual or intermediate surveys, as appropriate, are carried out so that the maximum intervals between the surveys prescribed by regulation 10 of this Annex are not exceeded.

(9)   A Certificate issued under regulation 11 of this Annex shall cease to be valid in any of the following cases:

(a)   if the relevant surveys are not completed within the periods specified under regulation 10(1) of this Annex;

(b)   if the Certificate is not endorsed in accordance with regulation 10(1)(c) or 10(1)(d) of this Annex;

(c)   upon transfer of the ship to the flag of another State. A new Certificate shall only be issued when the Government issuing the new Certificate is fully satisfied that the ship is in compliance with the requirements of regulation 10(4)(a) and 10(4)(b) of this Annex. In the case of a transfer between Parties, if requested within 3 months after the transfer has taken place, the Government of the Party whose flag the ship was formerly entitled to fly shall, as soon as possible, transmit to the Administration copies of the Certificate carried by the ship before the transfer and, if available, copies of the relevant survey reports.

Regulation 12A
Survey and certification of chemical tankers

Notwithstanding the provisions of regulations 10, 11 and 12 of this Annex, chemical tankers which have been surveyed and certified by States Parties to the present Convention in accordance with the provisions of the International Bulk Chemical Code or the Bulk Chemical Code, as applicable, shall be deemed to have complied with the provisions of the said regulations, and the certificate issued under that Code shall have the same force and receive the same recognition as the certificate issued under regulation 11 of this Annex.

Regulation 13
Requirements for minimizing accidental pollution

(1) The design, construction, equipment and operation of ships carrying noxious liquid substances of Category A, B or C in bulk, shall be such as to minimize the uncontrolled discharge into the sea of such substances.

(2) Chemical tankers constructed on or after 1 July 1986 shall comply with the requirements of the International Bulk Chemical Code.

(3) Chemical tankers constructed before 1 July 1986 shall comply with the following requirements:

    (a) The following chemical tankers shall comply with the requirements of the Bulk Chemical Code as applicable to ships referred to in 1.7.2 of that Code:

        (i) ships for which the building contract is placed on or after 2 November 1973 and which are engaged on voyages to ports or terminals under the jurisdiction of other States Parties to the Convention; and

        (ii) ships constructed on or after 1 July 1983 which are engaged solely on voyages between ports or terminals within the State the flag of which the ship is entitled to fly.

    (b) The following chemical tankers shall comply with the requirements of the Bulk Chemical Code as applicable to ships referred to in 1.7.3 of that Code:

        (i) ships for which the building contract is placed before 2 November 1973 and which are engaged on voyages to

Annex II

ports or terminals under the jurisdiction of other States Parties to the Convention; and

(ii) ships constructed before 1 July 1983 which are engaged on voyages between ports or terminals within the State the flag of which the ship is entitled to fly, except that for ships of less than 1,600 tons gross tonnage compliance with the Code in respect of construction and equipment shall take effect not later than 1 July 1994.

(4) In respect of ships other than chemical tankers carrying noxious liquid substances of Category A, B or C in bulk, the Administration shall establish appropriate measures based on the Guidelines developed by the Organization in order to ensure that the provisions of paragraph (1) of this regulation are complied with.

SEE INTERPRETATION 6A.1.1

Regulation 14
Carriage and discharge of oil-like substances

SEE INTERPRETATIONS 7.1 and 7.2

Notwithstanding the provisions of other regulations of this Annex, noxious liquid substances referred to in appendix II of this Annex as falling under Category C or D and identified by the Organization* as oil-like substances under the criteria developed by the Organization, may be carried on an oil tanker as defined in Annex I of the Convention and discharged in accordance with the provisions of Annex I of the present Convention, provided that all of the following conditions are complied with:

(a) the ship complies with the provisions of Annex I of the present Convention as applicable to product carriers as defined in that Annex;

(b) the ship carries an International Oil Pollution Prevention Certificate and its Supplement B and the certificate is endorsed to indicate that the ship may carry oil-like substances in conformity with this regulation and the endorsement includes a list of oil-like substances the ship is allowed to carry;

(c) in the case of Category C substances the ship complies with the ship type 3 damage stability requirements of;

(i) the International Bulk Chemical Code in the case of a ship constructed on or after 1 July 1986; or

---

* Refer to Interpretation 7.2.1 of the Unified Interpretations of Annex II.

(ii) the Bulk Chemical Code, as applicable under regulation 13 of this Annex, in the case of a ship constructed before 1 July 1986; and

SEE INTERPRETATION 7.3

(d) the oil content meter in the oil discharge monitoring and control system of the ship is approved by the Administration for use in monitoring the oil-like substances to be carried.

SEE INTERPRETATION 7.4

Annex II

Regulation 15
*Port State control on operational requirements**

(1) A ship when in a port of another Party is subject to inspection by officers duly authorized by such Party concerning operational requirements under this Annex, where there are clear grounds for believing that the master or crew are not familiar with essential shipboard procedures relating to the prevention of pollution by noxious liquid substances.

(2) In the circumstances given in paragraph (1) of this regulation, the Party shall take such steps as will ensure that the ship shall not sail until the situation has been brought to order in accordance with the requirements of this Annex.

(3) Procedures relating to the port State control prescribed in article 5 of the present Convention shall apply to this regulation.

(4) Nothing in this regulation shall be construed to limit the rights and obligations of a Party carrying out control over operational requirements specifically provided for in the present Convention.

Regulation 16
Shipboard marine pollution emergency plan
for noxious liquid substances

(1) Every ship of 150 gross tonnage and above certified to carry noxious liquid substances in bulk shall carry on board a shipboard marine pollution emergency plan for noxious liquid substances approved by

---

* Refer to the Procedures for port State control adopted by the Organization by resolution A.787(19) and amended by A.882(21); see IMO sales publication IMO-650E.

the Administration. This requirement shall apply to all such ships not later than 1 January 2003.

(2) Such a plan shall be in accordance with Guidelines* developed by the Organization and written in a working language or languages understood by the master and officers. The plan shall consist at least of:

    (a) the procedure to be followed by the master or other persons having charge of the ship to report a noxious liquid substances pollution incident, as required in article 8 and Protocol I of the present Convention, based on the Guidelines developed by the Organization;[†]

    (b) the list of authorities or persons to be contacted in the event of a noxious liquid substance pollution incident;

    (c) a detailed description of the action to be taken immediately by persons on board to reduce or control the discharge of noxious liquid substances following the incident; and

    (d) the procedures and point of contact on the ship for co-ordinating shipboard action with national and local authorities in combating the pollution.

(3) In the case of ships to which regulation 26 of Annex I of the Convention also applies, such a plan may be combined with the shipboard oil pollution emergency plan required under regulation 26 of Annex I of the Convention. In this case, the title of such a plan shall be "Shipboard marine pollution emergency plan".

---

* Refer to the Guidelines for the development of shipboard marine pollution emergency plans for oil and/or noxious liquid substances as adopted by the Organization by resolution MEPC.85(44); see IMO sales publication IMO-586E.

† Refer to the General principles for ship reporting systems and ship reporting requirements, including guidelines for reporting incidents involving dangerous goods, harmful substances and/or marine pollutants, adopted by the Organization by resolution A.851(20).

Appendices to Annex II

Appendix I
Guidelines for the categorization of noxious liquid substances

Category A   Substances which are bioaccumulated and liable to produce a hazard to aquatic life or human health, or which are highly toxic to aquatic life (as expressed by a Hazard Rating 4, defined by a TL_m less than 1 ppm); and additionally certain substances which are moderately toxic to aquatic life (as expressed by a Hazard Rating 3, defined by a TL_m of 1 ppm or more, but less than 10 ppm) when particular weight is given to additional factors in the hazard profile or to special characteristics of the substance.

Category B   Substances which are bioaccumulated with a short retention of the order of one week or less, or which are liable to produce tainting of the sea food, or which are moderately toxic to aquatic life (as expressed by a Hazard Rating 3, defined by a TL_m of 1 ppm or more, but less than 10 ppm); and additionally certain substances which are slightly toxic to aquatic life (as expressed by a Hazard Rating 2, defined by a TL_m of 10 ppm or more, but less than 100 ppm) when particular weight is given to additional factors in the hazard profile or to special characteristics of the substance.

Category C   Substances which are slightly toxic to aquatic life (as expressed by a Hazard Rating 2, defined by a TL_m of 10 ppm or more, but less than 100 ppm); and additionally certain substances which are practically non-toxic to aquatic life (as expressed by a Hazard Rating 1, defined by a TL_m of 100 ppm or more, but less than 1,000 ppm) when particular weight is given to additional factors in the hazard profile or to special characteristics of the substance.

Category D   Substances which are practically non-toxic to aquatic life (as expressed by a Hazard Rating 1, defined by a TL_m of 100 ppm or more, but less than 1,000 ppm); or causing deposits blanketing the sea floor with a high biochemical oxygen demand (BOD); or which are highly hazardous to human health, with an LD_{50} of less than 5 mg/kg; or which produce moderate reduction of amenities because of persistency, smell or poisonous or irritant characteristics, possibly interfering with use of beaches; or which are moderately hazardous to human health, with an LD_{50} of 5 mg/kg or more, but less than 50 mg/kg, and produce slight reduction of amenities.

Other Liquid Substances (for the purposes of regulation 4 of this Annex) Substances other than those categorized in Categories A, B, C, and D above.

Appendix II

List of noxious liquid substances carried in bulk

Noxious liquid substances carried in bulk and which are presently categorized as Category A, B, C or D and subject to the provisions of this Annex, are so indicated in the Pollution Category column of chapters 17 or 18 of the International Bulk Chemical Code.

Appendix III

List of other liquid substances

Liquid substances carried in bulk which are identified as falling outside Categories A, B, C and D and not subject to the provisions of this Annex and indicated as 'III' in the Pollution Category column of chapters 17 or 18 of the International Bulk Chemical Code.

Appendix IV

Form of Cargo Record Book for ships carrying noxious liquid substances in bulk

CARGO RECORD BOOK FOR SHIPS CARRYING NOXIOUS LIQUID SUBSTANCES IN BULK

Name of ship . . . .

Distinctive number or letters . . . .

Gross tonnage . . . .

Period from . . . . to . . . .

Note: Every ship carrying noxious liquid substances in bulk shall be provided with a Cargo Record Book to record relevant cargo/ballast operations.

Name of ship . . . .

Distinctive number or letters . . . .

PLAN VIEW OF CARGO AND SLOP TANKS
(to be completed on board)

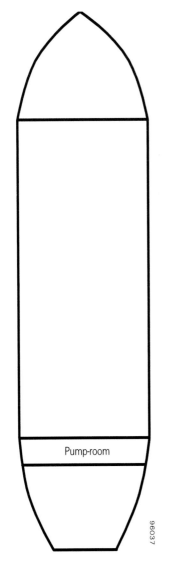

Identification of the tanks	Capacity

Pump-room

96037

(Give the capacity of each tank in cubic metres)

Introduction

The following pages show a comprehensive list of items of cargo and ballast operations which are, when appropriate, to be recorded in the Cargo Record Book on a tank-to-tank basis in accordance with paragraph 2 of regulation 9 of Annex II of the International Convention for the Prevention of Pollution from Ships, 1973, as modified by the Protocol of 1978 relating thereto, as amended. The items have been grouped into operational sections, each of which is denoted by a letter.

When making entries in the Cargo Record Book, the date, operational code and item number shall be inserted in the appropriate columns and the required particulars shall be recorded chronologically in the blank spaces.

Each completed operation shall be signed for and dated by the officer or officers in charge and, if applicable, by a surveyor authorized by the competent authority of the State in which the ship is unloading. Each completed page shall be countersigned by the master of the ship.

Entries in the Cargo Record Book are required only for operations involving Categories A, B, C and D substances.

For the category of a substance, refer to table 1 of the ship's Procedures and Arrangements Manual.

List of items to be recorded

Entries are required only for operations involving Categories A, B, C and D substances.

(A) Loading of cargo

1. Place of loading.

2. Identify tank(s), name of substance(s) and category(ies).

(B) Internal transfer of cargo

3. Name and category of cargo(es) transferred.

4. Identity of tanks:
   .1 from:
   .2 to:

5. Was (were) tank(s) in 4.1 emptied?

6. If not, quantity remaining in tank(s).

(C) Unloading of cargo

7. Place of unloading.

8. Identity of tank(s) unloaded.

9. Was (were) tank(s) emptied?
   .1 If yes, confirm that the procedure for emptying and stripping has been performed in accordance with the ship's Procedures and Arrangements Manual (i.e. list, trim, stripping temperature).
   .2 If not, quantity remaining in tank(s).

10. Does the ship's Procedures and Arrangements Manual require a prewash with subsequent disposal to reception facilities?

11. Failure of pumping and/or stripping system:
   .1 time and nature of failure;
   .2 reasons for failure;
   .3 time when system has been made operational.

(D) Mandatory prewash in accordance with the ship's Procedures and Arrangements Manual

12. Identify tank(s), substance(s) and category(ies).

13. Washing method:
    .1 number of washing machines per tank;
    .2 duration of wash/washing cycles;
    .3 hot/cold wash.

14. Prewash slops transferred to:
    .1 reception facility in unloading port (identify port);
    .2 reception facility otherwise (identify port).

(E) Cleaning of cargo tanks except mandatory prewash (other prewash operations, final wash, ventilation etc.)

15. State time, identify tank(s), substance(s) and category(ies) and state:
    .1 washing procedure used;
    .2 cleaning agent(s) (identify agent(s) and quantities);
    .3 dilution of cargo residues with water (state how much water used (only Category D substances));
    .4 ventilation procedure used (state number of fans used, duration of ventilation).

16. Tank washings transferred:
    .1 into the sea;
    .2 to reception facility (identify port);
    .3 to slops collecting tank (identify tank).

(F) Discharge into the sea of tank washings

17. Identify tank(s):
    .1 Were tank washings discharged during cleaning of tank(s)? If so at what rate?
    .2 Were tank washing(s) discharged from a slops collecting tank? If so, state quantity and rate of discharge.

18. Time pumping commenced and stopped.

19. Ship's speed during discharge.

(G) Ballasting of cargo tanks

20. Identity of tank(s) ballasted.

21. Time at start of ballasting.

(H) Discharge of ballast water from cargo tanks

22. Identity of tank(s).

Annex II

23. Discharge of ballast:
    .1 into the sea;
    .2 to reception facilities (identify port).

24. Time ballast discharge commenced and stopped.

25. Ship's speed during discharge.

(I) Accidental or other exceptional discharge

26. Time of occurrence.

27. Approximate quantity, substance(s) and category(ies).

28. Circumstances of discharge or escape and general remarks.

(J) Control by authorized surveyors

29. Identify port.

30. Identify tank(s), substance(s), category(ies) discharged ashore.

31. Have tank(s), pump(s), and piping system(s) been emptied?

32. Has a prewash in accordance with the ship's Procedures and Arrangements Manual been carried out?

33. Have tank washings resulting from the prewash been discharged ashore and is the tank empty?

34. An exemption has been granted from mandatory prewash.

35. Reasons for exemption.

36. Name and signature of authorized surveyor.

37. Organization, company, government agency for which surveyor works.

(K) Additional operational procedures and remarks

Name of ship . . . .

Distinctive number or letters . . . .

CARGO/BALLAST OPERATIONS

Date	Code (letter)	Item (number)	Record of operations/signature of officer in charge/name of and signature of authorized surveyor

Signature of master . . . .

Annex II

Appendix V
Form of NLS Certificate

INTERNATIONAL POLLUTION PREVENTION CERTIFICATE
FOR THE CARRIAGE OF NOXIOUS LIQUID SUBSTANCES IN BULK

Issued under the provisions of the International Convention for the Prevention of Pollution from Ships, 1973, as modified by the Protocol of 1978 relating thereto, and as amended by resolution MEPC.39(29) (hereinafter referred to as "the Convention") under the authority of the Government of:

. . . .
(full designation of the country)

by. . . . .
*(full designation of the competent person or organization
authorized under the provisions of the Convention)*

Particulars of ship*

Name of ship . . . .

Distinctive number or letters . . . .

Port of registry . . . .

Gross tonnage . . . .

IMO Number†. . . . .

---

* Alternatively, the particulars of the ship may be placed horizontally in boxes.

† In accordance with resolution A.600(15), IMO Ship Identification Number Scheme, this information may be included voluntarily.

THIS IS TO CERTIFY:

1   That the ship has been surveyed in accordance with regulation 10 of Annex II of the Convention.

2   That the survey showed that the structure, equipment, systems, fitting, arrangements and material of the ship and the condition thereof are in all respects satisfactory and that the ship complies with the applicable requirements of Annex II of the Convention.

3   That the ship has been provided with a Manual in accordance with the Standards for Procedures and Arrangements as called for by regulations 5, 5A and 8 of Annex II of the Convention, and that the arrangements and equipment of the ship prescribed in the Manual are in all respects satisfactory and comply with the applicable requirements of the said Standards.

4   That the ship is suitable for the carriage in bulk of the following noxious liquid substances, provided that all relevant operational provisions of Annex II of the Convention are observed.

Noxious liquid substances	Conditions of carriage (tank numbers etc.)
Continued on additional signed and dated sheets*	

This certificate is valid until . . . . †
subject to surveys in accordance with regulation 10 of Annex II of the Convention.

Issued at. . . . .
(Place of issue of certificate)

. . . .
(Date of issue)   *(Signature of authorized official issuing the certificate)*

(Seal or stamp of the authority, as appropriate)

---

* Delete as appropriate.

† Insert the date of expiry as specified by the Administration in accordance with regulation 12(1) of Annex II of the Convention. The day and the month of this date correspond to the anniversary date as defined in regulation 1(14) of Annex II of the Convention, unless amended in accordance with regulation 12(8) of Annex II of the Convention.

Annex II

ENDORSEMENT FOR ANNUAL AND INTERMEDIATE SURVEYS

THIS IS TO CERTIFY that, at a survey required by regulation 10 of Annex II of the Convention, the ship was found to comply with the relevant provisions of the Convention:

Annual survey: Signed . . . .
(Signature of authorized official)

Place . . . .

Date . . . .

(Seal or stamp of the authority, as appropriate)

Annual/Intermediate* survey: Signed . . . .
(Signature of authorized official)

Place . . . .

Date . . . .

(Seal or stamp of the authority, as appropriate)

Annual/Intermediate* survey: Signed . . . .
(Signature of authorized official)

Place . . . .

Date . . . .

(Seal or stamp of the authority, as appropriate)

Annual survey: Signed . . . .
(Signature of authorized official)

Place . . . .

Date . . . .

(Seal or stamp of the authority, as appropriate)

---

* Delete as appropriate.

ANNUAL/INTERMEDIATE SURVEY IN ACCORDANCE
WITH REGULATION 12(8)(c)

THIS IS TO CERTIFY that, at an annual/intermediate* survey in accordance with regulation 12(8)(c) of Annex II of the Convention, the ship was found to comply with the relevant provisions of the Convention:

Signed . . . .
(Signature of authorized official)

Place . . . .

Date . . . .

(Seal or stamp of the authority, as appropriate)

ENDORSEMENT TO EXTEND THE CERTIFICATE IF VALID
FOR LESS THAN 5 YEARS WHERE REGULATION 12(3) APPLIES

The ship complies with the relevant provisions of the Convention, and this Certificate shall, in accordance with regulation 12(3) of Annex II of the Convention, be accepted as valid until . . . .

Signed . . . .
(Signature of authorized official)

Place . . . .

Date . . . .

(Seal or stamp of the authority, as appropriate)

ENDORSEMENT WHERE THE RENEWAL SURVEY HAS BEEN
COMPLETED AND REGULATION 12(4) APPLIES

The ship complies with the relevant provisions of the Convention, and this Certificate shall, in accordance with regulation 12(4) of Annex II of the Convention, be accepted as valid until . . . .

Signed . . . .
(Signature of authorized official)

Place . . . .

Date . . . .

(Seal or stamp of the authority, as appropriate)

---

* Delete as appropriate.

Annex II

287

ENDORSEMENT TO EXTEND THE VALIDITY
OF THE CERTIFICATE UNTIL REACHING THE
PORT OF SURVEY OR FOR A PERIOD OF GRACE
WHERE REGULATION 12(5) OR 12(6) APPLIES

This Certificate shall, in accordance with regulation 12(5) or 12(6)* of Annex II of the Convention, be accepted as valid until . . . .

Signed . . . .
(Signature of authorized official)

Place . . . .

Date . . . .

(Seal or stamp of the authority, as appropriate)

ENDORSEMENT FOR ADVANCEMENT OF ANNIVERSARY DATE
WHERE REGULATION 12(8) APPLIES

In accordance with regulation 12(8) of Annex II of the Convention, the new anniversary date is . . . .

Signed . . . .
(Signature of authorized official)

Place . . . .

Date . . . .

(Seal or stamp of the authority, as appropriate)

In accordance with regulation 12(8) of Annex II of the Convention, the new anniversary date is . . . .

Signed . . . .
(Signature of authorized official)

Place . . . .

Date . . . .

(Seal or stamp of the authority, as appropriate)

---

* Delete as appropriate.

Unified Interpretations of Annex II

(Unless specified, regulations referred to are those of Annex II)

1 Definitions

Reg. 1(12)

1.1 *Conversion and modification of ships built before 1 July 1986*

1.1.1 An oil tanker or a chemical tanker previously not certified to carry safety hazard chemicals but which changes to a service of carrying these cargoes should be considered as having undergone a conversion. Safety hazard cargoes are identified in chapter VI of the Bulk Chemical Code (BCH Code) or chapter 17 of the International Bulk Chemical Code (IBC Code).

1.1.2 The last sentence of regulation 1(12) should apply only to modifications made on oil tankers and chemical tankers and the expression "modification" referred thereto should generally be those changes necessary to comply with Annex II and include the fitting of improved stripping systems and underwater discharge arrangements but do not include major structural changes such as those which might be necessary to comply with ship type requirements.

2 Application

Reg. 2(5) and 2(6)

2.1 *Equivalency for gas carriers*

2.1.1 With respect to liquefied gas carriers carrying Annex II substances listed in the Gas Carrier Code, equivalency may be permitted under the terms of regulation 2(5) on construction and equipment requirements contained in regulations 5, 5A and 13 when a gas carrier meets all the following conditions:

.1 hold a Certificate of Fitness in accordance with the appropriate Gas Carrier Code for ships carrying liquefied gases in bulk;

.2 hold an International Pollution Prevention Certificate for the Carriage of Noxious Liquid Substances in Bulk (NLS Certificate);

.3 be provided with segregated ballast arrangements;

.4 be provided with deepwell pumps and arrangements which minimize the amount of cargo residue remaining after discharge, to the extent that the Administration is satisfied on the basis of the design that the stripping requirements of regulation 5A(2)(b) or 5A(4)(b), without regard to the limiting date, are met and the cargo residue can be vented to the atmosphere through the approved venting arrangements;

.5 be provided with a Procedures and Arrangements Manual approved by the Administration. This manual should ensure that no operational mixing of cargo residues and water will occur and, after venting, no cargo residues will remain; and

.6 be certified in an NLS Certificate to carry only those Annex II noxious liquid substances listed in the appropriate Gas Carrier Code.

When such equivalency is granted, notification required by regulation 2(6) need not be made.

2A Categorization of substances

Reg. 3(4)

2A.1 When a substance which is not included in appendix II or III of MARPOL 73/78 is offered for bulk carriage, the provisional category should be established in accordance with the following procedure:*

.1 The Government of the State Party to MARPOL 73/78 shipping or producing the substance should check MEPC circulars to see whether the substance has been categorized by the Organization, or provisionally assessed by another State Party to MARPOL 73/78;

.2 if no information is found in the circulars, the Government of the Party should contact the Organization[†] to see if the substance has already been given a provisional assessment by the Organization or by another Government of a Party to MARPOL 73/78. If the latter is the case, the details should be obtained and, if satisfied, the Government of the Party may accept that provisional assessment;

.3 if there has been no previous provisional assessment, or the Government of the Party is not satisfied with the previous provisional assessment given, the Government of the Party shipping or producing the substance should carry out a provisional assessment in accordance with the attached guidelines;[‡]

---

* In carrying out the evaluation of substances, it will be necessary to establish minimum carriage requirements not only for Annex II purposes but also for safety purposes. Due regard should, therefore, be given to the "Criteria for Hazard Evaluation of Bulk Chemicals" approved by the MSC at its forty-second session (Annex 3 to the 1985 edition of the BCH Code, also included in the 1993 edition of the BCH Code and the 1994 and 1998 editions of the IBC Code).

† The enquiry should be addressed to: The Director, MED, IMO, 4 Albert Embankment, London SE1 7SR, United Kingdom; tel.: 44 020 7735 7611, telex: 23588 IMOLDN G, telefax: 44 020 7587 3210; and include the enquirer's mailing address, telex and telefax numbers. The latter, if available, would facilitate a quick reply.

‡ The Guidelines for the provisional assessment of liquid substances offered to be carried in bulk have been superseded by the Guidelines for the provisional assessment of liquids transported in bulk; see IMO sales publication IMO-653E.

.4   the Government of the Party should notify the Government of the State in whose port the cargo will be received and the Government of the flag State of their assessment along with information providing the basis for their pollution and safety hazard assessment, or the provisional assessment registered at the Organization, by the quickest means available;

.5   in the event of disagreement, the most severe conditions proposed should prevail;

.6   in the absence of an interim or final response to the notification from any of the other Parties involved within 14 days of the despatch, the provisional assessment made by the Government of the Party shipping or producing the substance should be deemed to have been accepted;

.7   the Organization should be notified and details provided of the provisional assessment made as required by regulation 3(4) (i.e. within 90 days, but preferably as soon as possible);

.8   the Organization should circulate the information as it is received through an MEPC circular and submit provisional assessments to the BCH Sub-Committee for review. The Organization should also maintain a register of all such substances and their provisional assessments until such time as the substances are formally included in the Annex II lists and the IBC and BCH Codes; and

.9   the Organization should forward to GESAMP all such information received, with a view to formal hazard evaluation and subsequent categorization and establishment of minimum carriage requirements by the BCH Sub-Committee, with a view to formal amendment of Annex II of MARPOL 73/78 and the IBC and BCH Codes.

2A.2   In the case that such provisionally assessed substances fall into Category A, B, C or D, amendment sheets to the ship's Certificate of Fitness, or to the NLS Certificate only in cases of Category D substances, and to the Ship's Procedures and Arrangements (P and A) Manual should be issued by the Administration before the ship sails, thus permitting their carriage. This authorization for the carriage of the substance may take the form of telex or equivalent means, which should be kept with the ship's Certificate of Fitness and P and A Manual until the substance is accepted as an amendment to the IBC/BCH Codes. Then the Certificate and the P and A Manual should formally be amended.

2A.3   In cases where it is necessary to provisionally assess pollutant-only mixtures, which contain substances for which the Organization has assigned a pollution category or a provisional pollution category and for which the Organization has assigned a ship type or provisional ship-type requirement, the following procedure may be applied in lieu of 2A.1, subparagraphs .4, .5 and .6:

.1   the mixtures need only be categorized by the Government of the Party shipping or producing the mixture, by the use of the

Annex II

Unified Interpretations

calculation procedure described in section 5 of the Guidelines for the Provisional Assessment of Liquids Transported in Bulk;

.2   the mixtures may contain up to 3% unassessed components which should be evaluated by the Government of a Party producing or shipping the mixture. Where evaluation by the use of GESAMP hazard profiles, data or assimilation by analogy from related substances is not possible for a component, classification of such a component should be taken as pollution category A, ship type 2;

.3   The mixture may contain components which have been identified as having safety hazards ('S' in column d of chapter 17 of the IBC Code) or which would justify inclusion in the Code, so long as it is judged that the dilution of these components results in a mixture that does not possess safety hazards. For purposes of this paragraph, mixtures meeting this characteristic are considered pollutant-only mixtures;

.4   the Government of the Party should notify the Government of the State in whose port the cargo will be received and the Government of the flag State of the assessment; and

.5   the Government of the Party may authorize the manufacturer to carry out the assignment (n.o.s. number, the appropriate shipping name, viscosity and melting point) on its behalf. In this case the obligation to inform flag States and receiving countries of the performed assignment falls on the authorized manufacturer. The manufacturer should also inform the authorizing Government of the assignment performed along with details of the assessment. The manufacturer should inform IMO if so requested by the Government of the shipping or producing country. Upon request, the manufacturer should provide the Government of the flag State or receiving State with full details of the mixture. Notification of the assignment by the manufacturer should be accompanied by the authorization letter indicating that the manufacturer acts under instruction and on behalf of the Government of the Party.

2A.4   Notwithstanding 2A.2, if a ship is certified fit to carry n.o.s. (not otherwise specified) substances of the provisionally assessed category and ship type, no amendment need be made to the ship's Certificate of Fitness or the ship's P and A Manual.

3   Discharge of residue

Reg. 5

3.1   *En route*

3.1.1   The term *"en route"* is taken to mean that the ship is under way at sea on a course, or courses, which so far as practicable for navigational purposes will cause any discharge to be spread over as great an area of the sea as is reasonably practicable.

4   Pumping, piping and unloading arrangements

**Reg. 5A
(6)(b)(iv)
and (7)(e)**

4.1 *Appropriate action in case of exemption*

4.1.1   With regard to the term "appropriate action, if any", any Party to the Convention that has an objection to the particulars of an exemption submitted by another Party should communicate this objection to the Organization and to the Party which issued the exemption within one year after the Organization circulates the particulars of the exemption to the Parties.

5   Reception facilities

Reg. 7(1)(b)

5.1 *Reception facilities in repair ports*

5.1.1   This regulation is taken to mean that ship repair ports undertaking repairs to chemical tankers should have facilities adequate for the reception of residues and mixtures containing noxious liquid substances as would remain for disposal from ships carrying them as a result of the application of this Annex.

5A   Measures of control

**Reg. 8(5)(a)(i)
and 8(7)(a)(i)**

5A.1   The wording "substance unloaded is identified in the standards developed by the Organization as resulting in a residue quantity exceeding the maximum quantity which may be discharged into the sea" of subparagraphs (5)(a)(i) and (7)(a)(i) of regulation 8 refers to high-viscosity or solidifying substances as defined in paragraphs 1.3.7 and 1.3.9 of the Standards for Procedures and Arrangements.

6   Survey and certification

**Reg. 10(1)
(c) and (d)**

6.1 *Intermediate and annual surveys for ships
      not required to hold NLS Certificate*

6.1.1   The applicability of regulation 10(1)(c) and (d) and/or corresponding requirements of the IBC and BCH Codes under regulation 12A to ships which are not required to hold an International Pollution Prevention Certificate for the Carriage of Noxious Liquid Substances in Bulk by virtue of regulation 11 should be determined by the Administration.

6A.1   Requirements for minimizing accidental pollution

Reg. 13(4)

6A.1.1 *Ships other than chemical tankers*

.1   For the purpose of regulation 13(4) of Annex II of MARPOL 73/78, the Organization has developed guidelines for offshore support vessels and ships engaged in dumping at sea.

.2   For the purpose of that regulation, for ships other than chemical tankers, which are not referred to in paragraph .1 above, all applicable requirements of the IBC or BCH Codes

Annex II

Unified Interpretations

should be complied with when such ships carry Category A, B or C noxious liquid substances in bulk.

7 Oil-like substances

Reg. 14

7.1 *List of oil-like substances*

Category C substances
Aviation alkylates
Cycloheptane
Cyclohexane
Cyclopentane
p-Cymene
Diethylbenzene
Dipentene
Ethylbenzene
Ethylcyclohexane
Heptene (all isomers)
Hexane (all isomers)
Hexene (all isomers)
Isopropylcyclohexane
Methylcyclohexane
2-Methyl-1-pentene
Nonane (all isomers)
Octane (all isomers)
Olefin mixtures (C_5–C_7)
Pentane (all isomers)
Pentene (all isomers)
1-Phenyl-1-xylylethane
Propylene dimer
Tetrahydronaphthalene
Toluene
Xylenes

Category D substances
Alkyl (C_9–C_{17}) benzenes
Diisopropyl naphthalene
Dodecane (all isomers)

For each of the above substances, compliance with the oil-like substance criterion 7.2.1.4 given below has to be demonstrated for the particular oil content meter installed.

7.2 *Selection criteria*

7.2.1 The following criteria define an oil-like Category C or D noxious liquid substance:

.1   the substance's mass density (specific gravity) is less than 1.0 at 20°C;

.2   the substance's solubility in seawater at 20°C is less than 0.1%;

.3   the substance is a hydrocarbon;

.4   the substance can be monitored by an oil content meter required by regulation 15 of Annex I of MARPOL 73/78;*

.5   in the case of Category C substances, the ship type requirement, as specified by the Bulk Chemical or International Bulk Chemical Codes, is type 3; and

.6   the substance is not regulated by the Bulk Chemical or International Bulk Chemical Codes for safety purposes as indicated in chapters VI and 17 of the Codes respectively.

Reg. 14(c)   7.3 *Damage stability calculation*

7.3.1 A new ship of 150 m or more in length under Annex I should be considered to comply with the requirements of regulation 14(c) if compliance with regulation 25 of Annex I has been demonstrated.

Reg. 14(d)   7.4 *Applicability of the waiver under regulation 15(5) of Annex I of MARPOL 73/78 to oil tankers carrying oil-like Annex II substances*

7.4.1 Since regulation 14 of Annex II applies to oil tankers, as defined in Annex I, which are allowed to carry oil-like substances and discharge these under the provisions of Annex I, any waiver granted to such oil tankers in respect of the requirement to be fitted with an oil discharge monitoring and control system encompasses the requirements contained in regulation 14(d) of Annex II. It should be noted, however, that in considering the issuance of a waiver under the provisions of Annex I, an Administration should determine that adequate reception facilities are available to receive the residues and mixtures at loading ports or terminals at which the tanker calls and that the facilities are also suitable for the treatment and ultimate disposal of the oil-like substances received.

---

* In approving an oil discharge monitoring and control system for the purpose of this regulation, the Administration should ensure through tests that the system can monitor concentrations of each oil-like substance in conformity with the Recommendation on international performance specifications for oily-water separating equipment and oil content meters adopted by the Organization by resolution A.393(X) or the Revised guidelines and specifications for oil discharge monitoring and control systems for oil tankers, resolution A.586(14). If it is necessary to readjust the monitor when changing from oil products to oil-like noxious substances, information on the readjustment should be provided and special operating procedures ensuring that discharges of oil-like noxious substances are measured accurately should be approved by the Administration. When the oil content meter is readjusted an entry should be made in the Oil Record Book.

Annex II

Unified Interpretations

Appendix to Unified Interpretations of Annex II

Guidelines for the application of amendments to the list of substances in Annex II of MARPOL 73/78 and in the IBC Code and the BCH Code with respect to pollution hazards

1   General

1.1   The present guidelines apply to amendments to the list of substances set out in appendices II and III to Annex II of MARPOL 73/78, in chapters 17 and 18 of the IBC Code and in chapters VI and VII of the BCH Code, namely the addition or deletion of substances, and changes of the pollution category or the ship-type requirements on existing substances.

1.2   Regulation 2(7)(a) of Annex II of MARPOL 73/78 stipulates that where an amendment to this Annex and the International Bulk Chemical Code and the Bulk Chemical Code involves changes to the structure or equipment and fittings due to the upgrading of the requirements for the carriage of certain substances, the Administration may modify or delay for a specified period the application of such an amendment to ships constructed before the date of entry into force of that amendment, if the immediate application of such an amendment is considered unreasonable or impracticable. Such relaxation shall be determined with respect to each substance, having regard to the Guidelines developed by the Organization. The present Guidelines have been developed to ensure uniform application of that regulation.

1.3   With respect to the preparation and circulation of proposed amendments to the list of substances, paragraphs 1 to 4 of the Guidelines for future amendments to the IBC Code and the BCH Code (MEPC 25/20, annex 7) should apply.

2   Definitions

For the purposes of the present Guidelines, the following definitions apply:

2.1   *New ship* means a ship the keel of which is laid or which is at a stage at which:

   .1   construction identifiable with the ship begins; and

   .2   assembly has commenced comprising at least 50 tonnes or 1% of the estimated mass of all structural material, whichever is less;

on or after the date of entry into force of the relevant amendment.

297

2.2 A ship, irrespective of the date of construction, which is converted to a chemical tanker on or after the date of entry into force of the relevant amendment, should be treated as a chemical tanker constructed on the date on which such conversion commences. This conversion provision does not apply to the modification of a ship referred to in regulation 1(12) of Annex II of MARPOL 73/78.

2.3 *Existing ship* means a ship which is not a new ship as defined in paragraph 2.1.

2.4 *Dedicated ship* means a ship built or converted and specifically fitted and certified for the carriage of:

.1   one named product only; or

.2   a restricted number of products each in a tank or group of tanks such that each tank or group of tanks is certified for the carriage of one named product only or compatible products not requiring cargo tank washing for change of cargo.

2.5 *Domestic trade* means a trade solely between ports or terminals within the State the flag of which the ship is entitled to fly, without entering into the territorial waters of other States.

2.6 *International trade* means a trade which is not a domestic trade as defined in paragraph 2.5.

2.7 *Structure of a ship* includes only the major structural elements, such as double bottom, longitudinal and transverse bulkheads, essential to the completeness of the hull necessary to meet the ship-type requirements. Piping systems, fittings and equipment such as underwater discharge outlets, stripping systems, high-level alarms, gauging devices etc., are not considered to be part of the structure of a ship.

2.8 *New substance* means a substance which was not previously carried in bulk. A substance which is not included in Annex II of MARPOL 73/78, the IBC Code or the BCH Code but which is transported in bulk may be treated as an existing substance, provided that such substance has been provisionally assessed under the provisions of regulation 3(4) of Annex II of MARPOL 73/78 or is carried in accordance with the provisions of Annex I of MARPOL 73/78.

2.9 *Existing substance* means a substance which is not a new substance.

3   Application of amendments to new and existing ships

3.1 All amendments which constitute the inclusion of new substances and those resulting in downgrading of requirements for existing substances should apply to new and existing ships as from the date of entry into force of the amendments.

3.2 Amendments resulting in upgrading of requirements for existing substances:

New ship

3.2.1 All amendments should apply to new ships as from the date of entry into force of the amendments.

Existing ships

3.2.2   All amendments involving only operational requirements should apply to existing ships as from the date of entry into force of the amendments.

3.2.3   The Administration may modify or delay for a specified period the application of amendments involving changes to the structure or equipment and fittings to existing ships, if the immediate application of such amendments is considered unreasonable or impracticable. Such relaxation should be determined with respect to each substance, having regard to such factors as volume of cargo shipped, whether or not dedicated ships are involved, types and ages of ships involved, types of trades (e.g. domestic or international trades), etc.

3.2.4   When allowing such relaxation the following guidelines should apply:

  .1   In the case of amendments affecting the structure of ships:

  .1.1   existing ships engaged in domestic trades should comply with the amended ship-type requirements not later than the end of the specified period which should not exceed ten years after the date of entry into force of the amendments;

  .1.2   existing ships engaged on restricted voyages in international trade as determined by the Administration should comply with the amended ship-type requirements not later than the end of the specified period which should not exceed ten years after the date of entry into force of amendments, provided that:

  .1.2.1   such relaxation is agreed among the Governments of the Parties concerned; and

  .1.2.2   the Certificate of Fitness is endorsed to the effect that the ship is solely engaged in such restricted voyages;

  .1.3   existing ships engaged in international trade other than the above should comply with the amended ship-type requirements as from the date of entry into force of the amendments.

  .2   In the case of amendments affecting the equipment and fittings:

  .2.1   if the amendments necessitate the provision of an underwater discharge outlet, the outlet should be fitted not later than two years following the entry into force of the amendments;

  .2.2   if the amendments necessitate the efficient stripping system:

  .2.2.1   until the end of the period of two years following the entry into force of the amendments or until 2 October 1994, whichever occurs later, the ship should comply with the requirements of regulation 5A(2)(b) or 5A(4)(b) of Annex II of MARPOL 73/78, as applicable;

  .2.2.2   after the above date, the efficient stripping system should be fitted to comply with the applicable requirements of regulation 5A.

  .2.3   The requirements for the discharge of effluent below the waterline need not apply until the underwater discharge outlet has been fitted.

3.2.5   As a general rule, the relaxation mentioned in paragraph 3.2.4.1 should be accepted only for existing dedicated ships. In exceptional cases, however, where the immediate application of such amendments to existing non-dedicated ships will create serious difficulties for clear and acceptable reasons, such as very large volumes of cargo being shipped, the application may be delayed for a limited period.

3.2.6   The Certificate of Fitness should be endorsed by the Administration specifying the relaxation allowed.

3.2.7   The Administration allowing a relaxation of the application of amendments should submit to the Organization a report giving details of the ship or ships concerned, the cargoes carried, the trade in which each ship is engaged and the justification for such relaxation.

3.2.8   A Member Government may notify the Organization that it does not accept the relaxation.

3.2.9   The notification made under 3.2.8 and 3.2.9 should be circulated to other Governments.

Standards for procedures and arrangements for the discharge of noxious liquid substances

(called for by Annex II of MARPOL 73/78, as amended)

Preamble

1   Annex II of the International Convention for the Prevention of Pollution from Ships, 1973, as modified by the Protocol of 1978 relating thereto (MARPOL 73/78) and as further amended by the Organization (hereafter referred to as Annex II), *inter alia* provides for the control of operational discharges of noxious liquid substances carried in bulk by ships. *Operational discharges* in this context means the discharges of noxious liquid substances or water contaminated by these substances which are the result of cargo tank and line washing, deballasting of unwashed cargo tanks or cargo pump-room bilge slops.

2   Annex II prohibits the discharge into the sea of noxious liquid substances except when the discharge is made under specified conditions. These conditions vary according to the degree of hazard a noxious liquid substance poses to the marine environment. For this purpose the noxious liquid substances have been divided into four categories, A, B, C and D.

3   Regulation 5 of Annex II specifies the conditions under which discharge of residues of Categories A, B, C and D substances may take place. These conditions, which are not reproduced in this document, include such parameters as: the maximum quantity which may be discharged into the sea, speed of ship, distance from nearest land, depth of water, maximum concentration of substance in ship's wake or dilution of substance prior to discharge.

4   For certain sea areas, referred to as "special areas", more stringent discharge criteria apply.

5   The standards for procedures and arrangements called for by Annex II (hereafter referred to as the Standards) have been developed in response to resolution 13 of the International Conference on Marine Pollution, 1973, and in compliance with regulations 5, 5A and 8 of Annex II. The Standards provide a uniform basis for the guidance of the Parties to MARPOL 73/78

in approving procedures and arrangements for the discharge of noxious liquid substances of a specific ship.

6   The Standards took effect on 6 April 1987, the date of implementation of Annex II, and apply to all ships which carry noxious liquid substances in bulk.

7   The Annex II requirements are not restated in the Standards. To ensure compliance with Annex II, the requirements of Annex II and those contained in the Standards should be considered together.

8   Annex II discharge requirements and certification requirements have been interpreted as requiring each ship to have a Procedures and Arrangements Manual approved by the Administration. The Manual should contain the information specified in the Standards and the requirements of Annex II. Compliance with the procedures and arrangements set out in a ship's Manual will ensure that the discharge requirements of Annex II are met.

9   Regulation 5A of Annex II requires that the efficiency of the cargo pumping system of a tank certified fit to carry Category B or C substances be tested in accordance with standards developed by the Organization. The test procedure is set out in the Standards. The pump stripping efficiency determined by the test will be assumed to be the stripping efficiency achieved when unloading the tank in accordance with the specified procedures.

10   The presence of a "sheen" resulting after discharges of some Category B, C and D substances should not be regarded as contrary to the principles of Annex II, provided that the discharges have been made in accordance with the Standards.

11   Throughout the Standards the word "discharge" is used to refer to the discharge of residues or residue/water mixtures either into the sea or to reception facilities, whilst the word "unloading" is used to refer to the unloading of cargo to receivers, terminals or ports.

Chapter 1
Introduction

1.1 *Purpose*

The purpose of the Standards is to provide a uniform international basis for approving procedures and arrangements by which ships carrying noxious liquid substances in bulk can satisfy the discharge provisions of Annex II. It is on the basis of these Standards that the Administration should approve the procedures and arrangements necessary for the issue of an International Pollution Prevention Certificate for the Carriage of Noxious Liquid Substances in Bulk or a Certificate of Fitness for the Carriage of Dangerous Chemicals in Bulk or an International Certificate of Fitness for the Carriage of Dangerous Chemicals in Bulk to each such ship. For that purpose the procedures and arrangements for each ship are to be laid out in an approved Procedures and Arrangements Manual (hereinafter called the Manual) for use on board the ship. It is not intended that these Standards be used by the ship's crew.

1.2 *Scope*

1.2.1   These Standards apply to all ships which carry Category A, B, C or D noxious liquid substances in bulk, including those provisionally assessed as such.

1.2.2   The Standards have been developed to ensure that the criteria for discharge of noxious liquid substances specified in regulations 5 and 8 will be met. For Category A substances, the Standards identify a prewash procedure which may be used in lieu of measuring the concentration of the effluent from a tank from which tank washings containing a Category A substance are discharged. For Category B and C substances, the Standards identify procedures and arrangements which will ensure that the maximum quantity of residue that may be discharged per tank and the maximum permitted concentration of the substance in the ship's wake are not exceeded. For Category B and C substances, the Standards identify procedures and arrangements for assessing compliance with regulation 5A. For Category A, B, C and D substances, the Standards identify ventilation procedures which may be used to remove residues from cargo tanks. The prewash procedures contained in appendix B to the Standards also enable Administrations to approve the prewash procedure referred to in regulation 5A(6)(b)(i).

1.2.3   The Standards do not cover the means by which the Administration ensures compliance with a ship's approved procedures and arrangements, and neither do they cover details of any constructions or materials used.

1.2.4 Regulation 13 requires, *inter alia*, chemical tankers carrying Category A, B or C noxious liquid substances to comply with the *International Code for the Construction and Equipment of Ships Carrying Dangerous Chemicals in Bulk** (hereinafter referred to as the IBC Code) or the *Code for the Construction and Equipment of Ships Carrying Dangerous Chemicals in Bulk** (hereinafter referred to as the BCH Code), as may be amended. All constructions, materials and equipment fitted as a requirement of Annex II and of the Standards shall therefore comply with the IBC Code or the BCH Code for all substances of Category A, B or C the chemical tanker is certified fit to carry in accordance with its Certificate of Fitness under that Code.

1.3 *Definitions*

1.3.1 *New ship* means a ship constructed on or after 1 July 1986.

1.3.2 *Existing ship* means a ship that is not a new ship.

1.3.3 *Residue* means any noxious liquid substance which remains for disposal.

1.3.4 *Residue/water mixture* means residue to which water has been added for any purpose (e.g. tank cleaning, ballasting, bilge slops).

1.3.5 *Miscible* means soluble with water in all proportions at washwater temperatures.

1.3.6 *Associated piping* means the pipeline from the suction point in a cargo tank to the shore connection used for unloading the cargo and includes all ship's piping, pumps and filters which are in open connection with the cargo unloading line.

1.3.7 *Solidifying substance* means a noxious liquid substance which:

 .1 in the case of substances with melting points less than 15°C, is at a temperature, at the time of unloading, of less than 5°C above its melting point; or

 .2 in the case of substances with melting points equal to or greater than 15°C, is at a temperature, at the time of unloading, of less than 10°C above its melting point.

1.3.8 *Non-solidifying substance* means a noxious liquid substance which is not a solidifying substance.

---

* The IBC and BCH Codes, extended to cover marine pollution aspects, were adopted by the Marine Environment Protection Committee of the Organization by resolution MEPC.19(22) and MEPC.20(22) respectively on 5 December 1985; see IMO sales publications IMO-100E and IMO-772E, respectively.

1.3.9 *High-viscosity substance* means:

    .1   in the case of Category A and B substances and in the case of Category C substances within special areas, a substance with a viscosity equal to or greater than 25 mPa·s at the unloading temperature; and

    .2   in the case of Category C substances outside special areas, a substance with a viscosity equal to or greater than 60 mPa·s at the unloading temperature.

1.3.10 *Low-viscosity substance* means a noxious liquid substance which is not a high-viscosity substance.

1.3.11 *Regulation* means a regulation of Annex II to MARPOL 73/78.

1.4 *Equivalents*

1.4.1 The equivalent provisions in regulation 2(5) and (6) are also applicable to the Standards.

1.5 *Certification*

1.5.1 Before issuing the appropriate Certificate referred to in section 1.1, the Administration should examine, and, if satisfied, approve:

    .1   the Manual for compliance with Annex II and the Standards; and

    .2   the equipment and arrangements provided for compliance with the Standards.

1.5.2 Reference to the approved Manual should be made by the Administration in the appropriate Certificate issued to the ship.

1.6 *Responsibilities of the master*

1.6.1 The master must ensure that no discharges into the sea of cargo residues or residue/water mixtures containing Category A, B, C or D substances shall take place, unless such discharges are made in full compliance with the operational procedures contained in the Manual and that the arrangements required by the Manual and needed for such discharges are used.

1.7 *Safety considerations*

1.7.1 The Standards are concerned with the marine environmental aspects of the cleaning of cargo tanks which have contained noxious liquid substances, and the discharge of residues and residue/water mixtures from

Annex II

these operations. Certain of these operations are potentially hazardous but no attempt is made in the Standards to lay down safety standards covering all aspects of these operations. For a description of potential hazards reference should be made to the IBC or BCH Codes and other documents as developed and published by the relevant associations or organizations, e.g. the Tanker Safety Guide (Chemicals) of the International Chamber of Shipping (ICS). Some potential safety hazards are mentioned below.

1.7.2 *Compatibility* – In mixing residue/water mixtures containing different substances, compatibility should be carefully considered.

1.7.3 *Electrostatic hazards* – The hazards associated with the generation of electrostatic charges during the cargo tank washing should be carefully considered.

1.7.4 *Tank entry hazards* – The safety of persons required to enter cargo tanks or slop tanks for any purpose should be carefully considered.

1.7.5 *Reactivity hazards* – The water washing of cargo tanks and slop tanks containing residues of certain substances may produce dangerous reactions and should be carefully considered.

1.7.6 *Ventilation hazards* – The hazards associated with tank ventilation identified in the ICS Tanker Safety Guide (Chemicals) should be carefully considered.

1.7.7 *Line clearing hazards* – The hazards associated with line clearing identified in the ICS Tanker Safety Guide (Chemicals) should be carefully considered.

1.7.8 *Fire hazards* – The fire hazards associated with the use of cleaning media other than water should be carefully considered.

1.8 *Cleaning agents or additives*

1.8.1 When a washing medium other than water, such as mineral oil or chlorinated solvent, is used instead of water to wash a tank, then its discharge is governed by the provisions of Annex I or Annex II, which would apply respectively if such a medium had been carried as cargo. Tank washing procedures involving the use of such mediums should be set out in the Procedures and Arrangements Manual and be approved by the Administration.

1.8.2 When small amounts of cleaning detergents are added to water in order to facilitate tank washing, no detergents containing pollution category A components should be used except those components that are readily biodegradable and present in a total concentration of less than 10%. No restrictions additional to those applicable to the tank due to the previous cargo should apply.

Chapter 2
Preparation of the Procedures and Arrangements Manual

2.1   Each ship which carries noxious liquid substances in bulk should be provided with a Manual as described in this chapter.

2.2   The main purpose of the Manual is to identify for the ship's officers the physical arrangements and all the operational procedures with respect to cargo handling, tank cleaning, slops handling, and cargo tank ballasting and deballasting which must be followed in order to comply with the requirements of Annex II.

2.3   The Manual should be based on the Standards. It should cover all noxious liquid substances which the ship is certified fit to carry.

2.4   The Manual should as a minimum contain the following information and operational instructions:

.1   a description of the main features of Annex II, including discharge requirements;

.2   a table of noxious liquid substances which the ship is certified fit to carry and which specifies information on these substances as detailed in appendix D;

.3   a description of the tanks carrying noxious liquid substances; and a table identifying in which cargo tanks each noxious liquid substance may be carried;

.4   a description of all arrangements and equipment including cargo heating and temperature control system, which are on board the ship and for which requirements are contained in chapters 3 or 8 including a list of all tanks that may be used as slop tanks, a description of the discharge arrangements, a schematic drawing of the cargo pumping and stripping systems showing the respective position of pumps and control equipment and identification of means for ensuring that the equipment is operating properly (check lists);

.5   details of the procedures set out in the Standards as applied to the individual ship which should, where appropriate, include instructions such as:

.5.1   methods of stripping cargo tanks and under what restrictions, such as minimum list and trim, the stripping system should be operated;

Annex II

.5.2   methods of draining cargo pumps, cargo lines and stripping lines;

.5.3   cargo tank prewash programmes;

.5.4   procedures for cargo tank ballasting and deballasting;

.5.5   procedures for discharge of residue/water mixtures; and

.5.6   procedures to be followed when a cargo tank cannot be unloaded in accordance with the required procedure;

.6   for existing ships operating under the provisions of regulation 5A(2)(b) or 5A(4)(b) a residue table developed in accordance with appendix A, which indicates for each tank in which Category B or C substances are to be carried the quantities of residue which will remain in the tank and associated piping system after unloading and stripping;

.7   a table which indicates the quantities measured as a result of carrying out the water test performed for assessing the "stripping quantity" referred to in paragraph 1.2.1 of appendix A; and

.8   the responsibility of the master in respect of operational procedures to be followed and the use of the arrangements. The master must ensure that no residues or residue/water mixtures are discharged into the sea, unless the arrangements listed in the Manual and needed for the discharge are used.

2.5   In the case of a ship engaged in international voyages, the Manual should be produced in the standard format as outlined in the attached appendix D. If the language used is neither English nor French, the text should include a translation into one of these languages.

2.6   The Administration may approve a Manual containing only those parts applicable to the substances the ship is certified fit to carry.

2.7   For a ship referred to in regulation 5A(6) or 5A(7), the format and the content of the Manual should be to the satisfaction of the Administration.

2.8   For a ship carrying only Category D substances, the format and the content of the Manual should be to the satisfaction of the Administration.

Chapter 3
Equipment and constructional standards for new ships

3.1 *General*

3.1.1 This chapter contains the standards for the equipment and constructional features enabling a new ship to comply with the residue discharge requirements of Annex II.

3.1.2 The equipment requirements in this chapter should be read in conjunction with the operating requirements in chapters 4, 5, 6 and 7 in order to determine what equipment is needed on the ship.

3.2 *Carriage requirements*

3.2.1 A Category B substance with a melting point equal to or greater than 15°C should not be carried in a cargo tank any boundary of which is formed by the ship's shell plating and should only be carried in a cargo tank fitted with a cargo heating system.

3.3 *Cargo unloading system*

3.3.1 The cargo unloading system for Category B and C substances should be capable of unloading the cargo to the residue quantities not in excess of the quantities specified in regulations 5 and 5A. The performance test required by regulation 5A(5) should be carried out in accordance with appendix A.

3.4 *Underwater discharge outlet location*

3.4.1 The underwater discharge outlet (or outlets) should be located within the cargo area in the vicinity of the turn of the bilge and should be so arranged as to avoid the re-intake of residue/water mixtures by the ship's seawater intakes.

3.5 *Underwater discharge outlet size*

3.5.1 The underwater discharge outlet arrangement should be such that the residue/water mixture discharged into the sea in accordance with the Standards will not pass through the ship's boundary layer.

To this end, when the discharge is made normal to the ship's shell plating, the minimum diameter of the discharge outlet is governed by the following equation:

$$D = \frac{Q_D}{5L}$$

where

D = minimum diameter of the discharge outlet (m)

L = distance from the forward perpendicular to the discharge outlet (m)

Q_D = the maximum rate selected at which the ship may discharge a residue/water mixture through the outlet (m^3/h).

3.5.2   When the discharge is directed at an angle to the ship's shell plating, the above relationship should be modified by substituting for Q_D the component of Q_D which is normal to the ship's shell plating.

3.6   Slop tanks

3.6.1   Although Annex II does not require the fitting of dedicated slop tanks, slop tanks may be needed for certain washing procedures. Cargo tanks may be used as slop tanks.

3.7   Ventilation equipment

3.7.1   If residues from cargo tanks are removed by means of ventilation, ventilation equipment meeting the requirements of appendix C should be provided.

Chapter 4
Operational standards for new ships carrying Category A substances

4.1 *General*

This chapter applies to any new ship certified fit to carry Category A substances.

4.2 *Pumping and stripping*

In unloading a cargo tank containing a Category A substance, the tank and its associated piping should be emptied to the maximum extent practicable by maintaining a positive flow of cargo to the tank's suction point and using the stripping procedure set out in the Manual.

4.3 *Prewash of Category A substances from cargo tanks*

4.3.1 Annex II requires that when a cargo tank that has contained a Category A substance is washed, the resulting residue/water mixtures be discharged to a reception facility until the concentration of the substance in the effluent is at or below a specified value and until the tank is empty. Where it is found to be impracticable to measure the concentration of the substance in the effluent, a prewash procedure in accordance with appendix B should be applied in conformity with regulation 8(4).

4.3.2 The residue/water mixture generated during the prewash should be discharged to a reception facility in accordance with regulation 8.

4.3.3 Any water subsequently introduced into the cargo tank may be discharged into the sea in accordance with the requirements of regulation 5(1) or regulation 5(7) in respect of the ship's position, speed and discharge outlet location.

4.4 *Ventilation of Category A substances from cargo tanks*

4.4.1 Ventilation procedures may be applied only to those substances having a vapour pressure greater than 5×10^3 Pa at 20°C.

4.4.2 The ventilation procedures set out in appendix C should be followed when a tank is to be ventilated.

4.4.3 In ventilating a tank the associated piping of the tank should be cleared of liquid and the tank should be ventilated until no visible remains of liquid can be observed in the tank. When direct observation is impossible or impracticable, means for detection of liquid remains should be provided.

Annex II

4.4.4   When the cargo tank has been ventilated dry in accordance with the Standards, any water subsequently introduced into the cargo tank for ballasting or for preparing the tank to receive the next cargo should be regarded as clean and should not be subject to the discharge requirements of Annex II.

Chapter 5
Operational standards for new ships carrying Category B substances

5.1 *General*

5.1.1   This chapter applies to any new ship certified fit to carry Category B substances.

5.1.2   If a cargo tank is to be washed or ballasted and some or all of the residue left in the tank is to be discharged into the sea, the requirements of sections 5.2 to 5.7 apply.

5.1.3   If the requirements of this chapter under which discharges into the sea of residues and residue/water mixtures containing Category B substances are allowed cannot be met, no such discharges may be made.

5.2 *Pumping and stripping*

5.2.1   In unloading a cargo tank containing a Category B substance, the tank and its associated piping should be emptied to the maximum extent practicable by maintaining a positive flow of cargo to the tank's suction point and using the stripping procedure set out in the Manual.

5.3 *Tank washing and residue discharge procedures outside special areas*

5.3.1   **High-viscosity or solidifying substances**

.1   A prewash procedure as specified in appendix B should be applied;

.2   the residue/water mixture generated during the prewash should be discharged to a reception facility in accordance with regulation 8; and

.3   any water subsequently introduced into the cargo tank may be discharged into the sea at a rate not exceeding the maximum rate for which the underwater discharge outlet(s) referred to in section 3.5 is(are) designed. The discharge must also be in accordance with the other discharge requirements of regulation 5(2) in respect of ship's position, speed, and discharge outlet location.

5.3.2   **Low-viscosity, non-solidifying substances**

.1   Any water introduced into the cargo tank may be discharged into the sea at a rate not exceeding the maximum rate for which the underwater discharge outlet(s) referred to in section 3.5 is(are) designed. The discharge must also be in accordance with the other discharge requirements of regulation 5(2) in respect of ship's position, speed and discharge outlet location.

5.4 Tank washing and residue discharge procedures within special areas

5.4.1   A prewash procedure as specified in appendix B should be applied.

5.4.2   The residue/water mixture generated during the prewash should be discharged to a reception facility in accordance with regulation 8.

5.4.3   Any water subsequently introduced into the cargo tank may be discharged into the sea at a rate not exceeding the maximum rate for which the underwater discharge outlet(s) referred to in section 3.5 is(are) designed. The discharge must also be in accordance with the other discharge requirements of regulation 5(8) in respect of ship's position, speed and discharge outlet location.

5.4.4   Notwithstanding the provisions of paragraphs 5.4.1 to 5.4.3, residues or residue/water mixtures containing only low-viscosity, non-solidifying substances may be retained on board and discharged into the sea outside special areas in accordance with the provisions of paragraphs 5.3.2 or 5.5.2.

5.5 Discharges from a slop tank

5.5.1   Residue/water mixtures in a slop tank should not be discharged into the sea within special areas.

5.5.2   Residue/water mixtures in a slop tank which contain only low-viscosity, non-solidifying substances may be discharged into the sea outside special areas at a rate not exceeding the maximum rate for which the underwater discharge outlet(s) referred to in section 3.5 is(are) designed. The discharge must also be in accordance with the other discharge requirements of regulation 5(2) in respect of ship's position, speed and discharge outlet location.

5.5.3   Residue/water mixtures in a slop tank which contain high-viscosity or solidifying substances, retained on board in accordance with regulation 8, should be discharged to a reception facility.

5.6 Ventilation of Category B substances from cargo tanks

5.6.1   When ventilation procedures are used to remove residue from cargo tanks, the requirements set out in section 4.3 apply.

5.7 *Ballasting and deballasting*

5.7.1 After unloading, and, if required, carrying out a prewash, a cargo tank may be ballasted. Procedures for the discharge of such ballast are set out in sections 5.3 and 5.4.

5.7.2 Ballast introduced into a cargo tank which has been washed to such an extent that the ballast contains less than 1 ppm of the substance previously carried, may be discharged into the sea without regard to the discharge rate, ship's speed and discharge outlet location, provided that the ship is not less than 12 miles from land and in water that is not less than 25 m deep. It is assumed this degree of cleanliness has been achieved when a prewash as specified in appendix B has been carried out and the tank has been subsequently washed with a complete cycle of the cleaning machine.

Annex II

Chapter 6
Operational standards for new ships carrying Category C substances

6.1 *General*

6.1.1   This chapter applies to any new ship certified fit to carry Category C substances.

6.1.2   If a cargo tank is to be washed or ballasted and some or all of the residue left in the tank is to be discharged into the sea, the requirements of sections 6.2 to 6.7 apply.

6.1.3   If the requirements of this chapter under which discharges into the sea of residues and residue/water mixtures containing Category C substances are allowed cannot be met, no such discharges may be made.

6.2 *Pumping and stripping*

6.2.1   In unloading a cargo tank containing a Category C substance, the tank and its associated piping should be emptied to the maximum extent practicable by maintaining a positive flow of cargo to the tank's suction point and using the stripping procedure set out in the Manual.

6.3 *Tank washing and residue discharge procedures outside special areas*

6.3.1   **High-viscosity or solidifying substances**

    .1   A prewash procedure as specified in appendix B should be applied;

    .2   the residue/water mixture generated during the prewash should be discharged to a reception facility in accordance with regulation 8; and

    .3   any water subsequently introduced into the cargo tank may be discharged into the sea at a rate not exceeding the maximum rate for which the underwater discharge outlet(s) referred to in section 3.5 is(are) designed. The discharge must also be in accordance with the other discharge requirements of regulation 5(3) in respect of ship's position, speed and discharge outlet location.

6.3.2   **Low-viscosity, non-solidifying substances**

    .1   Any water introduced into the cargo tank may be discharged into the sea at a rate not exceeding the maximum rate for which the underwater discharge outlet(s) referred to in section 3.5 is(are) designed. The discharge must also be in accordance with the other discharge requirements of regulation 5(3) in respect of ship's position, speed and discharge outlet location.

6.4 Tank washing and residue discharge procedures within special areas

6.4.1 High-viscosity* or solidifying substances

.1 A prewash procedure as specified in appendix B should be applied;

.2 the residue/water mixture generated during the prewash should be discharged to a reception facility in accordance with regulation 8;

.3 any water subsequently introduced into the cargo tank may be discharged into the sea at a rate not exceeding the maximum rate for which underwater discharge outlet(s) referred to in section 3.5 is(are) designed. The discharge must also be in accordance with the other discharge requirements of regulation 5(9) in respect of ship's position, speed and discharge outlet location; and

.4 notwithstanding the provisions of paragraphs 6.4.1.1 to 6.4.1.3, residue/water mixtures containing non-solidifying substances with a viscosity less than 60 mPa·s at the unloading temperature may be retained on board and discharged into the sea outside special areas in accordance with the provisions of paragraph 6.3.2.

6.4.2 Low-viscosity,† non-solidifying substances

.1 Any water introduced into the cargo tank may be discharged into the sea at a rate not exceeding the maximum rate for which the underwater discharge outlet(s) referred to in section 3.5 is(are) designed. The discharge must also be in accordance with the other discharge requirements of regulation 5(9) in respect of ship's position, speed and discharge outlet location.

---

* i.e. a substance with a viscosity equal to or greater than 25 mPa·s at the unloading temperature. See definition of a high-viscosity Category C substance discharged within special areas.

† i.e. a substance with a viscosity less than 25 mPa·s at the unloading temperature, within special areas.

6.5 Discharges from a slop tank

6.5.1   Residue/water mixtures in a slop tank which contains only low-viscosity,* non-solidifying substances may be discharged into the sea at a rate not exceeding the maximum rate for which the underwater discharge outlet(s) referred to in section 3.5 is(are) designed. The discharge must also be in accordance with the other discharge requirements of regulations 5(3) and 5(9) in respect of ship's position, speed and discharge outlet location.

6.5.2   Residue/water mixtures in a slop tank which contains high-viscosity or solidifying substances, retained on board in accordance with regulation 8, should be discharged to a reception facility.

6.6 Ventilation of Category C substances from cargo tanks

6.6.1   When ventilation procedures are used to remove residue from cargo tanks, the requirements set out in section 4.4 apply.

6.7 Ballasting and deballasting

6.7.1   After unloading and, if required, carrying out a prewash, a cargo tank may be ballasted. Procedures for the discharge of such ballast are set out in sections 6.3 and 6.4.

6.7.2   Ballast introduced into a cargo tank which has been washed to such an extent that the ballast contains less than 1 ppm of the substance previously carried, may be discharged into the sea without regard to the discharge rate, ship's speed and discharge outlet location, provided that the ship is not less than 12 miles from land and in water that is not less than 25 m deep. It is assumed this degree of cleanliness has been achieved when a prewash as specified in appendix B has been carried out and the tank has been subsequently washed with a complete cycle of the cleaning machine.

---

* i.e. a substance with a viscosity less than 25 mPa·s at the unloading temperature if discharged within special areas, or a substance with a viscosity less than 60 mPa·s at the unloading temperature if discharged outside special areas.

Chapter 7
Operational standards for new ships carrying Category D substances

7.1 *General*

7.1.1   This chapter applies to any new ship certified fit to carry Category D substances.

7.2 *Discharge of Category D residues*

7.2.1   Although residue(s) of Category D substances is(are) required to be discharged within and outside special areas in a diluted form in accordance with regulation 5(4), such residue(s) may also be discharged in accordance with the operational standards for low-viscosity, non-solidifying Category C substances as specified in chapter 6.

7.3 *Ventilation of Category D substances from cargo tanks*

7.3.1   When ventilation procedures are used to remove residue from cargo tanks the requirements set out in section 4.4 apply.

Chapter 8
Equipment and constructional standards for existing ships

8.1 *General*

8.1.1 This chapter contains the standards for the equipment and constructional features enabling an existing ship to comply with the residue discharge requirements of Annex II.

8.1.2 The equipment requirements in this chapter should be read in conjunction with the operating requirements in chapters 9, 10, 11 and 12 in order to determine what equipment is needed on the ship.

8.2 *Carriage requirements*

8.2.1   A Category B substance with a melting point equal to or greater than 15°C should not be carried in a cargo tank any boundary of which is formed by the ship's shell plating and should only be carried in a cargo tank fitted with a cargo heating system.

8.3 *Cargo unloading system*

8.3.1   The cargo unloading system for Category B and C substances should be capable of unloading the cargo to the residue quantities not in excess of the quantities specified in regulations 5 and 5A. The performance test required by regulation 5A(5) should be carried out in accordance with appendix A.

8.4 *Residue discharge system*

8.4.1   When for the purpose of discharging residues into the sea controlled pumping rates are needed to meet the requirements of chapter 10, one of the following systems should be used:

   .1   a variable rate pumping system in which:

   .1.1   the capacity is adjusted by varying the pump speed; or

   .1.2   the capacity is adjusted through the use of a throttling arrangement fitted on the discharge piping;

   .2   a fixed-rate pumping system with a capacity not exceeding the permissible discharge rate as set out under sections 10.5 and 10.6.

8.4.2 If the pumping rates are controlled in accordance with 8.4.1.1, a flow rate indicating device should be provided.

8.5 *Underwater discharge outlet location*

8.5.1 The underwater discharge outlet (or outlets) should be located within the cargo area in the vicinity of the turn of the bilge and should be so arranged as to avoid the re-intake of residue/water mixtures by the ship's seawater intakes.

8.5.2 If dual outlets are provided to achieve a higher permissible discharge rate, these should be located on opposite sides of the ship.

8.6 *Underwater discharge outlet size*

8.6.1 The underwater discharge outlet arrangement should be such that the residue/water mixture discharged into the sea in accordance with the Standards will not pass through the ship's boundary layer. To this end, when the discharge is made normal to the ship's shell plating, the minimum diameter of the discharge outlet is governed by the following equation:

$$D = \frac{Q_D}{5L}$$

where

D = minimum diameter of the discharge outlet (m)

L = distance from the forward perpendicular to the discharge outlet (m)

Q_D = the maximum rate selected at which the ship may discharge a residue/water mixture through the outlet (m^3/h).

8.6.2 When the discharge is directed at an angle to the ship's shell plating, the above relationship should be modified by substituting for Q_D the component of Q_D which is normal to the ship's shell plating.

8.7 *Recording devices*

8.7.1 When in accordance with chapter 10 it is necessary to record the discharge of residue/water mixtures, means should be provided for recording the start and stop time of the discharge with actual time (GMT or other standard time). The device should be in operation when there is a discharge into the sea which is to be recorded. The date should be recorded either manually or automatically. The record should be identifiable as to time and date and should be kept for at least three years.

Annex II

8.7.2   When in accordance with chapter 10 it is necessary to record the rate at which residue/water mixtures are discharged, means should be provided for measuring such flow rates. The accuracy of the flow recording unit should be within 15% of the actual flow.

8.7.3   If the recording units described in paragraphs 8.7.1 or 8.7.2 become defective, a manual alternative method should be used. The master should record such a defect in the Cargo Record Book. The defective unit should be made operable as soon as possible but at least within a period of 60 days.

8.8   *Slop tanks*

8.8.1   Although Annex II does not require the fitting of dedicated slop tanks, slop tanks may be needed for certain washing procedures. Cargo tanks may be used as slop tanks.

8.9   *Ventilation equipment*

8.9.1   If residues from cargo tanks are removed by means of ventilation, ventilation equipment meeting the requirements of appendix C should be provided.

Chapter 9
Operational standards for existing ships carrying Category A substances

9.1 *General*

9.1.1 This chapter applies to any existing ship certified fit to carry Category A substances.

9.2 *Prewash of a Category A substance from a cargo tank*

9.2.1 Annex II requires that when a tank that has contained a Category A substance is washed, the resulting residue/water mixtures be discharged to a reception facility until the concentration of the substance in the effluent is reduced below a specified value and until the tank is empty. Where it is found to be impracticable to measure the concentration of the substance in the effluent, a prewash procedure in accordance with appendix B should be applied in conformity with regulation 8(4).

9.2.2 The residue/water mixture generated during the prewash should be discharged to a reception facility in accordance with regulation 8.

9.2.3 Any water subsequently introduced into the cargo tank may be discharged into the sea in accordance with the requirements of regulation 5(1) or regulation 5(7) in respect of the ship's position, speed and discharge outlet location.

9.3 *Ventilation of Category A substances from cargo tanks*

9.3.1 Ventilation procedures may be applied only to those substances having a vapour pressure greater than 5×10^3 Pa at 20°C.

9.3.2 The ventilation procedures set out in appendix C should be followed when a tank is to be ventilated.

9.3.3 In ventilating a tank the associated piping of the tank should be cleared of liquid and the tank should be ventilated until no visible remains of liquid can be observed in the tank. When direct observation is impossible or impracticable, means for detection of liquid remains should be provided.

9.3.4 When the cargo tank has been ventilated dry in accordance with the Standards, any water subsequently introduced into the cargo tank for ballasting or for preparing the tank to receive the next cargo should be regarded as clean and should not be subject to the discharge requirements of Annex II.

Annex II

Chapter 10
Operational standards for existing ships carrying Category B substances

10.1 *General*

10.1.1 This chapter applies to any existing ship certified fit to carry Category B substances.

10.1.2 When a cargo tank on an existing ship is fitted with a cargo unloading system capable of unloading the cargo to the residue quantities not in excess of the quantity specified in regulation 5A(2)(a) and if the tank is to be washed or ballasted and some or all of the residue left in the tank is to be discharged into the sea, the requirements of chapter 5 apply.

10.1.3 If a tank other than that referred to in paragraph 10.1.2 is to be washed or ballasted and some or all of the residue left in the tank is to be discharged into the sea, the requirements of sections 10.2 to 10.8 apply.

10.1.4 If the requirements of this chapter under which discharges into the sea of residues and residue/water mixtures containing Category B substances are allowed cannot be met, no such discharges may be made.

10.2 *Pumping and stripping*

10.2.1 In unloading a cargo tank containing a Category B substance, the tank and its associated piping should be emptied to the maximum extent practicable by maintaining a positive flow of cargo to the tank's suction point and using the stripping procedure set out in the Manual.

10.3 *Tank washing and residue discharge procedures outside special areas*

10.3.1 **High-viscosity or solidifying substances**

.1 A prewash procedure as specified in appendix B should be applied;

.2 the residue/water mixture generated during the prewash should be discharged to a reception facility in accordance with regulation 8; and

.3 any water subsequently introduced into the cargo tank may be discharged into the sea at a rate not exceeding the maximum rate for which the underwater discharge outlet(s) referred to in section 8.6 is(are) designed. The discharge must also be in accordance with the other discharge requirements of regulation 5(2) in respect of ship's position, speed and discharge outlet location.

10.3.2 Low-viscosity, non-solidifying substances

.1 A prewash procedure as specified in appendix B should be applied;

.2 the residue/water mixture generated during the prewash should be discharged to a reception facility in accordance with regulation 8 or transferred to a slop tank for subsequent discharge into the sea in accordance with sections 10.5 or 10.6; and

.3 any water subsequently introduced into the cargo tank may be discharged into the sea at a rate not exceeding the maximum rate for which the underwater discharge outlet(s) referred to in section 8.6 is(are) designed. The discharge must also be in accordance with the other discharge requirements of regulation 5(2) in respect of ship's position, speed and discharge outlet location.

Annex II

10.4 *Tank washing and residue discharge procedures within special areas*

10.4.1 A prewash procedure as specified in appendix B should be applied.

10.4.2 The residue/water mixture generated during the prewash should be discharged to a reception facility in accordance with regulation 8.

10.4.3 Any water subsequently introduced into the tank may be discharged into the sea at a rate not exceeding the maximum rate for which the underwater discharge outlet(s) referred to in section 8.6 is(are) designed. The discharge must also be in accordance with the requirements of regulation 5(8) in respect of ship's position, speed and discharge outlet location.

10.4.4 Notwithstanding the provisions of paragraphs 10.4.1 to 10.4.3, residue or residue/water mixtures containing only low-viscosity, non-solidifying substances may be retained on board and discharged into the sea outside special areas in accordance with section 10.5 or 10.6.

10.5 *Discharge into the sea of a miscible residue/water mixture from a slop tank*

10.5.1 Prewash residue/water mixtures containing Category B substances should not be discharged into the sea within special areas.

10.5.2 Before a miscible residue/water mixture is discharged into the sea outside special areas, the composite concentration, C_s, should be determined as follows:

$$C_s = n/V_r$$

where n = number of tanks containing Category B residues which have been transferred to the slop tank. (For the sake of simplification, it is assumed that each tank contains 1 m³ of residue.)

V_r = volume of residue/water mixtures in the slop tank prior to discharge (determined from ullage tables) (m³).

10.5.3 The residue/water mixture may be discharged into the sea, provided that the rate does not exceed the maximum rate for which the underwater discharge outlet(s) referred to in section 8.6 is(are) designed or that defined by one of the equations below, whichever is smaller:

$$Q_D = \frac{KV^{1.4}L^{1.6}}{C_s} \qquad \text{when a single outlet is used; or}$$

$$Q_D = \frac{1.5\,KV^{1.4}L^{1.6}}{C_s} \qquad \text{when dual outlets are used.}$$

where Q_D = rate of discharge of residue/water mixture (m³/h)

V = ship's speed (knots)

L = ship's length (m)

K = 4.3×10^{-5}

C_s = composite concentration referred to in paragraph 10.5.2.

10.5.4 The discharge must also be in accordance with the other discharge requirements of regulation 5(2) in respect of ship's position, speed and discharge outlet location.

10.5.5 Residue/water mixtures discharged into the sea in accordance with this section should be recorded using the device referred to in paragraph 8.7.1. If a variable capacity pump is used for the discharge, the flow rate should so be recorded using the device referred to in paragraph 8.7.2.

10.6 Discharge into the sea of an immiscible residue/water mixture from a slop tank

10.6.1 Prewash residue/water mixtures containing Category B substances should not be discharged into the sea within special areas.

10.6.2 The residue/water mixture may be discharged into the sea outside special areas, provided that the rate does not exceed the maximum rate for which the underwater discharge outlet(s) referred to in section 8.6 is(are) designed or that defined by one of the equations below, whichever is smaller:

$$Q_D = KV^{1.4}L^{1.6} \qquad \text{when a single outlet is used; or}$$

$$Q_D = 1.5\,KV^{1.4}L^{1.6} \qquad \text{when dual outlets are used.}$$

10.6.3 The discharge must also be in accordance with the other discharge requirements of regulation 5(2) in respect of ship's position, speed and discharge outlet location.

10.6.4 Residue/water mixtures discharged into the sea in accordance with this section should be recorded using the device referred to in paragraph 8.7.1. If a variable capacity pump is used for the discharge, the flow rate should also be recorded using the device referred to in paragraph 8.7.2.

10.7 *Ventilation of Category B substances from cargo tanks*

10.7.1 When ventilation procedures are used to remove residue from cargo tanks, the requirements set out in section 9.3 apply.

10.8 *Ballasting and deballasting*

10.8.1 After unloading, and, if required, carrying out a prewash, a cargo tank may be ballasted. Procedures for the discharge of such ballast are set out in sections 10.3 to 10.6.

10.8.2 Ballast introduced into a cargo tank which has been washed to such an extent that the ballast contains less than 1 ppm of the substance previously carried, may be discharged into the sea without regard to the discharge rate, ship's speed and discharge outlet location, provided that the ship is not less than 12 miles from land and in water that is not less than 25 m deep. It is assumed this degree of cleanliness has been achieved when a prewash as specified in appendix B has been carried out and the tank has been subsequently washed with a complete cycle of the cleaning machine.

Annex II

Chapter 11
Operational standards for existing ships carrying Category C substances

11.1 *General*

11.1.1 This chapter applies to any existing ship certified fit to carry Category C substances.

11.1.2 When a cargo tank on an existing ship is fitted with a cargo unloading system capable of unloading the cargo to the residue quantities not in excess of the quantity specified in regulation 5A(4)(a) and if the tank is to be washed or ballasted and some or all of the residue left in the tank is to be discharged into the sea, the requirements of chapter 6 apply. However, an existing ship may only discharge residue/water mixtures containing Category C substances within special areas in accordance with paragraph 6.4.2.1 if the cargo unloading system meets the requirements as specified for new ships in regulation 5A(3). If the cargo unloading system does not meet these requirements, discharge of residue/water mixtures within special areas should be carried out in accordance with section 11.4 or 11.5.

11.1.3 If a cargo tank other than that referred to in paragraph 11.1.2 is to be washed or ballasted and some or all of the residue left in the tank is to be discharged into the sea, the requirements of sections 11.2 to 11.7 apply.

11.1.4 If the requirements of this chapter under which discharges into the sea of residues and residue/water mixtures containing Category C substances are allowed cannot be met, no such discharges may be made.

11.2 *Pumping and strippings*

11.2.1 In unloading a cargo tank containing a Category C substance, the tank and its associated piping should be emptied to the maximum extent practicable by maintaining a positive flow of cargo to the tank's suction point and using the stripping procedure set out in the Manual.

11.3 *Tank washing and residue discharge procedures outside special areas*

11.3.1 **High-viscosity or solidifying substances**

    .1   A prewash procedure as specified in appendix B should be applied;

    .2   the residue/water mixture generated during the prewash should be discharged to a reception facility in accordance with regulation 8; and

.3   any water subsequently introduced into the tank may be discharged
into the sea at a rate not exceeding the maximum rate for which
the underwater discharge outlet(s) referred to in section 8.6 is(are)
designed. The discharge must also be in accordance with the other
discharge requirements of regulation 5(3) in respect of ship's
position, speed and discharge outlet location.

11.3.2 **Low-viscosity, non-solidifying substances**

.1   any water introduced into the cargo tank may be discharged into
the sea at a rate not exceeding the maximum rate for which the
underwater discharge outlet(s) referred to in section 8.6 is(are)
designed. The discharge must also be in accordance with the
other discharge requirements of regulation 5(3) in respect of
ship's position, speed and discharge outlet location.

11.4 *Tank washing and residue discharge procedures within special areas*

11.4.1   A prewash procedure as specified in appendix B should be applied.

11.4.2   The residue/water mixture generated during the prewash should be
discharged to a reception facility in accordance with regulation 8.

11.4.3   Any water subsequently introduced into the cargo tank may be
discharged into the sea at a rate not exceeding the maximum rate for which
the underwater discharge outlet(s) referred to in section 8.6 is(are) designed.
The discharge must also be in accordance with the other discharge
requirements of regulation 5(9) in respect of ship's position, speed and
discharge outlet location.

11.4.4   Notwithstanding the provisions of paragraphs 11.4.1 to 11.4.3,
residue/water mixtures containing only non-solidifying substances with a
viscosity less than 60 mPa·s at the unloading temperature may be retained
on board and discharged into the sea outside special areas in accordance
with paragraph 11.5.2.

11.5 *Discharges from a slop tank*

11.5.1   Residue/water mixtures in a slop tank should not be discharged into
the sea within special areas.

11.5.2   Residue/water mixtures in a slop tank which contain only low-
viscosity, non-solidifying substances may be discharged into the sea outside
special areas at a rate not exceeding the maximum rate for which the
underwater discharge outlet(s) referred to in section 8.6 is(are) designed.
The discharge must also be in accordance with the other discharge
requirements of regulation 5(3) in respect of the ship's position, speed and
discharge outlet location.

Annex II

11.5.3 Residue/water mixtures in a slop tank which contain high-viscosity or solidifying substances, retained on board in accordance with regulation 8, should be discharged to a reception facility.

11.6 *Ventilation of Category C substances from cargo tanks*

11.6.1 When ventilation procedures are used to remove residue from cargo tanks, the requirements set out in section 9.3 apply.

11.7 *Ballasting and deballasting*

11.7.1 After unloading, and, if required, carrying out a prewash, a cargo tank may be ballasted. Procedures for the discharge of such ballast are set out in sections 11.3 to 11.4.

11.7.2 Ballast introduced into a cargo tank which has been washed to such an extent that the ballast contains less than 1 ppm of the substance previously carried, may be discharged into the sea without regard to the discharge rate, ship's speed and discharge outlet location, provided that the ship is not less than 12 miles from land and in water that is not less than 25 m deep. It is assumed this degree of cleanliness has been achieved when a prewash as specified in appendix B has been carried out and the tank has been subsequently washed with a complete cycle of the cleaning machine.

Chapter 12
Operational standards for existing ships carrying Category D substances

12.1 *General*

12.1.1 This chapter applies to any existing ship certified fit to carry Category D substances.

12.2 *Discharge of Category D residues*

12.2.1 Although residue(s) of Category D substances is(are) required to be discharged within and outside special areas in a diluted form in accordance with regulation 5(4), such residue(s) may also be discharged in accordance with the operational standards for low-viscosity, non-solidifying Category C substances as specified in chapter 11.

12.3 *Ventilation of Category D substances from cargo tanks*

12.3.1 When ventilation procedures are used to remove residue from cargo tanks the requirements set out in section 9.3 apply.

Appendix A
Assessment of residue quantities in cargo tanks, pumps and piping

1 Introduction

1.1 Purpose

1.1.1 The purpose of this appendix is:

    .1   to provide the procedure for testing the efficiency of cargo pumping systems; and

    .2   to provide the method for calculating the residue quantities on the cargo tank surfaces.

1.2 Background

1.2.1 The ability of the pumping system of a tank to comply with regulation 5A(1), (2), (3) or (4) is determined by performing a test in accordance with the procedure set out in section 3 of this appendix. The quantity measured is termed the "stripping quantity". The stripping quantity of each tank shall be recorded in the ship's Manual.

1.2.2 For tanks of existing ships not satisfying the appropriate pumping efficiency requirement of regulation 5A(2)(a) or (4)(a) it is necessary to calculate the quantity of residue remaining on tank surfaces. The method for calculating the clingage residue is given in section 4.

1.2.3 For tanks referred to in 1.2.2, it is necessary to calculate the total quantity of residue remaining in the cargo tanks and its associated piping. The total residue quantity is the sum of the water test result and the calculated clingage quantity.

1.2.4 After having determined the stripping quantity and calculated clingage quantity (when required) of one tank, the Administration may use the determined quantities for a similar tank, provided the Administration is satisfied that the pumping system in that tank is similar and operating properly.

2 Design criteria and performance test

2.1   The cargo pumping systems should be designed to meet the required maximum amount of residue per tank of 0.1 m³ and 0.3 m³ or 0.3 m³ and 0.9 m³ respectively for Category B or C substances as specified by regulation 5A to the satisfaction of the Administration.

2.2   In accordance with regulation 5A(5), the cargo pumping systems should be tested with water to prove their performance. Such water tests should, by measurement, show that the system meets the requirements of regulation 5A with the tolerance of 50 *l* per tank.

3   *Water test procedure*

3.1   Test condition

3.1.1   The ship's trim and list should be such as to provide favourable drainage to the suction point. During the water test the ship's trim should not exceed 3° by the stern, and the ship's list should not exceed 1°.

3.1.2   The trim and list chosen for the water test should be the minimum favourable trim and list as given in the ship's Manual for the stripping of the cargo tanks.

3.1.3   During the water test means should be provided to maintain a back-pressure of not less than 1 bar at the cargo tank's unloading manifold (see figures A-1 and A-2).

3.2   Test procedure

3.2.1   Ensure that the cargo tank to be tested and its associated piping have been cleaned and that the cargo tank is safe for entry.

3.2.2   Fill the cargo tank with water to a depth necessary to carry out normal end of unloading procedures.

3.2.3   Pump and strip the cargo tank and its associated piping in accordance with the ship's approved Manual.

3.2.4   Collect water remaining in the cargo tank and its associated piping into a calibrated container for measurement. Water residues should be collected from the following points:

    .1   the cargo tank suction and its vicinity;

    .2   any entrapped areas on the cargo tank bottom;

    .3   the low point drain of the cargo pump; and

    .4   all low point drains of piping associated with the cargo tank up to the manifold valve.

3.2.5   The total water volumes collected above determine the stripping quantity for the cargo tank.

3.2.6   Where a group of tanks is served by a common pump or piping, the water test residues associated with the common system(s) may be apportioned equally among the tanks provided that the following operational restriction is included in the ship's approved Manual: "For sequential unloading of tanks in this group, the pump or piping is not to be washed until all tanks in the group have been unloaded."

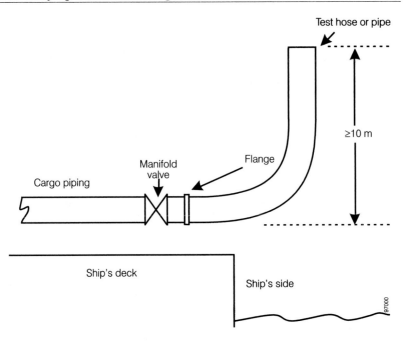

Figure A-1

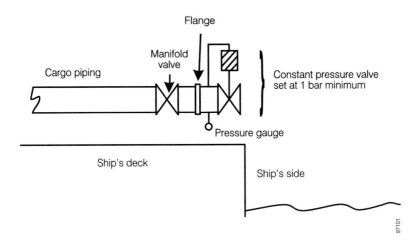

Figure A-2

The above figures illustrate test arrangements that would provide a backpressure of not less than 1 bar at the cargo tank's unloading manifold.

4 Calculation of clingage residues

4.1 Calculate the clingage residues using the following formula:

$$Q_{RES}^{(surf)} = 1.1 \times 10^{-4} A_d + 1.5 \times 10^{-5} A_w + 4.5 \times 10^{-4} L^{1/2} A_b$$

4.2 Symbols and units used in residue equation:

A_b = Area of tank bottom and horizontal components of tank structural members facing upwards (m^2)

A_d = Area underdecks and horizontal components of tank structural members facing downwards (m^2)

A_w = Surface area of tank walls and vertical components of tank structural members (m^2)

L = Length of tank (m)

$Q_{RES}^{(surf)}$ = Amount of clingage residue on tank surfaces (m^3).

Notes: 1. For purposes of calculating A_b, A_d and A_w, inclined (greater than 30° from the horizontal) and curved surfaces should be treated as vertical.

2. Methods of approximating A_b, A_d and A_w are permissible. (A method presented in BCH 15/INF.5 by Japan is an example.)

Appendix B
Prewash procedures [Ships built before 1 July 1994]*

In several sections of the Standards a prewash procedure is required in order to meet certain Annex II requirements. This appendix explains how these prewash procedures should be performed.

Prewash procedures for non-solidifying substances

1   Tanks should be washed by means of a rotary water jet, operated at sufficiently high water pressure. In the case of Category A substances washing machines should be operated in such locations that all tank surfaces are washed. In the case of Category B and C substances only one location need be used.

2   During washing the amount of water in the tank should be minimized by continuously pumping out slops and promoting flow to the suction point (positive list and trim). If this condition cannot be met the washing procedure should be repeated three times, with thorough stripping of the tank between washings.

3   Those substances which have a viscosity equal to or greater than 25 mPa·s at 20°C should be washed with hot water (temperature at least 60°C).

4   The number of cycles of the washing machine used should not be less than that specified in table B-1. A washing machine cycle is defined as the period between two consecutive identical orientations of the washing machine (rotation through 360°).

5   After washing, the washing machine(s) should be kept operating long enough to flush the pipeline, pump and filter.

Prewash procedures for solidifying substances

1   Tanks should be washed as soon as possible after unloading. If possible tanks should be heated prior to washing.

2   Residues in hatches and manholes should preferably be removed prior to the prewash.

---

* The Sub-Committee on Bulk Chemicals decided at its twenty-third session (September 1993) that the new appendix B† would be mandatory for new ships built on or after 1 July 1994 but could be applied to existing ships immediately on a voluntary basis, if so approved by the Administration. For ships built before 1 July 1994, therefore, the unamended text of appendix B still applies.

† See Revised appendix B.

3   Tanks should be washed by means of a rotary water jet operated at sufficiently high water pressure and in locations to ensure that all tank surfaces are washed.

4   During washing the amount of water in the tank should be minimized by pumping out slops continuously and promoting flow to the suction point (positive list and trim). If this condition cannot be met, the washing procedure should be repeated three times with thorough stripping of the tank between washings.

5   Tanks should be washed with hot water (temperature at least 60°C).

6   The number of cycles of the washing machine used should not be less than that specified in table B-1. A washing machine cycle is defined as the period between two consecutive identical orientations of the machine (rotation through 360°).

7   After washing, the washing machine(s) should be kept operating long enough to flush the pipeline, pump and filter.

Annex II

**Table B-1 – Number of washing machine cycles
to be used in each location**

Category of substance	Number of washing machine cycles	
	non-solidifying substances	solidifying substances
Category A (residual concentration 0.1% or 0.05%)	1	2
Category A (residual concentration 0.01% or 0.005%)	2	3
Category B	1/2	1
Category C	1/2	1

Note: For an explanation of "residual concentration" see regulation 5(1) and 5(7) of Annex II.

Revised appendix B
Prewash procedures for new ships*

In several sections of the Standards a prewash procedure is required in order to meet certain Annex II requirements. This appendix explains how these prewash procedures should be performed and how the minimum volumes of washing media to be used should be determined. Smaller volumes of washing media may be used based on actual verification testing to the satisfaction of the Administration. Where reduced volumes are approved an entry to that effect must be recorded in the Procedures and Arrangements Manual.

The applicable safety considerations listed in section 1.7 of the Standards should be taken into account when developing procedures employing recycling of wash water, or when washing is conducted with a medium other than water.

If a medium other than water is used for the prewash, the provisions of 1.8.1 of the Standards apply.

Prewash procedures for non-solidifying substances without recycling

1   Tanks should be washed by means of a rotary jet(s), operated at sufficiently high water pressure. In the case of Category A substances washing machines should be operated in such locations that all tank surfaces are washed. In the case of Category B and C substances only one location need be used.

2   During washing the amount of liquid in the tank should be minimized by continuously pumping out slops and promoting flow to the suction point. If this condition cannot be met, the washing procedure should be repeated three times, with thorough stripping of the tank between washings.

3   Those substances which have a viscosity equal to or greater than 25 mPa·s at 20°C should be washed with hot water (temperature at least 60°C), unless the properties of such substances make the washing less effective.

4   The quantities of wash water used should not be less than those specified in paragraph 20 or determined according to paragraph 21.

---

* The Sub-Committee on Bulk Chemicals decided at its twenty-third session (September 1993) that the new appendix B would be mandatory for new ships built on or after 1 July 1994 but could be applied to existing ships immediately on a voluntary basis, if so approved by the Administration. For ships built before 1 July 1994, therefore, the unamended text of appendix B still applies.

5   After prewashing the tanks and lines should be thoroughly stripped.

Prewash procedures for solidifying substances without recycling

6   Tanks should be washed as soon as possible after unloading. If possible, tanks should be heated prior to washing.

7   Residues in hatches and manholes should preferably be removed prior to the prewash.

8   Tanks should be washed by means of a rotary jet(s) operated at sufficiently high water pressure and in locations to ensure that all tank surfaces are washed.

9   During washing the amount of liquid in the tank should be minimized by pumping out slops continuously and promoting flow to the suction point. If this condition cannot be met, the washing procedure should be repeated three times with thorough stripping of the tank between washings.

10   Tanks should be washed with hot water (temperature at least 60°C), unless the properties of such substances make the washing less effective.

11   The quantities of wash water used should not be less than those specified in paragraph 20 or determined according to paragraph 21.

12   After prewashing the tanks and lines should be thoroughly stripped.

Prewash procedures with recycling of washing medium

13   Washing with a recycled washing medium may be adopted for the purpose of washing more than one cargo tank. In determining the quantity, due regard must be given to the expected amount of residues in the tanks and the properties of the washing medium and whether any initial rinse or flushing is employed. Unless sufficient data are provided, the calculated end concentration of cargo residues in the washing medium should not exceed 5% based on nominal stripping quantities.

14   The recycled washing medium should only be used for washing tanks having contained the same or similar substance.

15   A quantity of washing medium sufficient to allow continuous washing should be added to the tank or tanks to be washed.

16   All tank surfaces should be washed by means of a rotary jet(s) operated at sufficiently high pressure. The recycling of the washing medium may either be within the tank to be washed or via another tank, e.g. a slop tank.

17   The washing should be continued until the accumulated throughput is not less than that corresponding to the relevant quantities given in paragraph 20 or determined according to paragraph 21.

18   Solidifying substances and substances with viscosity equal to or greater than 25 mPa·s at 20°C should be washed with hot water (temperature at least 60°C) when water is used as the washing medium, unless the properties of such substances make the washing less effective.

19   After completing the tank washing with recycling to the extent specified in paragraph 17, the washing medium should be discharged and the tank thoroughly stripped. Thereafter, the tank should be subjected to a rinse, using clean washing medium, with continuous drainage and discharge. The rinse should as a minimum cover the tank bottom and be sufficient to flush the pipelines, pump and filter.

Minimum quantity of water to be used in a prewash

20   The minimum quantity of water to be used in a prewash is determined by the residual quantity of noxious liquid substance in the tank, the tank size, the cargo properties, the permitted concentration in any subsequent wash water effluent, and the area of operation. The minimum quantity is given by the following formula:

$$Q = k\left(15r^{0.8} + 5r^{0.7} \times V/1000\right)$$

where

Q =   the required minimum quantity in m^3

r =   the residual quantity per tank in m^3. The value of r shall be the value demonstrated in the actual stripping efficiency test, but should not be taken lower than 0.100 m^3 for a tank volume of 500 m^3 and above and 0.040 m^3 for a tank volume of 100 m^3 and below. For tank sizes between 100 m^3 and 500 m^3 the minimum value of r allowed to be used in the calculations is obtained by linear interpolation.

For Category A substances the value of r should either be determined based on stripping tests according to the Standards, observing the lower limits as given above, or be taken to be 0.9 m^3.

V =   tank volume in m^3

k =   a factor having values as follows:

Category A, non-solidifying, low-viscosity substance, outside special areas	$k = 1.0$
Category A, non-solidifying, low-viscosity substance, inside special areas	$k = 1.2$
Category A, solidifying or high-viscosity substance, outside special areas	$k = 2.0$

Category A, solidifying or high-viscosity
substance, inside special areas $k = 2.4$

Phosphorus, in all areas $k = 3.0$

Category B and C, non-solidifying,
low-viscosity substance $k = 0.5$

Category B and C, solidifying or
high-viscosity substance $k = 1.0$

The table below is calculated using the formula with a k factor of 1 and may
be used as an easy reference.

Stripping quantity (m^3)	Tank volume (m^3)		
	100	500	3,000
$\leqslant 0.04$	1.2	2.9	5.4
0.10	2.5	2.9	5.4
0.30	5.9	6.8	12.2
0.90	14.3	16.1	27.7

21   Verification testing for approval of prewash volumes lower than those
given in paragraph 20 may be carried out to the satisfaction of the
Administration to prove that the requirements of regulation 5 are met,
taking into account the substances the tanker is certified to carry. The
prewash volume so verified should be adjusted for other prewash conditions
by application of the factor k as defined in paragraph 20.

Annex II

Appendix C
Ventilation procedures

1   Cargo residues of substances with a vapour pressure greater than 5×10^3 Pa at 20°C may be removed from a cargo tank by ventilation.

2   Before residues of noxious liquid substances are ventilated from a tank the safety hazards relating to cargo flammability and toxicity should be considered. With regard to safety aspects, the operational requirements for openings in cargo tanks in the International Bulk Chemical Code, the Bulk Chemical Code, and the ventilation procedures in the ICS Tanker Safety Guide (Chemicals) should be consulted.

3   Port authorities may also have regulations on cargo tank ventilation.

4   The procedures for ventilation of cargo residues from a tank are as follows:

    .1   the pipelines should be drained and further cleared of liquid by means of ventilation equipment;

    .2   the list and trim should be adjusted to the minimum levels possible so that evaporation of residues in the tank is enhanced;

    .3   ventilation equipment producing an airjet which can reach the tank bottom shall be used. Figure C-1 could be used to evaluate the adequacy of ventilation equipment used for ventilating a tank of a given depth;

    .4   ventilation equipment should be placed in the tank opening closest to the tank sump or suction point;

    .5   ventilation equipment should, when practicable, be positioned so that the airjet is directed at the tank sump or suction point and impingement of the airjet on tank structural members is to be avoided as much as possible; and

    .6   ventilation shall continue until no visible remains of liquid can be observed in the tank. This shall be verified by a visual examination or an equivalent method.

Annex II

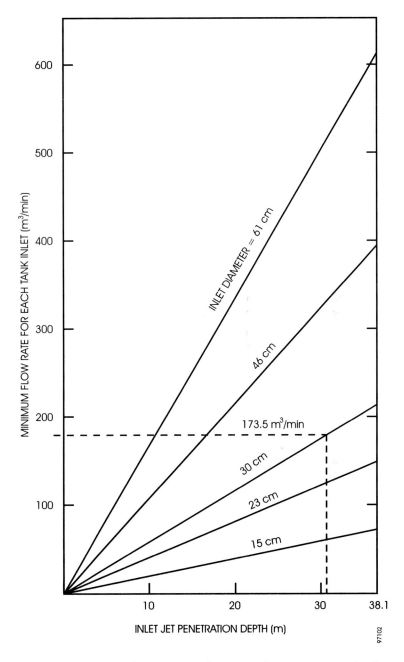

Figure C-1. Minimum flow rate as a function of jet penetration depth. Jet penetration depth should be compared against tank height.

Appendix D
Standard format for the Procedures and Arrangements Manual

Note 1: The standard format consists of a standardized text of an introduction, of an index and of the leading paragraphs to each section. This standardized text should be reproduced in the Manual provided for each ship followed by the information necessary to complete each section as applicable to the particular ship. The necessary information is indicated within with left-hand marking. When a section is not applicable, "NA" should be entered. It is recognized that the content of the Manual will vary depending on the design of the ship, the trade and the types of cargoes intended to be carried.

Note 2: If the Administration requires or accepts information and operational instructions in addition to those outlined in this Standard Format, they should be included in part 2 of the Manual. If no such additional information or operating instructions are required or accepted by the Administration, the Manual will consist of one part only.

STANDARD FORMAT

MARPOL 73/78 ANNEX II
PROCEDURES AND ARRANGEMENTS MANUAL

Name of ship:. . . . .

Distinctive number or letters: . . . .

Port of registry:. . . . .

Approval stamp of Administration:

INTRODUCTION

1.   The International Convention for the Prevention of Pollution from Ships, 1973, as modified by the Protocol of 1978 relating thereto (hereinafter referred to as MARPOL 73/78) was established in order to prevent the pollution of the marine environment by discharges into the sea from ships of harmful substances or effluents containing such substances. In order to achieve its aim, MARPOL 73/78 contains five annexes in which detailed regulations are given with respect to the handling on board ships and the discharge into the sea of five main groups of harmful substances, i.e. Annex I (mineral oils), Annex II (noxious liquid substances carried in bulk), Annex III (harmful substances carried in packaged forms), Annex IV (sewage) and Annex V (garbage).

2.   Regulation 5 of Annex II prohibits the discharge into the sea of noxious liquid substances of Categories A, B, C and D or of ballast water, tank washings or other residues or mixtures containing such substances, except in compliance with specified conditions including procedures and arrangements based upon standards developed by the International Maritime Organization (IMO) to ensure that the criteria specified for each Category will be met.

3.   The Standards for Procedures and Arrangements called for by Annex II of MARPOL 73/78 (as referred to above) require that each ship which is certified for the carriage of noxious liquid substances in bulk shall be provided with a Procedures and Arrangements Manual, hereinafter referred to as the Manual.

4.   This Manual has been written in accordance with chapter 2 of the Standards and is concerned with the marine environmental aspects of the cleaning of cargo tanks and the discharge of residues and mixtures from these operations. The Manual is not a safety guide and reference should be made to other publications specifically to evaluate safety hazards.

5.   The purpose of [Part 1 of]* the Manual is to identify the arrangements and equipment required to enable compliance with Annex II and to identify for the ship's officers all operational procedures with respect to cargo handling, tank cleaning, slops handling, residue discharging, ballasting and deballasting, which must be followed in order to comply with the requirements of Annex II. [Part 2 of the Manual contains additional information and operational instructions required or accepted by the Administration.]*

6.   In addition, this Manual, together with the ship's Cargo Record Book [and International Certificate for the Carriage of Noxious Liquid

---

* The parts in square brackets in the text or marked in the left margin are to be included only if a Part 2 is incorporated into the Manual.

Annex II

Substances in Bulk/Certificate of Fitness issued under the International Bulk Chemical Code/Certificate of Fitness issued under the Bulk Chemical Code],* will be used by Administrations for control purposes in order to ensure full compliance with the requirements of Annex II by this ship.

7.   The master shall ensure that no discharges into the sea of cargo residues or residue/water mixtures containing Category A, B, C or D substances shall take place, unless such discharges are made in full compliance with the operational procedures contained in this Manual and that the equipment required by this Manual and needed for such discharge is used.

8.   This Manual has been approved by the Administration and no alteration or revision shall be made to any part of it without the prior approval of the Administration.

---

* Include only the certificate issued to the particular ship.

INDEX OF SECTIONS

[Part 1]

Part 2

Additional information and operational instructions required or accepted by
the Administration (if necessary).

SECTION 1 – Main features of MARPOL 73/78, Annex II

1.1   The requirements of Annex II apply to all ships carrying noxious liquid substances in bulk. Substances posing a threat of harm to the marine environment are divided into four categories, A, B, C and D, and listed as such in appendix II to Annex II. Category A substances are those posing the greatest threat to the marine environment, whilst Category D substances are those posing the smallest threat.

1.2   Annex II prohibits the discharge into the sea of any effluent containing substances falling under these categories, except when the discharge is made under conditions which are specified in detail for each category. These conditions include, where applicable, such parameters as:

- the maximum quantity of substances per tank which may be discharged into the sea;
- the speed of the ship during the discharge;
- the minimum distance from the nearest land during discharge;
- the minimum depth of water at sea during discharge;
- the maximum concentration of substances in the ship's wake or the dilution of substances prior to discharge; and
- the need to effect the discharge below the waterline.

1.3   For certain sea areas identified as "special areas" more stringent discharge criteria are given. Under Annex II the special areas are the Baltic Sea area, the Black Sea area, and the Antarctic area.*

1.4   Annex II requires that every ship is provided with pumping and piping arrangements to ensure that each tank designated for the carriage of Category B and C substances does not retain after unloading a quantity of residue in excess of the quantity given in the Annex. For each tank intended for the carriage of such substances an assessment of the residue quantity has to be made. Only when the residue quantity as assessed is less than the quantity prescribed by the Annex may a tank be approved for the carriage of a Category B or a Category C substance.

1.5   In addition to the conditions referred to above, an important requirement contained in Annex II is that the discharge operations of certain cargo residues and certain tank cleaning and ventilation operations may only be carried out in accordance with approved procedures and arrangements based upon standards developed by the International Maritime Organization (IMO).

---

* MARPOL 73/78, Annex II defines these areas as follows:
  - The *Baltic Sea area* means the Baltic Sea proper with the Gulf of Bothnia, the Gulf of Finland and the entrance to the Baltic Sea bounded by the parallel of the Skaw in the Skagerrak at 57° 44.8′ N.
  - The *Black Sea area* means the Black Sea proper with the boundary between the Mediterranean and the Black Sea constituted by the parallel 41° N.
  - The *Antarctic area* means the sea south of latitude 60° S.

1.6   To enable this requirement to be complied with, this Manual contains in section 2 all particulars of the ship's equipment and arrangements, in section 3 operational procedures for cargo unloading and tank stripping and in section 4 procedures for discharge of cargo residues, tank washing, slops collection, ballasting and deballasting as may be applicable to the substances the ship is certified to carry.

1.7   By following the procedures as set out in this Manual, it will be ensured that the ship complies with all relevant requirements of Annex II to MARPOL 73/78.

SECTION 2 – Description of the ship's equipment and arrangements

2.1   This section contains all particulars of the ship's equipment and arrangements necessary to enable the crew to follow the operational procedures set out in sections 3 and 4.

2.2   *General arrangement of ship and description of cargo tanks*

> This section should contain a brief description of the cargo area of the ship with the main features of the cargo tanks and their positions.
>
> Line or schematic drawings showing the general arrangement of the ship and indicating the position and numbering of the cargo tanks and heating arrangements should be included. Identification of the cargo tanks certified fit to carry noxious liquid substances should be made in conjunction with table 1 of this Manual.

2.3   *Description of cargo pumping and piping arrangements and stripping system*

> This section should contain a description of the cargo pumping and piping arrangements and of the stripping system. Line or schematic drawings should be provided showing the following and be supported by textual explanation where necessary:
>
> – cargo piping arrangements with diameters;
>
> – cargo pumping arrangements with pump capacities;
>
> – piping arrangements of stripping system with diameters;
>
> – pumping arrangements of stripping system with pump capacities;
>
> – location of suction points of cargo lines and stripping lines inside every cargo tank;
>
> – if a suction well is fitted, the location and cubic capacity thereof;
>
> – line draining and stripping or blowing arrangements; and
>
> – quantity and pressure of nitrogen or air required for line blowing if applicable.

Annex II

2.4   *Description of ballast tanks and ballast pumping and piping arrangements*

This section should contain a description of the ballast tanks and ballast pumping and piping arrangements.

Line or schematic drawings and tables should be provided showing the following:

- a general arrangement showing the segregated ballast tanks and cargo tanks to be used as ballast tanks together with their capacities (cubic metres);
- ballast piping arrangement;
- pumping capacity for those cargo tanks which may also be used as ballast tanks; and
- any interconnection between the ballast piping arrangements and the underwater outlet system.

2.5   *Description of dedicated slop tanks with associated pumping and piping arrangements*

This section should contain a description of the dedicated slop tanks with the associated pumping and piping arrangements. Line or schematic drawings should be provided showing the following:

- which dedicated slop tanks are provided together with the capacities of such tanks;
- pumping and piping arrangements of dedicated slop tanks with piping diameters and their connection with the underwater discharge outlet.

2.6   *Description of underwater discharge outlet for effluents containing noxious liquid substances*

This section should contain information on position and maximum flow capacity of the underwater discharge outlet (or outlets) and the connections to this outlet from the cargo tanks and slop tanks. Line or schematic drawings should be provided showing the following:

- location and number of underwater discharge outlets;
- connections to underwater discharge outlet;
- location of all seawater intakes in relation to underwater discharge outlets.

2.7   *Description of flow rate indicating and recording devices*

This section, which applies only to ships operating under regulation 5A(2)(b), should contain a description of the means of measuring the flow rate, and if required also the means of recording the flow rate and time, and the methods of operation.

A line or schematic drawing showing the position and connections of these devices should be provided.

2.8　*Description of cargo tank ventilation system*

This section should contain a description of the cargo tank ventilation system.

Line or schematic drawings and tables should be provided showing the following and supported by textual explanation if necessary:

－ the noxious liquid substances the ship is certified fit to carry having a vapour pressure over 5×10^3 Pa at 20°C suitable for cleaning by ventilation to be listed in table 1;

－ ventilation piping and fans;

－ position of the ventilation openings;

－ the minimum flow rate of the ventilation system to adequately ventilate the bottom and all parts of the cargo tank;

－ the location of structures inside the tank affecting ventilation;

－ the method of ventilating the cargo pipeline system, pumps, filters, etc; and

－ means for ensuring that the tank is dry.

2.9　*Description of tank washing arrangements and washwater heating system*

This section should contain a description of the cargo tank washing arrangements, washwater heating system and all necessary tank washing equipment.

Line or schematic drawings and tables or charts showing the following:

－ arrangements of piping dedicated for tank washing with pipeline diameters;

－ type of tank washing machines with capacities and pressure rating;

－ maximum number of tank washing machines which can operate simultaneously;

－ position of deck openings for cargo tank washing;

－ the number of washing machines and their location required for ensuring complete coverage of the cargo tank walls;

－ maximum capacity of washwater which can be heated to 60°C by the installed heating equipment; and

－ maximum number of tank washing machines which can be operated simultaneously at 60°C.

SECTION 3 – Cargo unloading procedures and tank stripping

3.1　This section contains operational procedures in respect of cargo unloading and tank stripping which must be followed in order to ensure compliance with the requirements of Annex II.

3.2 *Cargo unloading*

This section should contain procedures to be followed including the pump and cargo unloading and suction line to be used for each tank. Alternative methods may be given.

The method of operation of the pump or pumps and the sequence of operation of all valves should be given.

The basic requirement is to unload the cargo to the maximum practicable extent.

3.3 *Cargo tank stripping*

This section should contain procedures to be followed during the stripping of each cargo tank.

The procedures should include the following:

– operation of stripping system;

– list and trim requirements;

– line draining and stripping or blowing arrangements if applicable.

3.4 *Cargo temperature*

This section should contain information on the heating requirements of cargoes which have been identified as being required to be at a certain minimum temperature during unloading.

Information should be given on control of the heating system and the method of temperature measurement.

3.5 *Procedures to be followed when a cargo tank cannot be unloaded in accordance with the required procedures*

This section should contain information on the procedures to be followed in the event that the requirements contained in sections 3.3 and/or 3.4 cannot be met due to circumstances such as the following:

– failure of cargo tank stripping system; and

– failure of cargo tank heating system.

3.6 *Cargo Record Book*

The Cargo Record Book should be completed in the appropriate places on completion of cargo unloading.

SECTION 4 – Procedures relating to the cleaning of cargo tanks, the discharge of residues, ballasting and deballasting

4.1   This section contains operational procedures in respect of tank cleaning, ballast and slops handling which must be followed in order to ensure compliance with the requirements of Annex II.

4.2   The following paragraphs outline the sequence of actions to be taken and contain the information essential to ensure that noxious liquid substances are discharged without posing a threat of harm to the marine environment.

4.3   Establish if the last cargo in the tank is included in the ship's approved list of noxious liquid substances (see table 1). If not included, no special tank cleaning, residue discharge, ballasting and deballasting procedures apply under the provisions of Annex II.

4.4   If the last cargo in the tank is included in the above-mentioned list, the information necessary to establish the procedures for discharging the residue of that cargo, cleaning, ballasting and deballasting the tank, should take into account the following:

4.4.1 *Category of substance*

Obtain the category of the substance from table 1.

4.4.2 *Stripping efficiency of tank pumping system*

> The contents of this section will depend on the design of the ship and whether it is a new ship or existing ship. (See flow diagrams pumping/ stripping requirements.)

4.4.3 *Vessel within or outside special area*

> This section should contain instructions on whether the tank washings can be discharged into the sea within a special area (as defined in section 1.3) or outside a special area. The different requirements should be made clear and will depend on the design and trade of the ship.

4.4.4 *Solidifying or high-viscosity substance*

The properties of the substance should be obtained from the shipping document.

4.4.5 *Miscibility in water*

This property of the substance should be obtained from table 1.

> *Note:* This section should be completed only for existing ships and only for Category B substances.

Annex II

4.4.6 *Compatibility with slops containing other substances*

This section should contain instructions on the permissible and non-permissible mixing of cargo slops. Reference should be made to compatibility guides.

4.4.7 *Discharge to reception facility*

This section should identify those substances the residues of which are required to be prewashed and discharged to a reception facility.

4.4.8 *Discharging into the sea*

This section should contain information on the factors to be considered in order to identify whether the residue/water mixtures are permitted to be discharged into the sea.

4.4.9 *Use of cleaning agents or additives*

This section should contain information on the use and disposal of cleaning agents (e.g. solvents used for tank cleaning) and additives to tank washing water (e.g. detergents).

4.4.10 *Use of ventilation procedures for tank cleaning*

This section should make reference to table 1 to ascertain the suitability of the use of ventilation procedures.

4.5   Having assessed the above information, the correct operational procedures to be followed should be identified using the instructions and flow diagrams in this section. Appropriate entries should be made in the Cargo Record Book indicating the procedure adopted.

This section should contain procedures, which will depend on the age of the ship and pumping efficiency, based on the Standards. Examples of flow diagrams referred to in this section are given at addendum A and incorporate comprehensive requirements applicable to both new and existing ships. The Manual for a particular ship should only contain those requirements specifically applicable to that ship. The Manual should contain the following information and procedures:

Table 1: List of noxious liquid substances allowed to be carried.

Table 2: Cargo tank information.

Addendum A: Flow diagrams.

Addendum B: Prewash procedures.

Addendum C: Ventilation procedures.

Addendum D: Determination of permitted residue discharge rates for Category B substances as required.

Outlines of the above tables and addenda follow.

Table 1 – List of noxious liquid substances allowed to be carried

Substance	Category	Tanks (tank groups)* fit for carriage	Melting point °C	Viscosity at 20°C mPa·s			Suitable for ventilation Yes/No	Miscible in water Yes/No
				<25	25-60	≥60		

Note: Information need only be inserted in the fourth and fifth columns, relating to melting point and viscosity, for those substances which have a melting point greater than 0°C or a viscosity greater than 25 mPa·s at 20°C. When more than one commercial grade is shipped and the viscosities or the melting points of those commercial grades differ, enter and note that other commercial grades may have lower viscosities or melting points or give the values for each commercial grade which will be shipped.

* Tank numbers (tank groups) should be identical to those in the ship's Certificate of Fitness.

Table 2 – Cargo tank information

Tank no.	Capacity (m³)	Stripping quantity in litres	Total residue* (m³)	Approved stripping level under reg. 5A

*For ships referred to in regulation 5A(2)(b) and 5A(4)(b) only.

Annex II

ADDENDUM A

Flow diagrams – Cleaning of cargo tanks and disposal of tank washings/ballast containing residues of Category A, B, C and D substances

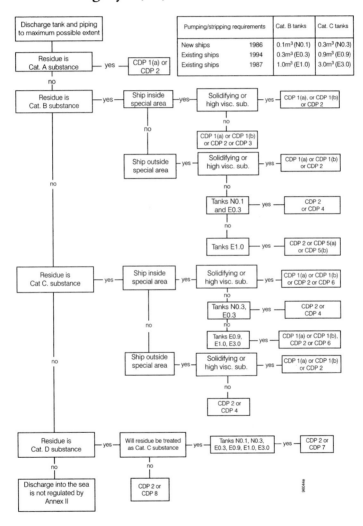

Note: This is a flow diagram giving comprehensive requirements applicable to new and existing ships. The flow diagram for a specific ship should only include parts applicable to that ship.

Cleaning and disposal procedures (CDP)	Sequence of procedures									
	1(a)	1(b)	2	3	4	5(a)	5(b)	6	7	8
Apply prewash in accordance with appendix B and discharge residue to reception facility	x	x								
Apply prewash in accordance with appendix B and transfer residues to slop tank for discharge to sea in accordance with chapter 10, section 10.5 or 10.6						x	x			
Apply subsequent wash of minimum one cycle		x					x			
Apply ventilation procedures in accordance with appendix C			x							
Residue may be retained on board and discharged outside special area				x						
Residues of substances with viscosities <60 mPa·s at the unloading temperature may be retained on board and discharged outside special area. Alternatively, tanks may be prewashed and slops discharged ashore								x		
Dilute residue in cargo tanks with water to obtain residue concentration in mixture of 10% or less										x
Ballast tank or, wash tank to commercial requirements	x			x	x	x		x	x	
Conditions for discharge of ballast/residue/water mixtures other than prewash: >12 miles from land	x	x		x	x	x	x	x	x	x
>7 knots ship's speed	x	x		x	x	x	x	x	x	x
>25 metres water depth	x	x		x	x	x	x	x		
Using underwater discharge	x	x		x	x	x	x	x	x	
Ballast added to tank		x					x			
Condition for discharge of ballasts: >12 miles from land		x					x			
>25 metres water depth		x					x			
Alternatively, residue/water mixtures may be discharged ashore (N.B. optional not MARPOL requirement)	x	x		x	x	x	x	x	x	x
Any water subsequently introduced into the tank may be discharged into the sea without restrictions	x	x	x	x	x	x	x	x	x	x

Note: Start at the top of the column under the CDP number specified and complete each procedure in sequence where marked x.

Annex II

Disposal of prewash or tank washings containing Category A, B, C or D substances from dedicated slop tanks or cargo tanks containing tank washings or slops

Annex II

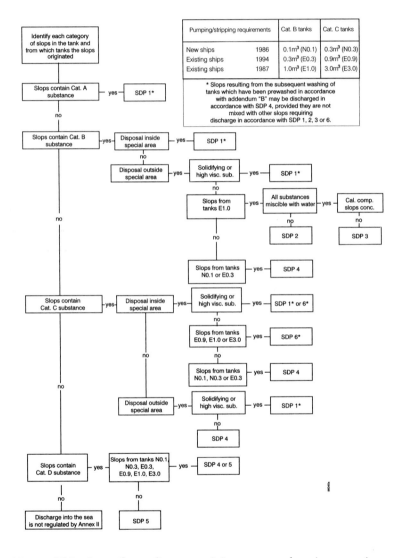

Pumping/stripping requirements		Cat. B tanks	Cat. C tanks
New ships	1986	0.1m³ (N0.1)	0.3m³ (N0.3)
Existing ships	1994	0.3m³ (E0.3)	0.9m³ (E0.9)
Existing ships	1987	1.0m³ (E1.0)	3.0m³ (E3.0)

* Slops resulting from the subsequent washing of tanks which have been prewashed in accordance with addendum "B" may be discharged in accordance with SDP 4, provided they are not mixed with other slops requiring discharge in accordance with SDP 1, 2, 3 or 6.

Note: This is a flow diagram giving comprehensive requirements applicable to new and existing ships. The flow diagram for a specific ship should only include parts applicable to that ship.

358

Slops disposal procedures (SDP)	Sequence of procedures					
	1	2	3	4	5	6
Slops must be discharged ashore	x					
Establish discharge rate of miscible residue/water mixture in accordance with addendum D		x				
Divide obtained discharge rate of pure product by composite slops concentration			x			
The figure obtained shows the rate at which discharge is permitted		x	x			
Residues of substances with viscosities < 60 mPa·s at the unloading temperature may be retained on board and discharged outside special area. Alternatively, tanks may be prewashed and slops discharged ashore						x
Dilute slops with water to obtain a solution of 10% or less no restrictions on discharge rate					x	
Discharge rate is maximum permitted by underwater discharge outlet				x		x
Additional discharge conditions						
– ship's speed at least 7 knots		x	x	x	x	x
– outside 12 miles from nearest land		x	x	x	x	x
– depth of water at least 25 m		x	x	x		x
– using underwater discharge		x	x	x		x

Note: Start at the top of the column under the SDP number specified and complete each procedure in sequence where marked x.

Annex II

359

ADDENDUM B – Prewash procedures

This addendum to the Manual should contain prewash procedures based on appendix B of the Standards. These procedures should contain specific requirements for the use of the tank washing arrangements and equipment provided on the particular ship and include the following:

– washing machine positions to be used;

– slops pumping out procedure;

– requirements for hot washing;

– number of cycles of washing machine (or time);

– minimum operating pressures.

ADDENDUM C – Ventilation procedures

This addendum to the Manual should contain ventilation procedures based on appendix C of the Standards. The procedures should contain specific requirements for the use of the cargo tank ventilation system, or equipment, fitted on the particular ship and should include the following:

– ventilation positions to be used;

– minimum flow or speed of fans;

– procedures for ventilating cargo pipeline, pumps, filters, etc.;

– procedures for ensuring that tanks are dry on completion.

ADDENDUM D – Determination of permitted residue discharge rates for Category B substances

This addendum to the Manual, which is required only by ships operating under regulation 5A(2)(b), should contain a method for the ship's crew to determine the permitted discharge rates for Category B substances. The method should be based on sections 10.5 and 10.6 of the Standards.

Annex III of MARPOL 73/78 (including amendments)

Regulations for the Prevention of Pollution by Harmful Substances Carried by Sea in Packaged Form

Annex III of MARPOL 73/78
(including amendments)

Regulations for the Prevention of Pollution by Harmful Substances Carried by Sea in Packaged Form

Regulation 1
Application

(1) Unless expressly provided otherwise, the regulations of this Annex apply to all ships carrying harmful substances in packaged form.

    (1.1) For the purpose of this Annex, "harmful substances" are those substances which are identified as marine pollutants in the International Maritime Dangerous Goods Code (IMDG Code).[*]

    (1.2) Guidelines for the identification of harmful substances in packaged form are given in the appendix to this Annex.

    (1.3) For the purposes of this Annex, "packaged form" is defined as the forms of containment specified for harmful substances in the IMDG Code.

(2) The carriage of harmful substances is prohibited, except in accordance with the provisions of this Annex.

(3) To supplement the provisions of this Annex, the Government of each Party to the Convention shall issue, or cause to be issued, detailed requirements on packing, marking, labelling, documentation, stowage, quantity limitations and exceptions for preventing or minimizing pollution of the marine environment by harmful substances.[*]

---

[*] Refer to the IMDG Code adopted by the Organization by resolution A.716(17), as it has been or may be amended by the Maritime Safety Committee; see IMO sales publications IMO-200E and IMO-210E.

(4) For the purposes of this Annex, empty packagings which have been used previously for the carriage of harmful substances shall themselves be treated as harmful substances unless adequate precautions have been taken to ensure that they contain no residue that is harmful to the marine environment.

(5) The requirements of this Annex do not apply to ship's stores and equipment.

Regulation 2
Packing

Packages shall be adequate to minimize the hazard to the marine environment, having regard to their specific contents.

Regulation 3
Marking and labelling

(1) Packages containing a harmful substance shall be durably marked with the correct technical name (trade names alone shall not be used) and, further, shall be durably marked or labelled to indicate that the substance is a marine pollutant. Such identification shall be supplemented where possible by any other means, for example, by use of the relevant United Nations number.

(2) The method of marking the correct technical name and of affixing labels on packages containing a harmful substance shall be such that this information will still be identifiable on packages surviving at least three months' immersion in the sea. In considering suitable marking and labelling, account shall be taken of the durability of the materials used and of the surface of the package.

(3) Packages containing small quantities of harmful substances may be exempted from the marking requirements.[*]

Regulation 4[†]
Documentation

(1) In all documents relating to the carriage of harmful substances by sea where such substances are named, the correct technical name of each such substance shall be used (trade names alone shall not be used) and

---

[*] Refer to the specific exemptions provided for in the IMDG Code; see IMO sales publications IMO-200E and IMO-210E.

[†] Reference to "documents" in this regulation does not preclude the use of electronic data processing (EDP) and electronic data interchange (EDI) transmission techniques as an aid to paper documentation.

the substance further identified by the addition of the words "MARINE POLLUTANT".

(2) The shipping documents supplied by the shipper shall include, or be accompanied by, a signed certificate or declaration that the shipment offered for carriage is properly packaged and marked, labelled or placarded as appropriate and in proper condition for carriage to minimize the hazard to the marine environment.

(3) Each ship carrying harmful substances shall have a special list or manifest setting forth the harmful substances on board and the location thereof. A detailed stowage plan which sets out the location of the harmful substances on board may be used in place of such special list or manifest. Copies of such documents shall also be retained on shore by the owner of the ship or his representative until the harmful substances are unloaded. A copy of one of these documents shall be made available before departure to the person or organization designated by the port State authority.

SEE INTERPRETATION 1.0

(4) When the ship carries a special list or manifest or a detailed stowage plan, required for the carriage of dangerous goods by the International Convention for the Safety of Life at Sea, 1974, as amended, the documents required by this regulation may be combined with those for dangerous goods. Where documents are combined, a clear distinction shall be made between dangerous goods and harmful substances covered by this Annex.

Annex III

Regulation 5
Stowage

Harmful substances shall be properly stowed and secured so as to minimize the hazards to the marine environment without impairing the safety of the ship and persons on board.

Regulation 6
Quantity limitations

Certain harmful substances may, for sound scientific and technical reasons, need to be prohibited for carriage or be limited as to the quantity which may be carried aboard any one ship. In limiting the quantity, due consideration shall be given to size, construction and equipment of the ship, as well as the packaging and the inherent nature of the substances.

Regulation 7
Exceptions

(1) Jettisoning of harmful substances carried in packaged form shall be prohibited, except where necessary for the purpose of securing the safety of the ship or saving life at sea.

(2) Subject to the provisions of the present Convention, appropriate measures based on the physical, chemical and biological properties of harmful substances shall be taken to regulate the washing of leakages overboard, provided that compliance with such measures would not impair the safety of the ship and persons on board.

Regulation 8
*Port State control on operational requirements**

(1) A ship when in a port of another Party is subject to inspection by officers duly authorized by such Party concerning operational requirements under this Annex, where there are clear grounds for believing that the master or crew are not familiar with essential shipboard procedures relating to the prevention of pollution by harmful substances.

(2) In the circumstances given in paragraph (1) of this regulation, the Party shall take such steps as will ensure that the ship shall not sail until the situation has been brought to order in accordance with the requirements of this Annex.

(3) Procedures relating to the port State control prescribed in article 5 of the present Convention shall apply to this regulation.

(4) Nothing in this regulation shall be construed to limit the rights and obligations of a Party carrying out control over operational requirements specifically provided for in the present Convention.

---

* Refer to the Procedures for port State control adopted by the Organization by resolution A.787(19) and amended by A.882(21); see IMO sales publication IMO-650E.

Appendix to Annex III
Guidelines for the identification of harmful substances in packaged form

For the purposes of this Annex, substances identified by any one of the following criteria are harmful substances:

- bioaccumulated to a significant extent and known to produce a hazard to aquatic life or to human health (Hazard Rating "+" in column A*); or

- bioaccumulated with attendant risk to aquatic organisms or to human health with a short retention of the order of one week or less (Hazard Rating "Z" in column A*); or

- highly toxic to aquatic life, defined by a $LC_{50}/96$ hour[†] less than 1 ppm (Hazard Rating "4" in column B*).

Unified Interpretation of Annex III

Reg. 4(3)   1.0   At any stopover, where any loading or unloading operations, even partial, are carried out, a revision of the documents listing the harmful substances taken on board, indicating their location on board or showing a detailed stowage plan, shall be made available before departure to the person or organization designated by the port State authority.

Annex III

Unified Interpretation

Annex IV of MARPOL 73/78

Regulations for the Prevention of Pollution by Sewage from Ships

Annex IV of MARPOL 73/78*

Regulations for the Prevention of Pollution by Sewage from Ships

Regulation 1
Definitions

For the purposes of the present Annex:

(1) *New ship* means a ship:

    (a) for which the building contract is placed, or in the absence of a building contract, the keel of which is laid, or which is at a similar stage of construction, on or after the date of entry into force of this Annex; or

    (b) the delivery of which is three years or more after the date of entry into force of this Annex.

(2) *Existing ship* means a ship which is not a new ship.

(3) *Sewage* means:

    (a) drainage and other wastes from any form of toilets, urinals, and WC scuppers;

    (b) drainage from medical premises (dispensary, sick bay, etc.) via wash basins, wash tubs and scuppers located in such premises;

    (c) drainage from spaces containing living animals; or

    (d) other waste waters when mixed with the drainages defined above.

(4) *Holding tank* means a tank used for the collection and storage of sewage.

(5) *Nearest land.* The term "from the nearest land" means from the baseline from which the territorial sea of the territory in question is established in accordance with international law except that, for the purposes of the present Convention, "from the nearest land" off the

---

* On the publication date of the 2002 Consolidated Edition, Annex IV had not met the conditions of entry into force. This Annex has been revised by the Marine Environment Protection Committee of the Organization. The revised text and resolution MEPC.88(44) are included in the Additional Information at the end of this publication.

north-eastern coast of Australia shall mean from a line drawn from a point on the coast of Australia in

> latitude 11°00′ S, longitude 142°08′ E
> to a point in latitude 10°35′ S, longitude 141°55′ E
> thence to a point latitude 10°00′ S, longitude 142°00′ E
> thence to a point latitude 9°10′ S, longitude 143°52′ E
> thence to a point latitude 9°00′ S, longitude 144°30′ E
> thence to a point latitude 13°00′ S, longitude 144°00′ E
> thence to a point latitude 15°00′ S, longitude 146°00′ E
> thence to a point latitude 18°00′ S, longitude 147°00′ E
> thence to a point latitude 21°00′ S, longitude 153°00′ E
> thence to a point on the coast of Australia in
> latitude 24°42′ S, longitude 153°15′ E.

Regulation 2
Application

The provisions of this Annex shall apply to:

(a)   (i)   new ships of 200 tons gross tonnage and above;

      (ii)   new ships of less than 200 tons gross tonnage which are certified to carry more than 10 persons;

      (iii)   new ships which do not have a measured gross tonnage and are certified to carry more than 10 persons; and

(b)   (i)   existing ships of 200 tons gross tonnage and above, 10 years after the date of entry into force of this Annex;

      (ii)   existing ships of less than 200 tons gross tonnage which are certified to carry more than 10 persons, 10 years after the date of entry into force of this Annex; and

      (iii)   existing ships which do not have a measured gross tonnage and are certified to carry more than 10 persons, 10 years after the date of entry into force of this Annex.

Regulation 3
Surveys

(1)   Every ship which is required to comply with the provisions of this Annex and which is engaged in voyages to ports or offshore terminals under the jurisdiction of other Parties to the Convention shall be subject to the surveys specified below:

  (a)   An initial survey before the ship is put in service or before the certificate required under regulation 4 of this Annex is issued for the first time, which shall include a survey of the ship which shall be such as to ensure:

Annex IV

   (i)   when the ship is fitted with a sewage treatment plant the plant shall meet operational requirements based on standards and the test methods developed by the Organization;*

   (ii)   when the ship is fitted with a system to comminute and disinfect the sewage, such a system shall be of a type approved by the Administration;

   (iii) when the ship is equipped with a holding tank the capacity of such tank shall be to the satisfaction of the Administration for the retention of all sewage having regard to the operation of the ship, the number of persons on board and other relevant factors. The holding tank shall have a means to indicate visually the amount of its contents; and

   (iv) that the ship is equipped with a pipeline leading to the exterior convenient for the discharge of sewage to a reception facility and that such a pipeline is fitted with a standard shore connection in compliance with regulation 11 of this Annex.

This survey shall be such as to ensure that the equipment, fittings, arrangements and material fully comply with the applicable requirements of this Annex.

(b)   Periodical surveys at intervals specified by the Administration but not exceeding five years which shall be such as to ensure that the equipment, fittings, arrangements and material fully comply with the applicable requirements of this Annex. However, where the duration of the International Sewage Pollution Prevention Certificate (1973) is extended as specified in regulation 7(2) or (4) of this Annex, the interval of the periodical survey may be extended correspondingly.

(2)   The Administration shall establish appropriate measures for ships which are not subject to the provisions of paragraph (1) of this regulation in order to ensure that the provisions of this Annex are complied with.

(3)   Surveys of the ship as regards enforcement of the provisions of this Annex shall be carried out by officers of the Administration. The Administration may, however, entrust the surveys either to surveyors nominated for the purpose or to organizations recognized by it. In every case the Administration concerned fully guarantees the completeness and efficiency of the surveys.

Annex IV

---

* Refer to the Recommendation on international effluent standards and guidelines for performance tests for sewage treatment plants adopted by the Marine Environment Protection Committee of the Organization by resolution MEPC.2(VI); see IMO sales publication IMO-592E.

(4) After any survey of the ship under this regulation has been completed, no significant change shall be made in the equipment, fittings, arrangements, or material covered by the survey without the approval of the Administration, except the direct replacement of such equipment or fittings.

Regulation 4
Issue of Certificate

(1) An International Sewage Pollution Prevention Certificate (1973) shall be issued, after survey in accordance with the provisions of regulation 3 of this Annex, to any ship which is engaged in voyages to ports or offshore terminals under the jurisdiction of other Parties to the Convention.

(2) Such Certificate shall be issued either by the Administration or by any persons or organization duly authorized by it. In every case the Administration assumes full responsibility for the Certificate.

Regulation 5
Issue of a Certificate by another Government

(1) The Government of a Party to the Convention may, at the request of the Administration, cause a ship to be surveyed and, if satisfied that the provisions of this Annex are complied with, shall issue or authorize the issue of an International Sewage Pollution Prevention Certificate (1973) to the ship in accordance with this Annex.

(2) A copy of the Certificate and a copy of the survey report shall be transmitted as early as possible to the Administration requesting the survey.

(3) A Certificate so issued shall contain a statement to the effect that it has been issued at the request of the Administration and it shall have the same force and receive the same recognition as the Certificate issued under regulation 4 of this Annex.

(4) No International Sewage Pollution Prevention Certificate (1973) shall be issued to a ship which is entitled to fly the flag of a State which is not a Party.

Regulation 6
Form of Certificate

The International Sewage Pollution Prevention Certificate (1973) shall be drawn up in an official language of the issuing country in the form corresponding to the model given in the appendix to this Annex. If the language used is neither English nor French, the text shall include a translation into one of these languages.

Regulation 7
Duration of Certificate

(1)   The International Sewage Pollution Prevention Certificate (1973) shall be issued for a period specified by the Administration, which shall not exceed five years from the date of issue, except as provided in paragraphs (2), (3) and (4) of this regulation.

(2)   If a ship at the time when the Certificate expires is not in a port or offshore terminal under the jurisdiction of the Party to the Convention whose flag the ship is entitled to fly, the Certificate may be extended by the Administration, but such extension shall be granted only for the purpose of allowing the ship to complete its voyage to the State whose flag the ship is entitled to fly or in which it is to be surveyed and then only in cases where it appears proper and reasonable to do so.

(3)   No Certificate shall be thus extended for a period longer than five months and a ship to which such extension is granted shall not on its arrival in the State whose flag it is entitled to fly or the port in which it is to be surveyed, be entitled by virtue of such extension to leave that port or State without having obtained a new certificate.

(4)   A Certificate which has not been extended under the provisions of paragraph (2) of this regulation may be extended by the Administration for a period of grace of up to one month from the date of expiry stated on it.

(5)   A Certificate shall cease to be valid if significant alterations have taken place in the equipment, fittings, arrangement or material required without the approval of the Administration, except the direct replacement of such equipment or fittings.

(6)   A Certificate issued to a ship shall cease to be valid upon transfer of such a ship to the flag of another State, except as provided in paragraph (7) of this regulation.

(7)   Upon transfer of a ship to the flag of another Party, the Certificate shall remain in force for a period not exceeding five months provided that it would not have expired before the end of that period, or until the Administration issues a replacement Certificate, whichever is earlier. As soon as possible after the transfer has taken place the Government of the Party whose flag the ship was formerly entitled to fly shall transmit to the Administration a copy of the Certificate carried by the ship before the transfer and, if available, a copy of the relevant survey report.

Annex IV

Regulation 8
Discharge of sewage

(1) Subject to the provisions of regulation 9 of this Annex, the discharge of sewage into the sea is prohibited, except when:

(a) the ship is discharging comminuted and disinfected sewage using a system approved by the Administration in accordance with regulation 3(1)(a) at a distance of more than 4 nautical miles from the nearest land, or sewage which is not comminuted or disinfected at a distance of more than 12 nautical miles from the nearest land, provided that in any case, the sewage that has been stored in holding tanks shall not be discharged instantaneously but at a moderate rate when the ship is *en route* and proceeding at not less than 4 knots; the rate of discharge shall be approved by the Administration based upon standards developed by the Organization; or

(b) the ship has in operation an approved sewage treatment plant which has been certified by the Administration to meet the operational requirements referred to in regulation 3(1)(a)(i) of this Annex, and

(i) the test results of the plant are laid down in the ship's International Sewage Pollution Prevention Certificate (1973);

(ii) additionally, the effluent shall not produce visible floating solids in, nor cause discoloration of, the surrounding water; or

(c) the ship is situated in the waters under the jurisdiction of a State and is discharging sewage in accordance with such less stringent requirements as may be imposed by such State.

(2) When the sewage is mixed with wastes or waste water having different discharge requirements, the more stringent requirements shall apply.

Regulation 9
Exceptions

Regulation 8 of this Annex shall not apply to:

(a) the discharge of sewage from a ship necessary for the purpose of securing the safety of a ship and those on board or saving life at sea; or

(b) the discharge of sewage resulting from damage to a ship or its equipment if all reasonable precautions have been taken before and after the occurrence of the damage, for the purpose of preventing or minimizing the discharge.

Regulation 10
Reception facilities

(1)   The Government of each Party to the Convention undertakes to ensure the provision of facilities at ports and terminals for the reception of sewage, without causing undue delay to ships, adequate to meet the needs of the ships using them.

(2)   The Government of each Party shall notify the Organization for transmission to the Contracting Governments concerned of all cases where the facilities provided under this regulation are alleged to be inadequate.

Regulation 11
Standard discharge connections

To enable pipes of reception facilities to be connected with the ship's discharge pipeline, both lines shall be fitted with a standard discharge connection in accordance with the following table:

Standard dimensions of flanges for discharge connections

Description	*Dimension*
Outside diameter	210 mm
Inner diameter	According to pipe outside diameter
Bolt circle diameter	170 mm
Slots in flange	4 holes 18 mm in diameter equidistantly placed on a bolt circle of the above diameter, slotted to the flange periphery. The slot width to be 18 mm
Flange thickness	16 mm
Bolts and nuts: quantity and diameter	4, each of 16 mm in diameter and of suitable length
The flange is designed to accept pipes up to a maximum internal diameter of 100 mm and shall be of steel or other equivalent material having a flat face. This flange, together with a suitable gasket, shall be suitable for a service pressure of 6 kg/cm^2	

For ships having a moulded depth of 5 m and less, the inner diameter of the discharge connection may be 38 mm.

Annex IV

Appendix to Annex IV
Form of Sewage Certificate

INTERNATIONAL SEWAGE POLLUTION PREVENTION CERTIFICATE (1973)

Issued under the Provisions of the International Convention for the Prevention of Pollution from Ships, 1973, under the authority of the Government of

. . . .

(full designation of the country)

by. . . . .

(full designation of the competent person or organization authorized under the provisions of the International Convention for the Prevention of Pollution from Ships, 1973)

Name of ship	Distinctive number or letters	Port of registry	Gross tonnage	Number of persons which the ship is certified to carry

New/existing ship*

    Date of building contract . . . .

    Date on which keel was laid or ship
    was at a similar stage of construction . . . .

    Date of delivery . . . .

Annex IV

---

* Delete as appropriate.

THIS IS TO CERTIFY:

(1)   The ship is equipped with a sewage treatment plant/comminuter/ holding tank* and a discharge pipeline in compliance with regulation 3(1)(a)(i) to (iv) of Annex IV of the Convention as follows:

    *(a)   Description of the sewage treatment plant:
          Type of sewage treatment plant . . . .
          Name of manufacturer . . . .
          The sewage treatment plant is certified by the Administration to meet the following effluent standards[†] . . . .

    *(b)   Description of comminuter:
          Type of comminuter . . . .
          Name of manufacturer . . . .
          Standard of sewage after disinfection . . . .

    *(c)   Description of holding tank equipment:
          Total capacity of the holding tank . . . . m^3
          Location . . . .

    (d)   A pipeline for the discharge of sewage to a reception facility, fitted with a standard shore connection.

(2)   The ship has been surveyed in accordance with regulation 3 of Annex IV of the International Convention for the Prevention of Pollution from Ships, 1973, concerning the prevention of pollution by sewage and the survey showed that the equipment of the ship and the condition thereof are in all respects satisfactory and the ship complies with the applicable requirements of Annex IV of the Convention.

This certificate is valid until . . . .

Issued at. . . . .
                    (place of issue of certificate)

. . . .
   (Date of issue)   *(Signature of official issuing the certificate)*

        (Seal or stamp of the issuing authority, as appropriate)

Under the provisions of regulation 7(2) and (4) of Annex IV of the Convention the validity of this certificate is extended until

. . . .

                 Signed . . . .
                   (Signature of duly authorized official)
                 Place . . . .
                 Date . . . .

        (Seal or stamp of the authority, as appropriate)

---

* Delete as appropriate.

[†] Parameters should be incorporated.

Annex V of MARPOL 73/78 (including amendments)

Regulations for the Prevention of Pollution by Garbage from Ships

Annex V of MARPOL 73/78
(including amendments)

Regulations for the Prevention
of Pollution by Garbage from Ships

Regulation 1
Definitions

For the purposes of this Annex:

(1) *Garbage* means all kinds of victual, domestic and operational waste excluding fresh fish and parts thereof, generated during the normal operation of the ship and liable to be disposed of continuously or periodically except those substances which are defined or listed in other Annexes to the present Convention.

(2) *Nearest land.* The term "from the nearest land" means from the baseline from which the territorial sea of the territory in question is established in accordance with international law, except that, for the purposes of the present Convention, "from the nearest land" off the north-eastern coast of Australia shall mean from a line drawn from a point on the coast of Australia in

> latitude 11°00′ S, longitude 142°08′ E
> to a point in latitude 10°35′ S, longitude 141°55′ E,
> thence to a point latitude 10°00′ S, longitude 142°00′ E,
> thence to a point latitude 09°10′ S, longitude 143°52′ E,
> thence to a point latitude 09°00′ S, longitude 144°30′ E,
> thence to a point latitude 10°41′ S, longitude 145°00′ E,
> thence to a point latitude 13°00′ S, longitude 145°00′ E,
> thence to a point latitude 15°00′ S, longitude 146°00′ E,
> thence to a point latitude 17°30′ S, longitude 147°00′ E,
> thence to a point latitude 21°00′ S, longitude 152°55′ E,
> thence to a point latitude 24°30′ S, longitude 154°00′ E,
> thence to a point on the coast of Australia in
> latitude 24°42′ S, longitude 153°15′ E.

(3) *Special area* means a sea area where for recognized technical reasons in relation to its oceanographical and ecological condition and to the particular character of its traffic the adoption of special mandatory methods for the prevention of sea pollution by garbage is required. Special areas shall include those listed in regulation 5 of this Annex.

Regulation 2
Application

Unless expressly provided otherwise, the provisions of this Annex shall apply to all ships.

Regulation 3
Disposal of garbage outside special areas

(1)   Subject to the provisions of regulations 4, 5 and 6 of this Annex:

    (a)   the disposal into the sea of all plastics, including but not limited to synthetic ropes, synthetic fishing nets, plastic garbage bags and incinerator ashes from plastic products which may contain toxic or heavy metal residues, is prohibited;

    (b)   the disposal into the sea of the following garbage shall be made as far as practicable from the nearest land but in any case is prohibited if the distance from the nearest land is less than:

        (i)   25 nautical miles for dunnage, lining and packing materials which will float;

        (ii)   12 nautical miles for food wastes and all other garbage including paper products, rags, glass, metal, bottles, crockery and similar refuse;

    (c)   disposal into the sea of garbage specified in subparagraph(b)(ii) of this regulation may be permitted when it has passed through a comminuter or grinder and made as far as practicable from the nearest land but in any case is prohibited if the distance from the nearest land is less than 3 nautical miles. Such comminuted or ground garbage shall be capable of passing through a screen with openings no greater than 25 mm.

(2)   When the garbage is mixed with other discharges having different disposal or discharge requirements the more stringent requirements shall apply.

Regulation 4
Special requirements for disposal of garbage

(1)   Subject to the provisions of paragraph (2) of this regulation, the disposal of any materials regulated by this Annex is prohibited from fixed or floating platforms engaged in the exploration, exploitation and associated offshore processing of sea-bed mineral resources, and from all other ships when alongside or within 500 m of such platforms.

(2)   The disposal into the sea of food wastes may be permitted when they have been passed through a comminuter or grinder from such fixed or floating platforms located more than 12 nautical miles from land and all other ships when alongside or within 500 m of such platforms. Such comminuted or ground food wastes shall be capable of passing through a screen with openings no greater than 25 mm.

Regulation 5
Disposal of garbage within special areas

(1)   For the purposes of this Annex the special areas are the Mediterranean Sea area, the Baltic Sea area, the Black Sea area, the Red Sea area, the "Gulfs area", the North Sea area, the Antarctic area and the Wider Caribbean Region, including the Gulf of Mexico and the Caribbean Sea, which are defined as follows:

(a)   The *Mediterranean Sea area* means the Mediterranean Sea proper including the gulfs and seas therein with the boundary between the Mediterranean and the Black Sea constituted by the 41° N parallel and bounded to the west by the Straits of Gibraltar at the meridian 5°36′ W.

(b)   The *Baltic Sea area* means the Baltic Sea proper with the Gulf of Bothnia and the Gulf of Finland and the entrance to the Baltic Sea bounded by the parallel of the Skaw in the Skagerrak at 57°44.8′ N.

(c)   The *Black Sea area* means the Black Sea proper with the boundary between the Mediterranean and the Black Sea constituted by the parallel 41° N.

(d)   The *Red Sea area* means the Red Sea proper including the Gulfs of Suez and Aqaba bounded at the south by the rhumb line between Ras si Ane (12°28.5′ N, 43°19.6′ E) and Husn Murad (12°40.4′ N, 43°30.2′ E).

(e)   The *Gulfs area* means the sea area located north-west of the rhumb line between Ras al Hadd (22°30′ N, 59°48′ E) and Ras al Fasteh (25°04′ N, 61°25′ E).

(f)   The *North Sea area* means the North Sea proper including seas therein with the boundary between:

(i)   the North Sea southwards of latitude 62° N and eastwards of longitude 4° W;

(ii)   the Skagerrak, the southern limit of which is determined east of the Skaw by latitude 57°44.8′ N; and

(iii)   the English Channel and its approaches eastwards of longitude 5° W and northwards of latitude 48°30′ N.

Annex V

387

(g)   The *Antarctic area* means the sea area south of latitude 60° S.

(h)   The *Wider Caribbean Region,* as defined in article 2, paragraph 1 of the Convention for the Protection and Development of the Marine Environment of the Wider Caribbean Region (Cartagena de Indias, 1983), means the Gulf of Mexico and Caribbean Sea proper including the bays and seas therein and that portion of the Atlantic Ocean within the boundary constituted by the 30° N parallel from Florida eastward to 77°30′ W meridian, thence a rhumb line to the intersection of 20° N parallel and 59° W meridian, thence a rhumb line to the intersection of 7°20′ N parallel and 50° W meridian, thence a rhumb line drawn south-westerly to the eastern boundary of French Guiana.

(2)   Subject to the provisions of regulation 6 of this Annex:

    (a)   disposal into the sea of the following is prohibited:

        (i)   all plastics, including but not limited to synthetic ropes, synthetic fishing nets, plastic garbage bags and incinerator ashes from plastic products which may contain toxic or heavy metal residues; and

        (ii)   all other garbage, including paper products, rags, glass, metal, bottles, crockery, dunnage, lining and packing materials;

    (b)   except as provided in subparagraph (c) of this paragraph, disposal into the sea of food wastes shall be made as far as practicable from land, but in any case not less than 12 nautical miles from the nearest land;

    (c)   disposal into the Wider Caribbean Region of food wastes which have been passed through a comminuter or grinder shall be made as far as practicable from land, but in any case not less than 3 nautical miles from the nearest land. Such comminuted or ground food wastes shall be capable of passing through a screen with openings no greater than 25 mm.

(3)   When the garbage is mixed with other discharges having different disposal or discharge requirements the more stringent requirements shall apply.

(4)   Reception facilities within special areas:

    (a)   The Government of each Party to the Convention, the coastline of which borders a special area, undertakes to ensure that as soon as possible in all ports within a special area adequate reception facilities are provided in accordance with regulation 7 of this Annex, taking into account the special needs of ships operating in these areas.

Annex V

(b) The Government of each Party concerned shall notify the Organization of the measures taken pursuant to subparagraph (a) of this regulation. Upon receipt of sufficient notifications the Organization shall establish a date from which the requirements of this regulation in respect of the area in question shall take effect. The Organization shall notify all Parties of the date so established no less than twelve months in advance of that date.

(c) After the date so established, ships calling also at ports in these special areas where such facilities are not yet available, shall fully comply with the requirements of this regulation.

(5) Notwithstanding paragraph 4 of this regulation, the following rules apply to the Antarctic area:

(a) The Government of each Party to the Convention at whose ports ships depart *en route* to or arrive from the Antarctic area undertakes to ensure that as soon as practicable adequate facilities are provided for the reception of all garbage from all ships, without causing undue delay, and according to the needs of the ships using them.

(b) The Government of each Party to the Convention shall ensure that all ships entitled to fly its flag, before entering the Antarctic area, have sufficient capacity on board for the retention of all garbage while operating in the area and have concluded arrangements to discharge such garbage at a reception facility after leaving the area.

Regulation 6
Exceptions

Regulations 3, 4 and 5 of this Annex shall not apply to:

(a) the disposal of garbage from a ship necessary for the purpose of securing the safety of a ship and those on board or saving life at sea; or

(b) the escape of garbage resulting from damage to a ship or its equipment provided all reasonable precautions have been taken before and after the occurrence of the damage, for the purpose of preventing or minimizing the escape; or

(c) the accidental loss of synthetic fishing nets, provided that all reasonable precautions have been taken to prevent such loss.

Annex V

389

Regulation 7
Reception facilities

(1)　The Government of each Party to the Convention undertakes to ensure the provision of facilities at ports and terminals for the reception of garbage, without causing undue delay to ships, and according to the needs of the ships using them.

(2)　The Government of each Party shall notify the Organization for transmission to the Parties concerned of all cases where the facilities provided under this regulation are alleged to be inadequate.

Regulation 8
*Port State control on operational requirements**

(1)　A ship when in a port of another Party is subject to inspection by officers duly authorized by such Party concerning operational requirements under this Annex, where there are clear grounds for believing that the master or crew are not familiar with essential shipboard procedures relating to the prevention of pollution by garbage.

(2)　In the circumstances given in paragraph (1) of this regulation, the Party shall take such steps as will ensure that the ship shall not sail until the situation has been brought to order in accordance with the requirements of this Annex.

(3)　Procedures relating to the port State control prescribed in article 5 of the present Convention shall apply to this regulation.

(4)　Nothing in this regulation shall be construed to limit the rights and obligations of a Party carrying out control over operational requirements specifically provided for in the present Convention.

Regulation 9
Placards, garbage management plans
and garbage record-keeping

(1)　(a)　Every ship of 12 m or more in length overall shall display placards which notify the crew and passengers of the disposal requirements of regulations 3 and 5 of this Annex, as applicable.

　　(b)　The placards shall be written in the working language of the ship's personnel and, for ships engaged in voyages to ports or

---

* Refer to the Procedures for port State control adopted by the Organization by resolution A.787(19) and amended by A.882(21); see IMO sales publication IMO-650E.

Annex V

offshore terminals under the jurisdiction of other Parties to the Convention, shall also be in English, French or Spanish.

(2) Every ship of 400 tons gross tonnage and above, and every ship which is certified to carry 15 persons or more, shall carry a garbage management plan which the crew shall follow. This plan shall provide written procedures for collecting, storing, processing and disposing of garbage, including the use of the equipment on board. It shall also designate the person in charge of carrying out the plan. Such a plan shall be in accordance with the guidelines developed by the Organization* and written in the working language of the crew.

(3) Every ship of 400 tons gross tonnage and above and every ship which is certified to carry 15 persons or more engaged in voyages to ports or offshore terminals under the jurisdiction of other Parties to the Convention and every fixed and floating platform engaged in exploration and exploitation of the sea-bed shall be provided with a Garbage Record Book. The Garbage Record Book, whether as a part of the ship's official log-book or otherwise, shall be in the form specified in the appendix to this Annex;

(a) each discharge operation, or completed incineration, shall be recorded in the Garbage Record Book and signed for on the date of the incineration or discharge by the officer in charge. Each completed page of the Garbage Record Book shall be signed by the master of the ship. The entries in the Garbage Record Book shall be at least in English, French or Spanish. Where the entries are also made in an official language of the State whose flag the ship is entitled to fly, these entries shall prevail in case of a dispute or discrepancy;

(b) the entry for each incineration or discharge shall include date and time, position of the ship, description of the garbage and the estimated amount incinerated or discharged;

(c) the Garbage Record Book shall be kept on board the ship and in such a place as to be available for inspection in a reasonable time. This document shall be preserved for a period of two years after the last entry is made on the record;

(d) in the event of discharge, escape or accidental loss referred to in regulation 6 of this Annex an entry shall be made in the Garbage Record Book of the circumstances of, and the reasons for, the loss.

---

* Refer to the Guidelines for the development of garbage management plans adopted by the Marine Environment Protection Committee of the Organization by resolution MEPC.71(38); see MEPC/Circ.317 and IMO sales publication IMO-656E.

Annex V

(4) The Administration may waive the requirements for Garbage Record Books for:

    (a) any ship engaged on voyages of 1 hour or less in duration which is certified to carry 15 persons or more; or

    (b) fixed or floating platforms while engaged in exploration and exploitation of the sea-bed.

(5) The competent authority of the Government of a Party to the Convention may inspect the Garbage Record Book on board any ship to which this regulation applies while the ship is in its ports or offshore terminals and may make a copy of any entry in that book, and may require the master of the ship to certify that the copy is a true copy of such an entry. Any copy so made, which has been certified by the master of the ship as a true copy of an entry in the ship's Garbage Record Book, shall be admissible in any judicial proceedings as evidence of the facts stated in the entry. The inspection of a Garbage Record Book and the taking of a certified copy by the competent authority under this paragraph shall be performed as expeditiously as possible without causing the ship to be unduly delayed.

(6) In the case of ships built before 1 July 1997, this regulation shall apply as from 1 July 1998.

Appendix to Annex V
Form of Garbage Record Book

GARBAGE RECORD BOOK

Name of ship: _____

Distinctive number or letters: _____

IMO No.: _____

Period:_____ From: _____ To: _____

1   Introduction
In accordance with regulation 9 of Annex V of the International Convention for the Prevention of Pollution from Ships, 1973, as modified by the Protocol of 1978 (MARPOL 73/78), a record is to be kept of each discharge operation or completed incineration. This includes discharges at sea, to reception facilities, or to other ships.

2   Garbage and garbage management
Garbage includes all kinds of food, domestic and operational waste excluding fresh fish and parts thereof, generated during the normal operation of the vessel and liable to be disposed of continuously or periodically except those substances which are defined or listed in other annexes to MARPOL 73/78 (such as oil, sewage or noxious liquid substances).

The Guidelines for the Implementation of Annex V of MARPOL 73/78[*] should also be referred to for relevant information.

3   Description of the garbage
The garbage is to be grouped into categories for the purposes of this record book as follows:

1   Plastics

2   Floating dunnage, lining, or packing material

3   Ground-down paper products, rags, glass, metal, bottles, crockery, etc.

4   Paper products, rags, glass, metal, bottles, crockery, etc.

---

[*] Refer to the Guidelines for the Implementation of Annex V of MARPOL 73/78; see IMO sales publication IMO-656E.

5   Food waste

6   Incinerator ash.

4   Entries in the Garbage Record Book

4.1   Entries in the Garbage Record Book shall be made on each of the following occasions:

(a)   When garbage is discharged into the sea:
   (i)   Date and time of discharge
   (ii) Position of the ship (latitude and longitude)
   (iii) Category of garbage discharged
   (iv) Estimated amount discharged for each category in cubic metres
   (v)   Signature of the officer in charge of the operation.

(b)   When garbage is discharged to reception facilities ashore or to other ships:
   (i)   Date and time of discharge
   (ii) Port or facility, or name of ship
   (iii) Category of garbage discharged
   (iv) Estimated amount discharged for each category in cubic metres
   (v)   Signature of officer in charge of the operation.

(c)   When garbage is incinerated:
   (i)   Date and time of start and stop of incineration
   (ii) Position of the ship (latitude and longitude)
   (iii) Estimated amount incinerated in cubic metres
   (iv) Signature of the officer in charge of the operation.

(d)   Accidental or other exceptional discharges of garbage
   (i)   Time of occurrence
   (ii) Port or position of the ship at time of occurrence
   (iii) Estimated amount and category of garbage
   (iv) Circumstances of disposal, escape or loss, the reason therefor and general remarks.

4.2   Receipts

The master should obtain from the operator of port reception facilities, or from the master of the ship receiving the garbage, a receipt or certificate specifying the estimated amount of garbage transferred. The receipts or certificates must be kept on board the ship with the Garbage Record Book for two years.

Annex V

4.3 Amount of garbage

The amount of garbage on board should be estimated in cubic metres, if possible separately according to category. The Garbage Record Book contains many references to estimated amount of garbage. It is recognized that the accuracy of estimating amounts of garbage is left to interpretation. Volume estimates will differ before and after processing. Some processing procedures may not allow for a usable estimate of volume, e.g. the continuous processing of food waste. Such factors should be taken into consideration when making and interpreting entries made in a record.

Annex V

RECORD OF GARBAGE DISCHARGES

Ship's name: _____ Distinctive No., or letters: _____ IMO No.: _____

Garbage categories:

1: Plastic.
2: Floating dunnage, lining, or packing materials.
3: Ground paper products, rags, glass, metal, bottles, crockery, etc.
4: Paper products, rags, glass, metal, bottles, crockery, etc.
5: Food waste.
6: Incinerator ash except from plastic products which may contain toxic or heavy metal residues.

NOTE: THE DISCHARGE OF ANY GARBAGE OTHER THAN FOOD WASTE IS PROHIBITED IN SPECIAL AREAS. ONLY GARBAGE DISCHARGED INTO THE SEA MUST BE CATEGORIZED. GARBAGE OTHER THAN CATEGORY 1 DISCHARGED TO RECEPTION FACILITIES NEED ONLY BE LISTED AS A TOTAL ESTIMATED AMOUNT.

Date/time	Position of the ship	Estimated amount discharged into sea (m³)						Estimated amount discharged to reception facilities or to other ship (m³)		Estimated amount incinerated (m³)	Certification/ Signature
		Cat. 2	Cat. 3	Cat. 4	Cat. 5	Cat.6		Cat. 1	Other		

Master's signature: _____ Date: _____

Annex VI of MARPOL 73/78

Regulations for the Prevention of Air Pollution from Ships

Annex VI of MARPOL 73/78

Regulations for the Prevention of Air Pollution from Ships

Chapter I – General

Regulation 1
Application

The provisions of this Annex shall apply to all ships, except where expressly provided otherwise in regulations 3, 5, 6, 13, 15, 18 and 19 of this Annex.

Regulation 2
Definitions

For the purpose of this Annex:

(1) *A similar stage of construction* means the stage at which:

    (a) construction identifiable with a specific ship begins; and

    (b) assembly of that ship has commenced comprising at least 50 tonnes or one per cent of the estimated mass of all structural material, whichever is less.

(2) *Continuous feeding* is defined as the process whereby waste is fed into a combustion chamber without human assistance while the incinerator is in nomal operating conditions with the combustion chamber operative temperature between 850°C and 1200°C.

(3) *Emission* means any release of substances subject to control by this Annex from ships into the atmosphere or sea.

(4) *New installations*, in relation to regulation 12 of this Annex, means the installation of systems, equipment, including new portable fire-extinguishing units, insulation, or other material on a ship after the date on which this Annex enters into force, but excludes repair or recharge of previously installed systems, equipment, insulation, or other material, or recharge of portable fire-extinguishing units.

(5) *NO$_x$ Technical Code* means the Technical Code on Control of Emission of Nitrogen Oxides from Marine Diesel Engines adopted by Conference resolution 2, as may be amended by the Organization,

provided that such amendments are adopted and brought into force in accordance with the provisions of article 16 of the present Convention concerning amendment procedures applicable to an appendix to an Annex.

(6) *Ozone-depleting substances* means controlled substances defined in paragraph 4 of article 1 of the Montreal Protocol on Substances that Deplete the Ozone Layer, 1987, listed in Annexes A, B, C or E to the said Protocol in force at the time of application or interpretation of this Annex.

Ozone-depleting substances that may be found on board ship include, but are not limited to:

Halon 1211	Bromochlorodifluoromethane
Halon 1301	Bromotrifluoromethane
Halon 2402	1,2-Dibromo-1,1,2,2-tetrafluoroethane (also known as Halon 114B2)
CFC-11	Trichlorofluoromethane
CFC-12	Dichlorodifluoromethane
CFC-113	1,1,2-Trichloro-1,2,2-trifluoroethane
CFC-114	1,2-Dichloro-1,1,2,2-tetrafluoroethane
CFC-115	Chloropentafluoroethane

(7) *Sludge oil* means sludge from the fuel or lubricating oil separators, waste lubricating oil from main or auxiliary machinery, or waste oil from bilge water separators, oil filtering equipment or drip trays.

(8) *Shipboard incineration* means the incineration of wastes or other matter on board a ship, if such wastes or other matter were generated during the normal operation of that ship.

(9) *Shipboard incinerator* means a shipboard facility designed for the primary purpose of incineration.

(10) *Ships constructed* means ships the keels of which are laid or which are at a similar stage of construction.

(11) SO_x *emission control area* means an area where the adoption of special mandatory measures for SO_x emissions from ships is required to prevent, reduce and control air pollution from SO_x and its attendant adverse impacts on land and sea areas. SO_x emission control areas shall include those listed in regulation 14 of this Annex.

(12) *Tanker* means an oil tanker as defined in regulation 1(4) of Annex I or a chemical tanker as defined in regulation 1(1) of Annex II of the present Convention.

(13) *The Protocol of 1997* means the Protocol of 1997 to amend the International Convention for the Prevention of Pollution from Ships, 1973, as amended by the Protocol of 1978 relating thereto.

Regulation 3
General exceptions

Regulations of this Annex shall not apply to:

    (a)   any emission necessary for the purpose of securing the safety of a ship or saving life at sea; or

    (b)   any emission resulting from damage to a ship or its equipment:

        (i)   provided that all reasonable precautions have been taken after the occurrence of the damage or discovery of the emission for the purpose of preventing or minimizing the emission; and

        (ii)   except if the owner or the master acted either with intent to cause damage, or recklessly and with knowledge that damage would probably result.

Regulation 4
Equivalents

(1)   The Administration may allow any fitting, material, appliance or apparatus to be fitted in a ship as an alternative to that required by this Annex if such fitting, material, appliance or apparatus is at least as effective as that required by this Annex.

(2)   The Administration which allows a fitting, material, appliance or apparatus as an alternative to that required by this Annex shall communicate to the Organization for circulation to the Parties to the present Convention particulars thereof, for their information and appropriate action, if any.

Chapter II – Survey, certification and means of control

Regulation 5

Surveys and inspections

(1)   Every ship of 400 gross tonnage or above and every fixed and floating drilling rig and other platforms shall be subject to the surveys specified below:

(a)   an initial survey before the ship is put into service or before the certificate required under regulation 6 of this Annex is issued for the first time. This survey shall be such as to ensure that the equipment, systems, fittings, arrangements and material fully comply with the applicable requirements of this Annex;

(b)   periodical surveys at intervals specified by the Administration, but not exceeding five years, which shall be such as to ensure that the equipment, systems, fittings, arrangements and material fully comply with the requirements of this Annex; and

(c)   a minimum of one intermediate survey during the period of validity of the certificate which shall be such as to ensure that the equipment and arrangements fully comply with the require-ments of this Annex and are in good working order. In cases where only one such intermediate survey is carried out in a single certificate validity period, and where the period of the certificate exceeds $2\frac{1}{2}$ years, it shall be held within six months before or after the halfway date of the certificate's period of validity. Such intermediate surveys shall be endorsed on the certificate issued under regulation 6 of this Annex.

(2)   In the case of ships of less than 400 gross tonnage, the Administration may establish appropriate measures in order to ensure that the applicable provisions of this Annex are complied with.

(3)   Surveys of ships as regards the enforcement of the provisions of this Annex shall be carried out by officers of the Administration. The Administration may, however, entrust the surveys either to surveyors nominated for the purpose or to organizations recognized by it. Such organizations shall comply with the guidelines adopted by the Organization.* In every case the Administration concerned shall fully guarantee the completeness and efficiency of the survey.

(4)   The survey of engines and equipment for compliance with regulation 13 of this Annex shall be conducted in accordance with the NO_x Technical Code.

---

* Refer to the Guidelines for the authorization of organizations acting on behalf of the Administration, adopted by the Organization by resolution A.739(18), and the Specifications on the survey and certification functions of recognized organizations acting on behalf of the Administration, adopted by the Organization by resolution A.789(19).

(5)   The Administration shall institute arrangements for unscheduled inspections to be carried out during the period of validity of the certificate. Such inspections shall ensure that the equipment remains in all respects satisfactory for the service for which the equipment is intended. These inspections may be carried out by their own inspection service, nominated surveyors, recognized organizations, or by other Parties upon request of the Administration. Where the Administration, under the provisions of paragraph (1) of this regulation, establishes mandatory annual surveys, the above unscheduled inspections shall not be obligatory.

(6)   When a nominated surveyor or recognized organization determines that the condition of the equipment does not correspond substantially with the particulars of the certificate, they shall ensure that corrective action is taken and shall in due course notify the Administration. If such corrective action is not taken, the certificate should be withdrawn by the Administration. If the ship is in a port of another Party, the appropriate authorities of the port State shall also be notified immediately. When an officer of the Administration, a nominated surveyor or recognized organization has notified the appropriate authorities of the port State, the Government of the port State concerned shall give such officer, surveyor or organization any necessary assistance to carry out their obligations under this regulation.

(7)   The equipment shall be maintained to conform with the provisions of this Annex and no changes shall be made in the equipment, systems, fittings, arrangements, or material covered by the survey, without the express approval of the Administration. The direct replacement of such equipment and fittings with equipment and fittings that conform with the provisions of this Annex is permitted.

(8)   Whenever an accident occurs to a ship or a defect is discovered, which substantially affects the efficiency or completeness of its equipment covered by this Annex, the master or owner of the ship shall report at the earliest opportunity to the Administration, a nominated surveyor, or recognized organization responsible for issuing the relevant certificate.

Regulation 6
Issue of International Air Pollution Prevention Certificate

(1)   An International Air Pollution Prevention Certificate shall be issued, after survey in accordance with the provisions of regulation 5 of this Annex, to:

(a)   any ship of 400 gross tonnage or above engaged in voyages to ports or offshore terminals under the jurisdiction of other Parties; and

(b) platforms and drilling rigs engaged in voyages to waters under the sovereignty or jurisdiction of other Parties to the Protocol of 1997.

(2) Ships constructed before the date of entry into force of the Protocol of 1997 shall be issued with an International Air Pollution Prevention Certificate in accordance with paragraph (1) of this regulation no later than the first scheduled drydocking after entry into force of the Protocol of 1997, but in no case later than three years after entry into force of the Protocol of 1997.

(3) Such Certificate shall be issued either by the Administration or by any person or organization duly authorized by it. In every case the Administration assumes full responsibility for the Certificate.

Regulation 7
Issue of a Certificate by another Government

(1) The Government of a Party to the Protocol of 1997 may, at the request of the Administration, cause a ship to be surveyed and, if satisfied that the provisions of this Annex are complied with, issue or authorize the issuance of an International Air Pollution Prevention Certificate to the ship in accordance with this Annex.

(2) A copy of the Certificate and a copy of the survey report shall be transmitted as soon as possible to the requesting Administration.

(3) A Certificate so issued shall contain a statement to the effect that it has been issued at the request of the Administration and it shall have the same force and receive the same recognition as a Certificate issued under regulation 6 of this Annex.

(4) No International Air Pollution Prevention Certificate shall be issued to a ship which is entitled to fly the flag of a State which is not a Party to the Protocol of 1997.

Regulation 8
Form of Certificate

The International Air Pollution Prevention Certificate shall be drawn up in an official language of the issuing country in the form corresponding to the model given in appendix I to this Annex. If the language used is not English, French, or Spanish, the text shall include a translation into one of these languages.

Regulation 9
Duration and validity of Certificate

(1) An International Air Pollution Prevention Certificate shall be issued for a period specified by the Administration, which shall not exceed five years from the date of issue.

(2) No extension of the five-year period of validity of the International Air Pollution Prevention Certificate shall be permitted, except in accordance with paragraph (3).

(3) If the ship, at the time when the International Air Pollution Prevention Certificate expires, is not in a port of the State whose flag it is entitled to fly or in which it is to be surveyed, the Administration may extend the Certificate for a period of no more than five months. Such extension shall be granted only for the purpose of allowing the ship to complete its voyage to the State whose flag it is entitled to fly or in which it is to be surveyed, and then only in cases where it appears proper and reasonable to do so. After arrival in the State whose flag it is entitled to fly or in which it is to be surveyed, the ship shall not be entitled by virtue of such extension to leave the port or State without having obtained a new International Air Pollution Prevention Certificate.

(4) An International Air Pollution Prevention Certificate shall cease to be valid in any of the following circumstances:

   (a) if the inspections and surveys are not carried out within the periods specified under regulation 5 of this Annex;

   (b) if significant alterations have taken place to the equipment, systems, fittings, arrangements or material to which this Annex applies without the express approval of the Administration, except the direct replacement of such equipment or fittings with equipment or fittings that conform with the requirements of this Annex. For the purpose of regulation 13, significant alteration shall include any change or adjustment to the system, fittings, or arrangement of a diesel engine which results in the nitrogen oxide limits applied to that engine no longer being complied with; or

   (c) upon transfer of the ship to the flag of another State. A new Certificate shall be issued only when the Government issuing the new Certificate is fully satisfied that the ship is in full compliance with the requirements of regulation 5 of this Annex. In the case of a transfer between Parties, if requested within three months after the transfer has taken place, the Government of the Party whose flag the ship was formerly entitled to fly shall, as soon as possible, transmit to the Administration of the other

Party a copy of the International Air Pollution Prevention Certificate carried by the ship before the transfer and, if available, copies of the relevant survey reports.

Regulation 10
Port State control on operational requirements

(1) A ship, when in a port or an offshore terminal under the jurisdiction of another Party to the Protocol of 1997, is subject to inspection by officers duly authorized by such Party concerning operational requirements under this Annex, where there are clear grounds for believing that the master or crew are not familiar with essential shipboard procedures relating to the prevention of air pollution from ships.

(2) In the circumstances given in paragraph (1) of this regulation, the Party shall take such steps as will ensure that the ship shall not sail until the situation has been brought to order in accordance with the requirements of this Annex.

(3) Procedures relating to the port State control prescribed in article 5 of the present Convention shall apply to this regulation.

(4) Nothing in this regulation shall be construed to limit the rights and obligations of a Party carrying out control over operational requirements specifically provided for in the present Convention.

Regulation 11
Detection of violations and enforcement

(1) Parties to this Annex shall co-operate in the detection of violations and the enforcement of the provisions of this Annex, using all appropriate and practicable measures of detection and environmental monitoring, adequate procedures for reporting and accumulation of evidence.

(2) A ship to which the present Annex applies may, in any port or offshore terminal of a Party, be subject to inspection by officers appointed or authorized by that Party for the purpose of verifying whether the ship has emitted any of the substances covered by this Annex in violation of the provision of this Annex. If an inspection indicates a violation of this Annex, a report shall be forwarded to the Administration for any appropriate action.

(3) Any Party shall furnish to the Administration evidence, if any, that the ship has emitted any of the substances covered by this Annex in violation of the provisions of this Annex. If it is practicable to do so,

the competent authority of the former Party shall notify the master of the ship of the alleged violation.

(4) Upon receiving such evidence, the Administration so informed shall investigate the matter, and may request the other Party to furnish further or better evidence of the alleged contravention. If the Administration is satisfied that sufficient evidence is available to enable proceedings to be brought in respect of the alleged violation, it shall cause such proceedings to be taken in accordance with its law as soon as possible. The Administration shall promptly inform the Party which has reported the alleged violation, as well as the Organization, of the action taken.

(5) A Party may also inspect a ship to which this Annex applies when it enters the ports or offshore terminals under its jurisdiction, if a request for an investigation is received from any Party together with sufficient evidence that the ship has emitted any of the substances covered by the Annex in any place in violation of this Annex. The report of such investigation shall be sent to the Party requesting it and to the Administration so that the appropriate action may be taken under the present Convention.

(6) The international law concerning the prevention, reduction, and control of pollution of the marine environment from ships, including that law relating to enforcement and safeguards, in force at the time of application or interpretation of this Annex, applies, *mutatis mutandis*, to the rules and standards set forth in this Annex.

Chapter III – Requirements for control of emissions from ships

Regulation 12
Ozone-depleting substances

(1)   Subject to the provisions of regulation 3, any deliberate emissions of ozone-depleting substances shall be prohibited. Deliberate emissions include emissions occurring in the course of maintaining, servicing, repairing or disposing of systems or equipment, except that deliberate emissions do not include minimal releases associated with the recapture or recycling of an ozone-depleting substance. Emissions arising from leaks of an ozone-depleting substance, whether or not the leaks are deliberate, may be regulated by Parties to the Protocol of 1997.

(2)   New installations which contain ozone-depleting substances shall be prohibited on all ships, except that new installations containing hydrochlorofluorocarbons (HCFCs) are permitted until 1 January 2020.

(3)   The substances referred to in this regulation, and equipment containing such substances, shall be delivered to appropriate reception facilities when removed from ships.

Regulation 13
Nitrogen oxides (NO$_x$)

(1)   (a)   This regulation shall apply to:

      (i)   each diesel engine with a power output of more than 130 kW which is installed on a ship constructed on or after 1 January 2000; and

      (ii)   each diesel engine with a power output of more than 130 kW which undergoes a major conversion on or after 1 January 2000.

   (b)   This regulation does not apply to:

      (i)   emergency diesel engines, engines installed in lifeboats and any device or equipment intended to be used solely in case of emergency; and

      (ii)   engines installed on ships solely engaged in voyages within waters subject to the sovereignty or jurisdiction of the State the flag of which the ship is entitled to fly, provided that such engines are subject to an alternative NO$_x$ control measure established by the Administration.

(c) Notwithstanding the provisions of sub-paragraph (a) of this paragraph, the Administration may allow exclusion from the application of this regulation to any diesel engine which is installed on a ship constructed, or on a ship which undergoes a major conversion, before the date of entry into force of the present Protocol, provided that the ship is solely engaged in voyages to ports or offshore terminals within the State the flag of which the ship is entitled to fly.

(2) (a) For the purpose of this regulation, *major conversion* means a modification of an engine where:

(i) the engine is replaced by a new engine built on or after 1 January 2000, or

(ii) any substantial modification, as defined in the NO_x Technical Code, is made to the engine, or

(iii) the maximum continuous rating of the engine is increased by more than 10%.

(b) The NO_x emission resulting from modifications referred to in the sub-paragraph (a) of this paragraph shall be documented in accordance with the NO_x Technical Code for approval by the Administration.

(3) (a) Subject to the provision of regulation 3 of this Annex, the operation of each diesel engine to which this regulation applies is prohibited, except when the emission of nitrogen oxides (calculated as the total weighted emission of NO_2) from the engine is within the following limits:

(i) 17.0 g/kW h when *n* is less than 130 rpm

(ii) $45.0 \times n^{-0.2}$ g/kW h when *n* is 130 or more but less than 2000 rpm

(iii) 9.8 g/kW h when *n* is 2000 rpm or more

where *n* = rated engine speed (crankshaft revolutions per minute).

When using fuel composed of blends from hydrocarbons derived from petroleum refining, test procedure and measurement methods shall be in accordance with the NO_x Technical Code, taking into consideration the test cycles and weighting factors outlined in appendix II to this Annex.

(b) Notwithstanding the provisions of sub-paragraph (a) of this paragraph, the operation of a diesel engine is permitted when:

(i) an exhaust gas cleaning system, approved by the Administration in accordance with the NO_x Technical Code, is applied to the engine to reduce onboard NO_x emissions at least to the limits specified in sub-paragraph (a), or

(ii)   any other equivalent method, approved by the Adminis-tration taking into account relevant guidelines to be developed by the Organization, is applied to reduce onboard NO_x emissions at least to the limit specified in sub-paragraph (a) of this paragraph.

Regulation 14
Sulphur oxides (SO_x)

General requirements

(1)   The sulphur content of any fuel oil used on board ships shall not exceed 4.5% m/m.

(2)   The world-wide average sulphur content of residual fuel oil supplied for use on board ships shall be monitored taking into account guidelines to be developed by the Organization.[*]

Requirements within SO_x emission control areas

(3)   For the purpose of this regulation, SO_x emission control areas shall include:

(a)   the Baltic Sea area as defined in regulation 10(1)(b) of Annex I; and

(b)   any other sea area, including port areas, designated by the Organization in accordance with criteria and procedures for designation of SO_x emission control areas with respect to the prevention of air pollution from ships contained in appendix III to this Annex.

(4)   While ships are within SO_x emission control areas, at least one of the following conditions shall be fulfilled:

(a)   the sulphur content of fuel oil used on board ships in a SO_x emission control area does not exceed 1.5% m/m;

(b)   an exhaust gas cleaning system, approved by the Administration taking into account guidelines to be developed by the Organization, is applied to reduce the total emission of sulphur oxides from ships, including both auxiliary and main propulsion engines, to 6.0 g SO_x/kW h or less calculated as the total weight of sulphur dioxide emission. Waste streams from the use of such equipment shall not be discharged into enclosed ports, harbours and estuaries unless it can be thoroughly documented by the ship that such waste streams have no adverse impact on the ecosystems of such enclosed ports, harbours and estuaries, based

---

[*] Refer to resolution MEPC.82(43), Guidelines for monitoring the world-wide average sulphur content of residual fuel oils supplied for use on board ships; see item 9 of the additional information.

upon criteria communicated by the authorities of the port State to the Organization. The Organization shall circulate the criteria to all Parties to the Convention; or

(c) any other technological method that is verifiable and enforceable to limit SO_x emissions to a level equivalent to that described in sub-paragraph (b) is applied. These methods shall be approved by the Administration taking into account guidelines to be developed by the Organization.

(5) The sulphur content of fuel oil referred to in paragraph (1) and paragraph (4)(a) of this regulation shall be documented by the supplier as required by regulation 18 of this Annex.

(6) Those ships using separate fuel oils to comply with paragraph (4)(a) of this regulation shall allow sufficient time for the fuel oil service system to be fully flushed of all fuels exceeding 1.5% m/m sulphur content prior to entry into a SO_x emission control area. The volume of low-sulphur fuel oils (less than or equal to 1.5% sulphur content) in each tank as well as the date, time, and position of the ship when any fuel-changeover operation is completed, shall be recorded in such log-book as prescribed by the Administration.

(7) During the first 12 months immediately following entry into force of the present Protocol, or of an amendment to the present Protocol designating a specific SO_x emission control area under paragraph (3)(b) of this regulation, ships entering a SO_x emission control area referred to in paragraph (3)(a) of this regulation or designated under paragraph (3)(b) of this regulation are exempted from the require-ments in paragraphs (4) and (6) of this regulation and from the requirements of paragraph (5) of this regulation insofar as they relate to paragraph (4)(a) of this regulation.

Regulation 15
Volatile organic compounds

(1) If the emissions of volatile organic compounds (VOCs) from tankers are to be regulated in ports or terminals under the jurisdiction of a Party to the Protocol of 1997, they shall be regulated in accordance with the provisions of this regulation.

(2) A Party to the Protocol of 1997 which designates ports or terminals under its jurisdiction in which VOCs emissions are to be regulated shall submit a notification to the Organization. This notification shall include information on the size of tankers to be controlled, on cargoes requiring vapour emission control systems, and the effective date of such control. The notification shall be submitted at least six months before the effective date.

(3) The Government of each Party to the Protocol of 1997 which designates ports or terminals at which VOCs emissions from tankers

are to be regulated shall ensure that vapour emission control systems, approved by that Government taking into account the safety standards developed by the Organization,* are provided in ports and terminals designated, and are operated safely and in a manner so as to avoid undue delay to the ship.

(4)   The Organization shall circulate a list of the ports and terminals designated by the Parties to the Protocol of 1997 to other Parties to the Protocol of 1997 and Member States of the Organization for their information.

(5)   All tankers which are subject to vapour emission control in accordance with the provisions of paragraph (2) of this regulation shall be provided with a vapour collection system approved by the Administration taking into account the safety standards developed by the Organization,* and shall use such system during the loading of such cargoes. Terminals which have installed vapour emission control systems in accordance with this regulation may accept existing tankers which are not fitted with vapour collection systems for a period of three years after the effective date identified in paragraph (2).

(6)   This regulation shall only apply to gas carriers when the type of loading and containment systems allow safe retention of non-methane VOCs on board, or their safe return ashore.

Regulation 16
Shipboard incineration

(1)   Except as provided in paragraph (5), shipboard incineration shall be allowed only in a shipboard incinerator.

(2)   (a)   Except as provided in sub-paragraph (b) of this paragraph, each incinerator installed on board a ship on or after 1 January 2000 shall meet the requirements contained in appendix IV to this Annex. Each incinerator shall be approved by the Administration taking into account the standard specifications for shipboard incinerators developed by the Organization.†

        (b)   The Administration may allow exclusion from the application of sub-paragraph (a) of this paragraph to any incinerator which is installed on board a ship before the date of entry into force of the Protocol of 1997, provided that the ship is solely engaged in voyages within waters subject to the sovereignty or jurisdiction of the State the flag of which the ship is entitled to fly.

---

* Refer to MSC/Circ.585, Standards for vapour emission control systems.

† Refer to resolution MEPC.76(40), Standard specification for shipboard incinerators, and resolution MEPC.93(45), Amendments to the standard specification for shipboard incinerators.

(3)   Nothing in this regulation affects the prohibition in, or other requirements of, the Convention on the Prevention of Marine Pollution by Dumping of Wastes and Other Matter, 1972, as amended, and the 1996 Protocol thereto.

(4)   Shipboard incineration of the following substances shall be prohibited:

(a)   Annex I, II and III cargo residues of the present Convention and related contaminated packing materials;

(b)   polychlorinated biphenyls (PCBs);

(c)   garbage, as defined in Annex V of the present Convention, containing more than traces of heavy metals; and

(d)   refined petroleum products containing halogen compounds.

(5)   Shipboard incineration of sewage sludge and sludge oil generated during the normal operation of a ship may also take place in the main or auxiliary power plant or boilers, but in those cases, shall not take place inside ports, harbours and estuaries.

(6)   Shipboard incineration of polyvinyl chlorides (PVCs) shall be prohibited, except in shipboard incinerators for which IMO Type Approval Certificates have been issued.

(7)   All ships with incinerators subject to this regulation shall possess a manufacturer's operating manual which shall specify how to operate the incinerator within the limits described in paragraph (2) of appendix IV to this Annex.

(8)   Personnel responsible for operation of any incinerator shall be trained and capable of implementing the guidance provided in the manufacturer's operating manual.

(9)   Monitoring of combustion flue gas outlet temperature shall be required at all times and waste shall not be fed into a continuous-feed shipboard incinerator when the temperature is below the minimum allowed temperature of 850°C. For batch-loaded shipboard incinerators, the unit shall be designed so that the temperature in the combustion chamber shall reach 600°C within five minutes after start-up.

(10)   Nothing in this regulation precludes the development, installation and operation of alternative design shipboard thermal waste treatment devices that meet or exceed the requirements of this regulation.

Regulation 17
Reception facilities

(1)   The Government of each Party to the Protocol of 1997 undertakes to ensure the provision of facilities adequate to meet the:

(a) needs of ships using its repair ports for the reception of ozone-depleting substances and equipment containing such substances when removed from ships;

(b) needs of ships using its ports, terminals or repair ports for the reception of exhaust gas cleaning residues from an approved exhaust gas cleaning system when discharge into the marine environment of these residues is not permitted under regulation 14 of this Annex;

without causing undue delay to ships, and

(c) needs in ship breaking facilities for the reception of ozone-depleting substances and equipment containing such substances when removed from ships.

(2) Each Party to the Protocol of 1997 shall notify the Organization for transmission to the Members of the Organization of all cases where the facilities provided under this regulation are unavailable or alleged to be inadequate.

Regulation 18
Fuel oil quality

(1) Fuel oil for combustion purposes delivered to and used on board ships to which this Annex applies shall meet the following requirements:

(a) except as provided in sub-paragraph (b):

(i) the fuel oil shall be blends of hydrocarbons derived from petroleum refining. This shall not preclude the incorporation of small amounts of additives intended to improve some aspects of performance;

(ii) the fuel oil shall be free from inorganic acid;

(iii) the fuel oil shall not include any added substance or chemical waste which either:

(1) jeopardizes the safety of ships or adversely affects the performance of the machinery, or

(2) is harmful to personnel, or

(3) contributes overall to additional air pollution; and

(b) fuel oil for combustion purposes derived by methods other than petroleum refining shall not:

(i) exeed the sulphur content set forth in regulation 14 of this Annex;

(ii) cause an engine to exceed the NO_x emission limits set forth in regulation 13(3)(a) of this Annex;

(iii) contain inorganic acid; and

          (iv)   (1)   jeopardize the safety of ships or adversely affect the performance of the machinery, or

                  (2)   be harmful to personnel, or

                  (3)   contribute overall to additional air pollution.

(2)   This regulation does not apply to coal in its solid form or nuclear fuels.

(3)   For each ship subject to regulations 5 and 6 of this Annex, details of fuel oil for combustion purposes delivered to and used on board shall be recorded by means of a bunker delivery note which shall contain at least the information specified in appendix V to this Annex.

(4)   The bunker delivery note shall be kept on board the ship in such a place as to be readily available for inspection at all reasonable times. It shall be retained for a period of three years after the fuel oil has been delivered on board.

(5)   (a)   The competent authority* of the Government of a Party to the Protocol of 1997 may inspect the bunker delivery notes on board any ship to which this Annex applies while the ship is in its port or offshore terminal, may make a copy of each delivery note, and may require the master or person in charge of the ship to certify that each copy is a true copy of such bunker delivery note. The competent authority may also verify the contents of each note through consultations with the port where the note was issued.

        (b)   The inspection of the bunker delivery notes and the taking of certified copies by the competent authority under this paragraph shall be performed as expeditiously as possible without causing the ship to be unduly delayed.

(6)   The bunker delivery note shall be accompanied by a representative sample of the fuel oil delivered, taking into account guidelines to be developed by the Organization. The sample is to be sealed and signed by the supplier's representative and the master or officer in charge of the bunker operation on completion of bunkering operations and retained under the ship's control until the fuel oil is substantially consumed, but in any case for a period of not less than 12 months from the time of delivery.

(7)   Parties to the Protocol of 1997 undertake to ensure that appropriate authorities designated by them:

        (a)   maintain a register of local suppliers of fuel oil;

---

* Refer to resolution A.787(19), Procedures for port State control, as amended by A.882(21); see IMO sales publication IMO-650E.

(b)   require local suppliers to provide the bunker delivery note and sample as required by this regulation, certified by the fuel oil supplier that the fuel oil meets the requirements of regulations 14 and 18 of this Annex;

(c)   require local suppliers to retain a copy of the bunker delivery note for at least three years for inspection and verification by the port State as necessary;

(d)   take action as appropriate against fuel oil suppliers that have been found to deliver fuel oil that does not comply with that stated on the bunker delivery note;

(e)   inform the Administration of any ship receiving fuel oil found to be non-compliant with the requirements of regulations 14 or 18 of this Annex; and

(f)   inform the Organization for transmission to Parties to the Protocol of 1997 of all cases where fuel oil suppliers have failed to meet the requirements specified in regulations 14 or 18 of this Annex.

(8)   In connection with port State inspections carried out by Parties to the Protocol of 1997, the Parties further undertake to:

(a)   inform the Party or non-Party under whose jurisdiction a bunker delivery note was issued of cases of delivery of non-compliant fuel oil, giving all relevant information; and

(b)   ensure that remedial action as appropriate is taken to bring non-compliant fuel oil discovered into compliance.

Regulation 19
Requirements for platforms and drilling rigs

(1)   Subject to the provisions of paragraphs (2) and (3) of this regulation, fixed and floating platforms and drilling rigs shall comply with the requirements of this Annex.

(2)   Emissions directly arising from the exploration, exploitation and associated offshore processing of sea-bed mineral resources are, consistent with article 2(3)(b)(ii) of the present Convention, exempt from the provisions of this Annex. Such emissions include the following:

(a)   emissions resulting from the incineration of substances that are solely and directly the result of exploration, exploitation and associated offshore processing of sea-bed mineral resources, including but not limited to the flaring of hydrocarbons and the burning of cuttings, muds, and/or stimulation fluids during well completion and testing operations, and flaring arising from upset conditions;

(b)   the release of gases and volatile compounds entrained in drilling fluids and cuttings;

(c)   emissions associated solely and directly with the treatment, handling, or storage of sea-bed minerals; and

(d)   emissions from diesel engines that are solely dedicated to the exploration, exploitation and associated offshore processing of sea-bed mineral resources.

(3)   The requirements of regulation 18 of this Annex shall not apply to the use of hydrocarbons which are produced and subsequently used on site as fuel, when approved by the Administration.

Appendices to Annex VI

Appendix I
Form of IAPP Certificate
(Regulation 8)

INTERNATIONAL AIR POLLUTION PREVENTION CERTIFICATE

Issued under the provisions of the Protocol of 1997 to amend the International Convention for the Prevention of Pollution from Ships, 1973, as modified of the Protocol of 1978 related thereto (hereinafter referred to as "the Convention") under the authority of the Government of:

....

(full designation of the country)

by ....

(full designation of the competent person or organization
authorized under the provisions of the Convention)

Name of ship	Distinctive number or letters	IMO number	Port of registry	Gross tonnage

Type of ship: ☐ tanker
                  ☐ ships other than a tanker

THIS IS TO CERTIFY:

1.   That the ship has been surveyed in accordance with regulation 5 of Annex VI of the Convention; and

2.   That the survey shows that the equipment, systems, fittings, arrangements and materials fully comply with the applicable requirements of Annex VI of the Convention.

This certificate is valid until .... subject to surveys in accordance with regulation 5 of Annex VI of the Convention.

Issued at ....

(Place of issue of certificate)

....

(Date of issue)   *(Signature of duly authorized official issuing the certificate)*

(Seal or stamp of the authority, as appropriate)

ENDORSEMENT FOR ANNUAL AND INTERMEDIATE SURVEYS

THIS IS TO CERTIFY that at a survey required by regulation 5 of Annex VI of the Convention the ship was found to comply with the relevant provisions of the Convention:

Annual survey:

Signed ....

(Signature of duly authorized official)

Place ....

Date ....

(Seal or stamp of the authority, as appropriate)

Annual*/Intermediate* survey:

Signed ....

(Signature of duly authorized official)

Place ....

Date ....

(Seal or stamp of the authority, as appropriate)

Annual*/Intermediate* survey:

Signed ....

(Signature of duly authorized official)

Place ....

Date ....

(Seal or stamp of the authority, as appropriate)

Annual survey:

Signed ....

(Signature of duly authorized official)

Place ....

Date ....

(Seal or stamp of the authority, as appropriate)

---

* Delete as appropriate.

SUPPLEMENT TO
INTERNATIONAL AIR POLLUTION PREVENTION CERTIFICATE
(IAPP CERTIFICATE)

RECORD OF CONSTRUCTION AND EQUIPMENT

In respect of the provisions of Annex VI of the International Convention for the Prevention of Pollution from Ships, 1973, as modified by the Protocol of 1978 relating thereto (hereinafter referred to as "the Convention").

Notes:

1 This Record shall be permanently attached to the IAPP Certificate. The IAPP Certificate shall be available on board the ship at all times.

2 If the language of the original Record is not English, French or Spanish, the text shall include a translation into one of these languages.

3 Entries in boxes shall be made by inserting either a cross (**x**) for the answer "yes" and "applicable" or a (–) for the answers "no" and "not applicable" as appropriate.

4 Unless otherwise stated, regulations mentioned in this Record refer to regulations of Annex VI of the Convention and resolutions or circulars refer to those adopted by the International Maritime Organization.

---

1   Particulars of ship

1.1   Name of ship ....

1.2   Distinctive number or letters ....

1.3   IMO number...

1.4   Port of registry..

1.5   Gross tonnage..

1.6   Date on which keel was laid or ship was at a similar stage of construction...

1.7   Date of commencement of major engine conversion (if applicable) (regulation 13):...

422

2   Control of emissions from ships

2.1 *Ozone-depleting substances (regulation 12)*

2.1.1 The following fire-extinguishing systems and equipment containing halons may continue in service: .... □

System equipment	Location on board

2.1.2 The following systems and equipment containing CFCs may continue in service: .... □

System equipment	Location on board

2.1.3 The following systems containing hydro-chlorofluorocarbons (HCFCs) installed before 1 January 2020 may continue in service: .... □

System equipment	Location on board

2.2 *Nitrogen oxides (NO_x) (regulation 13)*

2.2.1 The following diesel engines with power output greater than 130 kW, and installed on a ship constructed on or after 1 January 2000, comply with the emission standards of regulation 13(3)(a) in accordance with the NO_x Technical Code: .... □

Manufacturer and model	Serial number	Use	Power output (kW)	Rated speed (rpm)

2.2.2 The following diesel engines with power output greater than 130 kW, and which underwent major conversion per regulation 13(2) on or after 1 January 2000, comply with the emission standards of regulation 13(3)(a) in accordance with the NO_x Technical Code: .... ☐

Manufacturer and model	Serial number	Use	Power output (kW)	Rated speed (rpm)

2.2.3 The following diesel engines with a power output greater than 130 kW and installed on a ship constructed on or after 1 January 2000, or with a power output greater than 130 kW and which underwent major conversion per regulation 13(2) on or after 1 January 2000, are fitted with an exhaust gas cleaning system or other equivalent methods in accordance with regulation 13(3), and the NO_x Technical Code: ...... ☐

Manufacturer and model	Serial number	Use	Power output (kW)	Rated speed (rpm)

2.2.4 The following diesel engines from 2.2.1, 2.2.2 and 2.2.3 above are fitted with NO_x emission monitoring and recording devices in accordance with the NO_x Technical Code: .... ☐

Manufacturer and model	Serial number	Use	Power output (kW)	Rated speed (rpm)

2.3 *Sulphur oxides (SO$_x$) (regulation 14)*

2.3.1 When the ship operates within an SO_x emission control area specified in regulation 14(3), the ship uses:

.1 fuel oil with a sulphur content that does not exceed 1.5% m/m as documented by bunker delivery notes; or .... ☐

  .2 an approved exhaust gas cleaning system to reduce SO_x emissions below 6.0 g SO_x/kW h; or .... ☐

  .3 other approved technology to reduce SO_x emissions below 6.0 g SO_x/kW h .... ☐

2.4 *Volatile organic compounds (VOCs) (regulation 15)*

2.4.1 The tanker has a vapour collection system installed and approved in accordance with MSC/Circ.585 .... ☐

2.5 The ship has an incinerator:

  .1 which complies with resolution MEPC.76(40) as amended .... ☐

  .2 installed before 1 January 2000 which does not comply with resolution MEPC.76(40) as amended .... ☐

THIS IS TO CERTIFY that this Record is correct in all respects.

Issued at ....

(Place of issue of the Record)

....

Date of issue   *(Signature of duly authorized official issuing the Record)*

(Seal or stamp of the authority, as appropriate)

Appendix II

Test cycles and weighting factors (Regulation 13)

The following test cycles and weighting factors should be applied for verification of compliance of marine diesel engines with the NO_x limits in accordance with regulation 13 of this Annex using the test procedure and calculation method as specified in the NO_x Technical Code.

.1 For constant-speed marine engines for ship main propulsion, including diesel-electric drive, test cycle E2 should be applied.

.2 For variable-pitch propeller sets test cycle E2 should be applied.

.3 For propeller-law-operated main and propeller-law-operated auxiliary engines the test cycle E3 should be applied.

.4 For constant-speed auxiliary engines test cycle D2 should be applied.

.5 For variable-speed, variable-load auxiliary engines, not included above, test cycle C1 should be applied.

Test cycle for *constant-speed main propulsion* application
(including diesel-electric drive or variable-pitch propeller installations)

Test cycle type E2	Speed	100%	100%	100%	100%
	Power	100%	75%	50%	25%
	Weighting factor	0.2	0.5	0.15	0.15

Test cycle for *propeller-law-operated main and propeller-law-operated auxiliary engine* application

Test cycle type E3	Speed	100%	91%	80%	63%
	Power	100%	75%	50%	25%
	Weighting factor	0.2	0.5	0.15	0.15

Test cycle for *constant-speed auxiliary engine* application

Test cycle type D2	Speed	100%	100%	100%	100%	100%
	Power	100%	75%	50%	25%	10%
	Weighting factor	0.05	0.25	0.3	0.3	0.1

Test cycle for *variable-speed and -load auxiliary engine* application

Test cycle type C1	Speed	Rated				Intermediate			Idle
	Torque	100%	75%	50%	10%	100%	75%	50%	0%
	Weighting factor	0.15	0.15	0.15	0.1	0.1	0.1	0.1	0.15

Appendix III

Criteria and procedures for designation of SO$_x$ emission control areas (Regulation 14)

1 *Objectives*

1.1   The purpose of this appendix is to provide the criteria and procedures for the designation of SO$_x$ emission control areas. The objective of SO$_x$ emission control areas is to prevent, reduce, and control air pollution from SO$_x$ emissions from ships and their attendant adverse impacts on land and sea areas.

1.2   A SO$_x$ emission control area should be considered for adoption by the Organization if supported by a demonstrated need to prevent, reduce, and control air pollution from SO$_x$ emissions from ships.

2 *Proposal criteria for designation of a SO$_x$ emission control area*

2.1   A proposal to the Organization for designation of a SO$_x$ emission control area may be submitted only by Contracting States to the Protocol of 1997. Where two or more Contracting States have a common interest in a particular area, they should formulate a co-ordinated proposal.

2.2   The proposal shall include:

.1   a clear delineation of the proposed area of application of controls on SO$_x$ emissions from ships, along with a reference chart on which the area is marked;

.2   a description of the land and sea areas at risk from the impacts of ship SO$_x$ emissions;

.3   an assessment that SO$_x$ emissions from ships operating in the proposed area of application of the SO$_x$ emission controls are contributing to air pollution from SO$_x$, including SO$_x$ deposition, and their attendant adverse impacts on the land and sea areas under consideration. Such assessment shall include a description of the impacts of SO$_x$ emissions on terrestrial and aquatic ecosystems, areas of natural productivity, critical habitats, water quality, human health, and areas of cultural and scientific significance, if applicable. The sources of relevant data, including methodologies used, shall be identified;

.4   relevant information pertaining to the meteorological conditions in the proposed area of application of the SO_x emission controls and the land and sea areas at risk, in particular prevailing wind patterns, or to topographical, geological, oceanographic, morphological, or other conditions that may lead to an increased probability of higher localized air pollution or levels of acidification;

.5   the nature of the ship traffic in the proposed SO_x emission control area, including the patterns and density of such traffic; and

.6   a description of the control measures taken by the proposing Contracting State or Contracting States addressing land-based sources of SO_x emissions affecting the area at risk that are in place and operating concurrent with the consideration of measures to be adopted in relation to provisions of regulation 14 of Annex VI of the present Convention.

2.3   The geographical limits of an SO_x emission control area will be based on the relevant criteria outlined above, including SO_x emission and deposition from ships navigating in the proposed area, traffic patterns and density, and wind conditions.

2.4   A proposal to designate a given area as an SO_x emission control area should be submitted to the Organization in accordance with the rules and procedures established by the Organization.

3   *Procedures for the assessment and adoption of SO_x emission control areas by the Organization*

3.1   The Organization shall consider each proposal submitted to it by a Contracting State or Contracting States.

3.2   A SO_x emission control area shall be designated by means of an amendment to this Annex, considered, adopted and brought into force in accordance with article 16 of the present Convention.

3.3   In assessing the proposal, the Organization shall take into account the criteria which are to be included in each proposal for adoption as set forth in section 2 above, and the relative costs of reducing sulphur depositions from ships when compared with land-based controls. The economic impacts on shipping engaged in international trade should also be taken into account.

4   *Operation of SO_x emission control areas*

4.1   Parties which have ships navigating in the area are encouraged to bring to the Organization any concerns regarding the operation of the area.

Appendix IV

Type approval and operating limits for shipboard incinerators (Regulation 16)

(1) Shipboard incinerators described in regulation 16(2) shall possess an IMO type approval certificate for each incinerator. In order to obtain such certificate, the incinerator shall be designed and built to an approved standard as described in regulation 16(2). Each model shall be subject to a specified type approval test operation at the factory or an approved test facility, and under the responsibility of the Administration, using the following standard fuel/waste specification for the type approval test for determining whether the incinerator operates within the limits specified in paragraph (2) of this appendix:

Sludge oil consisting of:	75% sludge oil from HFO; 5% waste lubricating oil; and 20% emulsified water

Solid waste consisting of:	50% food waste
	50% rubbish containing
	    approx. 30% paper,
	    approx. 40% cardboard,
	    approx. 10% rags,
	    approx. 20% plastic

	The mixture will have up to 50% moisture and 7% incombustible solids.

(2) Incinerators described in regulation 16(2) shall operate within the following limits:

O_2 in combustion chamber:	6–12%
CO in flue gas maximum average:	200 mg/MJ
Soot number maximum average:	Bacharach 3 or Ringelman 1 (20% opacity) (A higher soot number is acceptable only during very short periods such as starting up)
Unburned components in ash residues:	maximum 10% by weight
Combustion chamber flue gas outlet temperature range:	850–1200°C

Appendix V

Information to be included in the bunker delivery note (Regulation 18(3))

Name and IMO number of receiving ship

Port

Date of commencement of delivery

Name, address, and telephone number of marine fuel oil supplier

Product name(s)

Quantity (metric tons)

Density at 15°C (kg/m^3)*

Sulphur content (% m/m)†

A declaration signed and certified by the fuel oil supplier's representative that the fuel oil supplied is in conformity with regulation 14 (1) or (4)(a) and regulation 18(1) of this Annex.

---

* Fuel oil should be tested in accordance with ISO 3675.
† Fuel oil should be tested in accordance with ISO 8754.

Additional information

1

List of unified interpretations of Annexes I, II and III of MARPOL 73/78

1   List of unified interpretations of Annex I of MARPOL 73/78

MEPC/Circ.97, annex 2 and Corr. 1	Uniform interpretation of provisions of Annex I
MEPC 17/21, paragraph 5.17	Revision of unified interpretation of regulation 10(3)(b)(vi)
MEPC 18/18, annex 5	Unified interpretations of regulations 1(8), 3, 16(1) and 16(2)(b), 25(1) and 25(2)
MEPC 19/18, annex 3	Unified interpretations of regulations 1(4) and 8
MEPC 20/19, annex 5	Unified interpretations of regulations 4, 5 and others, 15(5) and 16(3)(a) and 21
MEPC 21/19, annex 11	Unified interpretation of regulations 9(1) and 10(3) of Annex I and interpretation of Assembly resolution A.541(13) (*not included in this book*)
MEPC 25/20, paragraph 5.7	Uniform interpretation of "all oily mixtures" in regulations 15(5)(a) and 15(5)(b)(ii)(3)
MEPC 26/25, annex 5	Agreed amendment to and interpretation of regulation 17 of Annex I
MEPC 27/16, annex 7	Unified interpretation of regulation 17 of Annex I
MEPC 30/24, annex 7	Unified interpretation of regulation 1(1) of Annex I
MEPC 31/21, annex 5	Unified interpretation of regulation 1(17) of Annex I
MEPC 32/20, paragraph 5.2 and annex 3	Unified interpretation of regulation 26 of Annex I
MEPC 33/20, paragraph 4.5 and annex 5	Unified interpretations of regulations 9(4), 10(3), 16(1) and 16(2) of Annex I
MEPC 34/23, paragraph 7.2.2 and annex 6	Unified interpretations of regulations 7, 12(2), 13, 13G, 13F(3)(d), 13G(4), 15(7), 16(6) and 21 of Annex I

MEPC 35/21, paragraph 8.10 and annex 5	Unified interpretation of regulation 15(7) of Annex I
MEPC 36/22, paragraph 9.38 and annex 7	Unified interpretation of regulation 15(5) of Annex I
MEPC 38/20, paragraphs 3.9 and 8.14	Unified interpretations of regulations 13F(3)(d) and 13F of Annex I
MEPC 40/21, paragraphs 8.2 and 9.3 and annex 4	Unified interpretations of regulation 13(3)(b) and of regulation 25A(2) of Annex I
MEPC 43/21, paragraphs 11.23 to 11.25	IACS Unified Interpretation MPC 7 on hydrostatic balance loading, within MEPC/Circ.365

2 List of unified interpretations of Annex II of MARPOL 73/78

MEPC 22/21, annex 7	Texts of agreed unified interpretations of Annex II
MEPC 23/22, annex 6	Unified interpretations of the provisions of Annex II
MEPC 23/22, annex 7	Interpretation of Annex II in respect of incinerator ships (*not included in this book*)
MEPC 24/19, annex 2	Unified interpretations of the provisions of Annex II
MEPC 24/19, annex 3	Extension of the unified interpretation of regulation 3(4) of Annex II
MEPC 25/20, annex 4	Unified interpretation of the provisions of Annex II
MEPC 25/20, annex 5	Interpretation of Annex II in respect of ships engaged in dumping operations and explanatory notes thereto (*not included in this book*)
MEPC 29/22, annex 2	Unified interpretation of regulations of Annex II
MEPC 30/24, annex 11	Unified interpretation of regulation 3(4) of Annex II
MEPC 33/20, paragraph 3.19	Unified interpretation of regulation 3(4) of Annex II

3 List of unified interpretations of Annex III of MARPOL 73/78

| MEPC 36/22, paragraph 9.42 and annex 7 | Unified interpretation of regulation 4(3) of Annex III |

2
List of related documents

1   The following is a list of related documents which have been incorporated into this book.

Reference   *Document*

International Convention for the Prevention
of Pollution from Ships, 1973

Protocol of 1978 relating to the International
Convention for the Prevention of Pollution
from Ships, 1973

Protocol of 1997 to amend the International
Convention for the Prevention of Pollution
from Ships, 1973, as modified by the Protocol
of 1978 relating thereto

Protocol I: Provisions concerning reports
on incidents involving harmful substances

–	1985 amendments to Protocol	MEPC 22/21, annex 10
–	1996 amendment to article II(1)	MEPC 38/20, annex 2

Protocol II: Arbitration

Annex I

–	1984 Amendments	MEPC 20/19, annex 4
–	1987 Amendments	MEPC 25/20, annex 9
–	1990 Amendments	MEPC 30/24, annex 5
–	1991 Amendments	MEPC 31/21, annex 6
–	1992 Amendments	MEPC 32/20, annexes 5 & 6
–	1994 Amendments	MP/CONF.2/8
–	1997 Amendments	MEPC 40/21, annex 5
–	1999 Amendments	MEPC 43/21, annex 3
–	2001 Amendments	MEPC 46/23, annex 3
	– Condition assessment scheme	MEPC 46/23, annex 2

Annex II

–	1985 Amendments	MEPC 22/21, annex 2
–	1989 Amendments	MEPC 27/16, annex 5
–	1992 Amendments	MEPC 33/20, annex 8
–	1994 Amendments	MP/CONF.2/8

– 1999 Amendments	MEPC 43/21, annex 3
– Standards for procedures and arrangements for the discharge of noxious liquid substances	MEPC 22/21, annex 4
– 1992 Amendments	MEPC 33/20, annex 3
– 1994 Amendments	MEPC 35/21, annex 2
– 1995 Amendments	MEPC 37/22/Add.1, annex 4

Annex III

– Draft revised Annex III of MARPOL 73/78	MEPC 26/25, annex 6
– 1992 Amendments	MEPC 33/20, annex 9
– 1994 Amendments	MP/CONF.2/8
– 2000 Amendments	MEPC 44/20, annex 3

Annex IV

– Revised Annex IV of MARPOL 73/78	MEPC 44/20, annex 10
– Implementation of Annex IV	MEPC 44/20, annex 11

Annex V

– 1989 Amendments	MEPC 28/4, annex 2
– 1994 Amendments	MP/CONF.2/8
– 1995 Amendments	MEPC 37/22/Add.1, annex 13
– 2000 Amendments	MEPC 45/20, annex 3

Annex VI

2   **The following is a list of related documents which have not been included in this book.**

Reference	*Document or IMO publication sales number*

Protocol I

– Resolution A.851(20): General principles for ship reporting systems and ship reporting requirements, including guidelines for reporting incidents involving dangerous goods, harmful substances and/ or marine pollutants	IMO-516E
– Provisions concerning the reporting of incidents involving harmful substances under MARPOL 73/78 (1999 edition)	IMO-516E

Annex I

- Guidelines for surveys under Annex I of   IMO-526E
  MARPOL 73/78 (1983 edition)
- Guidelines for the development of   IMO-586E
  shipboard marine pollution emergency
  plans (2001 edition)
- Crude oil washing systems   IMO-617E
  (2000 edition)
- Dedicated clean ballast tanks   IMO-619E
  (1982 edition)
- Inert gas systems (1990 edition)   IMO-860E
- Guidelines on enhanced programme of   IMO-265E
  inspections during surveys of bulk carriers
  and oil tankers (2001 edition)

Annex II

- Guidelines for the provisional assessment of   IMO-653E
  liquids transported in bulk
    - Annex 1 – Flow chart for provisional
      assessment of liquids transported in
      bulk
    - *Annex 2 is included within this
      publication*
    - Annex 3 – Example of an
      amendment sheet to the ship's
      Certificate of Fitness and Procedures
      and Arrangements Manual
    - Annex 4 – Interpretation of the
      Guidelines for the categorization of
      noxious liquid substances
    - Annex 5 – Abbreviated legend to the
      hazard profiles
    - Annex 6 – Criteria for establishing
      ship type requirements from the
      marine pollution point of view
    - Annex 7 – Telex/Telefax format for
      proposing tripartite agreement for
      provisional assessment of liquid
      substances
    - Annex 8 – Format for assessment of
      liquid chemicals
    - Annex 9 – Examples of the
      calculation method

Additional Information

      – Annex 10 – Interpretation for assigning the minimum carriage requirements for mixtures involving products included in the IBC/BCH Codes for safety reasons

– International Code for the Construction and Equipment of Ships Carrying Dangerous Chemicals in Bulk (IBC Code) (1998 edition)   IMO-100E

– Guidelines for surveys under Annex II of MARPOL 73/78 (1987 edition)   IMO-508E

– Code for the Construction and Equipment of Ships Carrying Dangerous Chemicals in Bulk (BCH Code) (1993 edition)   IMO-772E

– Guidelines for the development of shipboard marine pollution emergency plans (2001 edition)   IMO-586E

Annex III

  – International Maritime Dangerous Goods Code (IMDG Code) (2000 edition)   IMO-200E

Annex V

– Guidelines for the implementation of Annex V   IMO-656E

       – Appendix 1 – Form for reporting alleged inadequacy of port reception facilities for garbage

       – Appendix 2 – Standard specification for shipboard incinerators (MEPC.59(33))

General

– Control of ships and discharges (1986 edition)   IMO-601E

– Procedures for port State control (2000 edition)   IMO-650E

– Comprehensive manual on port reception facilities (1999 edition)   IMO-597E

– Pollution prevention equipment required under MARPOL 73/78 (1997 edition)   IMO-646E

– MARPOL – How to do it   IMO-636E

3
List of MEPC resolutions

MEPC.1(II)*	Resolution on establishment of the list of substances to be annexed to the Protocol relating to Intervention on the High Seas in Cases of Marine Pollution by Substances other than Oil	–
MEPC.2(VI)	Recommendation on international effluent standards and guidelines for performance tests for sewage treatment plants	IMO-592E
MEPC.3(XII)	Recommendation on the standard format for the crude oil washing operations and equipment manual	IMO-617E
MEPC.4(XIII)	Recommendation regarding acceptance of oil content meters in oil tankers	–
MEPC.5(XIII)	Specification for oil/water interface detectors	IMO-608E (Oily-Water Separators and Monitoring Equipment)
MEPC.6(XIV)	Application of the provisions of Annex I of the International Convention for the Prevention of Pollution from Ships, 1973, as modified by the Protocol of 1978 relating thereto on the discharge of oil in the Baltic Sea area	–

Additional Information

---

* Roman or Arabic figures in brackets show the session number, and the texts of these resolutions are annexed to the MEPC report of that session.

MEPC.7(XV)	Entries in oil record books on methods of disposal of residue	–
MEPC.8(XVI)	Discharge of oils not specified by the International Convention for the Prevention of Pollution of the Sea by Oil, 1954, as amended in 1962 and 1969	–
MEPC.9(17)	Application of the provisions of Annex V of MARPOL 73/78 on the discharge of garbage in the Baltic Sea area	–
MEPC.10(18)	Application scheme for oil discharge monitoring and control systems	–
MEPC.11(18)	Guidelines for surveys under Annex I of the International Convention for the Prevention of Pollution from Ships, 1973, as modified by the Protocol of 1978 relating thereto	IMO-526E
MEPC.12(18)	Regional arrangements for combating major incidents of marine pollution	–
MEPC.13(19)	Guidelines for plan approval and installation survey of oil discharge monitoring and control systems for oil tankers and environmental testing of control sections thereof	IMO-608E (Oily-Water Separators and Monitoring Equipment)
MEPC.14(20)	Adoption of amendments to Annex I of MARPOL 73/78	–
MEPC.15(21)	Installation of oil discharge monitoring and control systems in existing oil tankers	–
MEPC.16(22)	Adoption of amendments to Annex II of MARPOL 73/78	–
MEPC.17(22)	Implementation of Annex II of MARPOL 73/78	–

MEPC.18(22)	Adoption of the standards for procedures and arrangements for the discharge of noxious liquid substances	–
MEPC.19(22)	Adoption of the International Code for the Construction and Equipment of Ships Carrying Dangerous Chemicals in Bulk (IBC Code)	IMO-100E
MEPC.20(22)	Adoption of the Code for the Construction and Equipment of Ships Carrying Dangerous Chemicals in Bulk (BCH Code)	IMO-772E
MEPC.21(22)	Adoption of amendments to Protocol I to MARPOL 73/78 and the text of the Protocol, as amended, annexed thereto	–
MEPC.22(22)	Adoption of guidelines for reporting incidents involving harmful substances and the text of guidelines annexed thereto	–
MEPC.23(22)	The application of Annex II of MARPOL 73/78 on the discharge of noxious liquid substances in the Baltic Sea area	–
MEPC.24(22)	Adoption of amendments to the Revised guidelines and specifications for oil discharge monitoring and control systems for oil tankers as adopted by the Organization by resolution A.586(14) and to the Recommendation on international performance and test specifications for oily-water separating equipment and oil content meters adopted by the Organization by resolution A.393(X)	IMO-608E (Oily-Water Separators and Monitoring Equipment)

Additional Information

MEPC.25(23)	Guidelines for surveys under Annex II of the International Convention for the Prevention of Pollution from Ships, 1973, as modified by the Protocol of 1978 relating thereto (MARPOL 73/78)	IMO-508E
MEPC.26(23)	Procedures for the control of ships and discharges under Annex II of the International Convention for the Prevention of Pollution from Ships, 1973, as modified by the Protocol of 1978 relating thereto (MARPOL 73/78)	IMO-601E (Control of Ships and Discharges)
MEPC.27(23)	Categorization of liquid substances	–
MEPC.28(24)	Compliance with Annex II of MARPOL 73/78	–
MEPC.29(25)	Adoption of amendments to the Annex of the Protocol of 1978 relating to the International Convention for the Prevention of Pollution from Ships, 1973 (designation of the Gulf of Aden as a special area)	–
MEPC.30(25)	Guidelines for reporting incidents involving harmful substances	–
MEPC.31(26)	Establishment of the date of application of the provisions of regulation 5 of Annex V of the International Convention for the Prevention of Pollution from Ships, 1973 as modified by the Protocol of 1978 relating thereto on the discharge of garbage in the Baltic Sea area	–

MEPC.32(27)	Adoption of amendments to the International Code for the Construction and Equipment of Ships Carrying Dangerous Chemicals in Bulk (IBC Code)	IMO-100E
MEPC.33(27)	Adoption of amendments to the Code for the Construction and Equipment of Ships Carrying Dangerous Chemicals in Bulk (BCH Code)	IMO-772E
MEPC.34(27)	Adoption of amendments to the Annex of the Protocol of 1978 relating to the International Convention for the Prevention of Pollution from Ships, 1973 (Appendices II and III of Annex II of MARPOL 73/78)	–
MEPC.35(27)	Implementation of Annex III of MARPOL 73/78	–
MEPC.36(28)	Adoption of amendments to the Annex of the Protocol of 1978 relating to the International Convention for the Prevention of Pollution from Ships, 1973 (Amendments to Annex V of MARPOL 73/78)	–
MEPC.37(28)	Establishment of the date of application of the provisions of regulation 5 of Annex V of the International Convention for the Prevention of Pollution from Ships, 1973 as modified by the Protocol of 1978 relating thereto on the discharge of garbage in the North Sea area	–
MEPC.38(29)	Application of the provisions of Annex IV of the International Convention for the Prevention of Pollution from Ships, 1973 as modified by the Protocol of 1978 relating thereto on the discharge of sewage in the Baltic Sea area	–

Additional Information

MEPC.39(29)   Adoption of amendments to   –
the Annex of the Protocol of
1978 relating to the International
Convention for the Prevention
of Pollution from Ships, 1973
(Introduction of the Harmonized
System of Survey and Certification
to Annexes I and II of
MARPOL 73/78)

MEPC.40(29)   Adoption of amendments to   –
the International Code for the
Construction and Equipment
of Ships Carrying Dangerous
Chemicals in Bulk (IBC Code)
(Harmonized System of Survey
and Certification)

MEPC.41(29)   Adoption of amendments to   –
the Code for the Construction
and Equipment of Ships Carrying
Dangerous Chemicals in Bulk
(BCH Code) (Harmonized System
of Survey and Certification)

MEPC.42(30)   Adoption of amendments to   –
the Annex of the Protocol of
1978 relating to the International
Convention for the Prevention of
Pollution from Ships, 1973
(Designation of Antarctic area as
a special area under Annexes I and V
of MARPOL 73/78)

MEPC.43(30)   Prevention of pollution by   –
garbage in the Mediterranean

MEPC.44(30)   Identification of the Great Barrier   –
Reef region as a particularly
sensitive area

MEPC.45(30)   Protection of the Great Barrier   –
Reef region

MEPC.46(30)   Measures to control potential adverse   –
impacts associated with use of
tributyl tin compounds in
anti-fouling paints

448

MEPC.47(31)	Adoption of amendments to the Annex of the Protocol of 1978 relating to the International Convention for the Prevention of Pollution from Ships, 1973 (new regulation 26 and other amendments to Annex I of MARPOL 73/78)	–
MEPC.48(31)	Adoption of amendments to the Annex of the Protocol of 1978 relating to the International Convention for the Prevention of Pollution from Ships, 1973 (designation of the Wider Caribbean area as a special area under Annex V of MARPOL 73/78)	–
MEPC.49(31)	Revision of the list of substances to be annexed to the Protocol relating to Intervention on the High Seas in Cases of Marine Pollution by Substances Other Than Oil, 1973	–
MEPC.50(31)	Guidelines for preventing the introduction of unwanted aquatic organisms and pathogens from ships' ballast water and sediment discharges	–
MEPC.51(32)	Adoption of amendments to the annex of the Protocol of 1978 relating to the International Convention for the Prevention of Pollution from Ships, 1973 (Discharge criteria of Annex I of MARPOL 73/78)	IMO-520E
MEPC.52(32)	Adoption of amendments to the annex of the Protocol of 1978 relating to the International Convention for the Prevention of Pollution from Ships, 1973 (New regulations 13F and 13G and related amendments to Annex I of MARPOL 73/78)	IMO-520E

Additional Information

449

MEPC.53(32)	Development of the capacity of ship scrapping for the smooth implementation of the amendments to Annex I of MARPOL 73/78	–
MEPC.54(32)	Guidelines for the development of shipboard oil pollution emergency plans	IMO-586E
MEPC.55(33)	Adoption of amendments to the International Code for the Construction and Equipment of Ships Carrying Dangerous Chemicals in Bulk (IBC Code)	–
MEPC.56(33)	Adoption of amendments to the Code for the Construction and Equipment of Ships Carrying Dangerous Chemicals in Bulk (BCH Code)	–
MEPC.57(33)	Adoption of amendments to the annex of the Protocol of 1978 relating to the International Convention for the Prevention of Pollution from Ships, 1973 (Designation of the Antarctic area as a special area and lists of liquid substances in Annex II)	–
MEPC.58(33)	Adoption of amendments to the annex of the Protocol of 1978 relating to the International Convention for the Prevention of Pollution from Ships, 1973 (Revised Annex III)	–
MEPC.59(33)	Revised guidelines for the implementation of Annex V of MARPOL 73/78	IMO-656E
MEPC.60(33)	Guidelines and specifications for pollution prevention equipment for machinery space bilges of ships	IMO-646E
MEPC.61(34)	Visibility limits of oil discharges of Annex I of MARPOL 73/78	–
MEPC.62(35)	Amendments to the standards for procedures and arrangements for the discharge of noxious liquid substances	–

MEPC.63(36)   Oil tanker stability, operational   —
              safety and protection of the marine
              environment

MEPC.64(36)   Guidelines for approval of   —
              alternative structural or
              operational arrangements as
              called for in regulation 13G(7)
              of Annex I of MARPOL 73/78

MEPC.65(37)   Amendments to the annex of   —
              the Protocol of 1978 relating
              to the International Convention
              for the Prevention of Pollution
              from Ships, 1973 (Amendments
              to regulation 2 and new
              regulation 9 of Annex V)

MEPC.66(37)   Interim Guidelines for the approval   —
              of alternative methods of design
              and construction of oil tankers
              under regulation 13F(5) of
              Annex I of MARPOL 73/78

MEPC.67(37)   Guidelines on application of the   —
              precautionary approach

MEPC.68(38)   Amendments to the annex of   —
              the Protocol of 1978 relating
              to the International Convention
              for the Prevention of Pollution
              from Ships, 1973 (Amendments
              to Protocol I)

MEPC.69(38)   Amendments to the International   —
              Code for the Construction and
              Equipment of Ships Carrying
              Dangerous Chemicals in Bulk
              (IBC Code)

MEPC.70(38)   Amendments to the Code for   —
              the Construction and Equipment
              of Ships Carrying Dangerous
              Chemicals in Bulk (BCH Code)

MEPC.71(38)   Guidelines for the development   IMO-656E
              of garbage management plans

Additional Information

MEPC.72(38)	Revision of the list of substances to be annexed to the Protocol relating to the Intervention on the High Seas in Cases of Marine Pollution by Substances other than Oil	–
MEPC.73(39)	Amendments to the International Code for the Construction and Equipment of Ships Carrying Dangerous Chemicals in Bulk (IBC Code) (Vague expressions)	–
MEPC.74(40)	Identification of the Archipelago of Sabana-Camagüey as a particularly sensitive sea area	–
MEPC.75(40)	Amendments to Annex I of the Protocol of 1978 relating to the International Convention for the Prevention of Pollution from Ships, 1973	–
MEPC.76(40)	Standard specification for shipboard incinerators	–
MEPC.77(41)	Establishment of the date on which the amendments to regulation 10 of Annex I of MARPOL 73/78 in respect of the North-West European Waters special area shall take effect	–
MEPC.78(43)	Amendments to the annex of the Protocol of 1978 relating to the International Convention for the Prevention of Pollution from Ships, 1973	–
MEPC.79(43)	Amendments to the International Code for the Construction and Equipment of Ships Carrying Dangerous Chemicals in Bulk (IBC Code)	–
MEPC.80(43)	Amendments to the Code for the Construction and Equipment of Ships Carrying Dangerous Chemicals in Bulk (BCH Code)	–
MEPC.81(43)	Amendments to section 9 of the Standard Format for the COW Manual (resolution MEPC.3(XII))	IMO-617E

MEPC.82(43)	Guidelines for monitoring the world-wide average sulphur content of residual fuel oils supplied for use on board ships	–
MEPC.83(44)	Guidelines for ensuring the adequacy of port waste reception facilities	–
MEPC.84(44)	Amendments to the Annex of the Protocol of 1978 Relating to the International Convention for the Prevention of Marine Pollution from Ships, 1973	–
MEPC.85(44)	Guidelines for the development of shipboard marine pollution emergency plans for oil and/or noxious liquid substances	IMO-586E
MEPC.86(44)	Amendments to the Guidelines for the development of shipboard oil pollution emergency plans	IMO-586E
MEPC.87(44)	Use of Spanish under IMO conventions relating to pollution prevention	–
MEPC.88(44)	Implementation of Annex IV of MARPOL 73/78	–
MEPC.89(45)	Amendments to the Annex of the Protocol of 1978 Relating to the International Convention for the Prevention of Pollution from Ships, 1973	–
MEPC.90(45)	Amendments to the International Code for the Construction and Equipment of Ships Carrying Dangerous Chemicals in Bulk (IBC Code)	–
MEPC.91(45)	Amendments to the Code for the Construction and Equipment of Ships Carrying Dangerous Chemicals in Bulk (BCH Code)	–
MEPC.92(45)	Amendments to the Revised Guidelines for the implementation of Annex V of MARPOL 73/78 (resolution MEPC.59(33))	–

Additional Information

MEPC.93(45)	Amendments to the Standard Specification for Shipboard Incinerators	–
MEPC.94(46)	Condition Assessment Scheme	–
MEPC.95(46)	Amendments to the Annex of the Protocol of 1978 Relating to the International Convention for the Prevention of Pollution from Ships, 1973	–

Additional Information

4
Status of MARPOL 73/78, amendments and related instruments

This list show the dates of entry into force of MARPOL 73/78, its protocols, annexes and amendments as at 1 January 2002.

Details of the amendments may be found in the list of MEPC resolutions.

International Convention for the Prevention of Pollution from Ships, 1973, as modified by the Protocol of 1978 relating thereto (MARPOL (amended) 73/78).

Entry into force:	2 October 1983
Annex I	2 October 1983
Annex II	6 April 1987
Annex III	1 July 1992
Annex IV	*not yet in force*
Annex V	31 December 1988
Annex VI	*not yet in force*
1984 amendments (MEPC.14(20)) (extensive amendments to Annex I which had been agreed over the years)	7 January 1986
1985 amendments (MEPC.16(22)) (extensive amendments to Annex II in preparation for its implementation – pumping, piping, control, etc.)	6 April 1987
1985 (Protocol I) amendments (MEPC.21(22)) (Reporting Protocol)	6 April 1987
1987 (Annex I) amendments (MEPC.29(25)) (designation of the Gulf of Aden as a special area)	1 April 1989
1989 (IBC Code) amendments (MEPC.32(27)) (list of chemicals)	13 October 1990
1989 (BCH Code) amendments (MEPC.33(27)) (list of chemicals)	13 October 1990
1989 (Annex II) amendments (MEPC.34(27)) (list of chemicals)	13 October 1990
1989 (Annex V) amendments (MEPC.36(28)) (designation of the North Sea as a special area)	18 February 1991
1990 (Annexes I and II) amendments (MEPC.39(29)) (harmonized system of survey and certification)	3 February 2000

1990 (IBC Code) amendments (MEPC.40(29)) (harmonized system of survey and certification)	3 February 2000
1990 (BCH Code) amendments (MEPC.41(29)) (harmonized system of survey and certification)	3 February 2000
1990 (Annexes I and V) amendments (MEPC.42(30)) (designation of the Antarctic area as a special area)	17 March 1992
1991 (Annex I) amendments (MEPC.47(31)) (new regulation 26 (Shipboard Oil Pollution Emergency Plan) and other amendments)	4 April 1993
1991 (Annex V) amendments (MEPC.48(31)) (designation of the Wider Caribbean area as a special area)	4 April 1993
1992 (Annex I) amendments (MEPC.51(32)) (discharge criteria)	6 July 1993
1992 (Annex I) amendments (MEPC.52(32)) (oil tanker design)	6 July 1993
1992 (IBC Code) amendments (MEPC.55(33)) (lists of chemicals, cargo tank venting and gas-freeing arrangements and other amendments)	1 July 1994
1992 (BCH Code) amendments (MEPC.56(33)) (lists of chemicals and other amendments)	1 July 1994
1992 (Annex II) amendments (MEPC.57(33)) (lists of chemicals and the designation of the Antarctic area as a special area)	1 July 1994
1992 (Annex III) amendments (MEPC.58(33)) (total revision of Annex III with the IMDG Code as a vehicle for its implementation)	28 February 1994
1994 (Annexes I, II, III and V) amendments (Conference resolutions 1-3) (Port State control on operational requirements)	3 March 1996
1995 (Annex V) amendments (MEPC.65(37)) (Application, placards, management plans and record keeping)	1 July 1997
1996 (Protocol I) amendments (MEPC.68(38)) (article II – when to make reports)	1 January 1998
1996 (IBC Code) amendments (MEPC.69(38)) (lists of chemicals)	1 July 1998
1996 (BCH Code amendments (MEPC.70(38)) (lists of chemicals)	1 July 1998

1997 (IBC Code) amendments (MEPC.73(39)) (vague expressions)	10 July 1998
1997 (Annex I) amendments (MEPC.75(40)) (designation of North-West European waters as a special area; new regulation 25A)	1 February 1999
1999 (Annexes I and II) amendments (MEPC.78(43)) (amendments to regulations 13G and 26 and IOPP Certificate of Annex I and addition of new regulation 16 to Annex II)	1 January 2001
1999 (IBC Code) amendments (MEPC.79(43)) (cargo-tank venting and gas-freeing arrangements)	[1 July 2002]
1999 (BCH Code) amendments (MEPC.80(43)) (cargo containment)	[1 July 2002]
2000 (Annex III) amendments (MEPC.84(44)) (amendments to appendix)	1 January 2002
2000 (Annex V) amendments (MEPC.89(45)) (amendments to regulations 1, 3, 5 and 9 and Record of Garbage Discharges)	[1 March 2002]
2000 (IBC Code) amendments (MEPC.90(45)) (amendments to chapters 5, 14, 15, 16)	[1 July 2002]
2000 (BCH Code) amendments (MEPC.91(45)) (amendments to chapters II, III, IV, V)	[1 July 2002]
2001 (Annex I) amendments (MEPC.95(46)) (amendments to regulation 13G and to Supplement to IOPP Certificate)	[1 September 2002*]

Additional Information

---

* Conditional upon acceptance on 1 March 2002.

457

5
Implementation of Annex IV

Resolution MEPC.88(44)

Implementation of Annex IV of MARPOL 73/78

adopted on 13 March 1990

THE MARINE ENVIRONMENT PROTECTION COMMITTEE,

RECALLING Article 38(a) of the Convention on the International Maritime Organization concerning the function of the Committee conferred upon it by international conventions for the prevention and control of marine pollution,

RECOGNIZING that approximately twenty-seven years have elapsed since the adoption of Annex IV of MARPOL 73/78,

RECOGNIZING ALSO that seventy-seven States, the combined merchant fleets of which constitute approximately forty-three per cent of the gross tonnage of the world's merchant shipping, have accepted Annex IV of MARPOL 73/78 as at 13 March 2000,

RECOGNIZING FURTHER that, since the Annex IV of MARPOL 73/78 will enter into force twelve months after the date on which not less than fifteen States, the combined merchant fleets of which constitute not less than fifty per cent of the gross tonnage of the world's merchant shipping, have become Parties to it, acceptance by States which represent approximately seven per cent of the world's tonnage is still needed to bring it into force,

NOTING that a large number of Member Governments, who have not yet accepted Annex IV of MARPOL 73/78, have expressed the view that they will face problems in complying with several provisions in Annex IV of MARPOL 73/78 if they accept Annex IV of MARPOL 73/78 in its present form,

NOTING ALSO that these provisions need to be amended to assist the entry into force of Annex IV of MARPOL 73/78, while maintaining the same level of protection of the marine environment,

NOTING FURTHER that the text of the revised Annex IV of MARPOL 73/78, as set out in annex 10 to MEPC 44/20,* was approved by the Committee at its forty-fourth session,

1.   AGREES to request the Secretary-General to circulate the text of the revised Annex IV of MARPOL 73/78 to all Members of the Organization and to all Parties to MARPOL 73/78 which are not Members of the Organization after conditions for entry into force of the existing Annex IV of MARPOL 73/78 have been met with a view to future adoption upon the entry into force of the existing Annex IV in accordance with article 16 of MARPOL 73/78;

2.   RESOLVES that the Parties to Annex IV of MARPOL 73/78 should implement the revised Annex IV of MARPOL 73/78 immediately after entry into force of the existing Annex IV of MARPOL 73/78, with a view to avoiding the creation of a dual treaty regime between the existing and the revised Annex IV of MARPOL 73/78;

3.   URGES States which have not yet accepted the existing Annex IV of MARPOL 73/78 to do so as soon as possible with the understanding that only the provisions of the revised Annex IV of MARPOL 73/78 will be implemented upon the entry into force of the existing Annex IV of MARPOL 73/78.

---

* The text of this document is item 6 of this additional information.

Regulations for the Prevention of Pollution by Sewage from Ships

Chapter I – General

Regulation 1
Definitions

For the purposes of this Annex:

1   *New ship* means a ship:

    .1   for which the building contract is placed, or in the absence of a building contract, the keel of which is laid, or which is at a similar stage of construction, on or after the date of entry into force of this Annex; or

    .2   the delivery of which is three years or more after the date of entry into force of this Annex.

2   *Existing ship* means a ship which is not a new ship.

3   *Sewage* means:

    .1   drainage and other wastes from any form of toilets and urinals;

    .2   drainage from medical premises (dispensary, sick bay, etc.) via wash basins, wash tubs and scuppers located in such premises;

    .3   drainage from spaces containing living animals; or

    .4   other waste waters when mixed with the drainages defined above.

4   *Holding tank* means a tank used for the collection and storage of sewage.

5   *Nearest land.* The term "from the nearest land" means from the baseline from which the territorial sea of the territory in question is established in accordance with international law except that, for the purposes of the present Convention, "from the nearest land" off the north-eastern

coast of Australia shall mean from a line drawn from a point on the coast of Australia in:

> latitude 11°00′ S, longitude 142°08′ E
> to a point in latitude 10°35′ S, longitude 141°55′ E,
> thence to a point latitude 10°00′ S, longitude 142°00′ E,
> thence to a point latitude 09°10′ S, longitude 143°52′ E,
> thence to a point latitude 09°00′ S, longitude 144°30′ E,
> thence to a point latitude 10°41′ S, longitude 145°00′ E,
> thence to a point latitude 13°00′ S, longitude 145°00′ E,
> thence to a point latitude 15°00′ S, longitude 146°00′ E,
> thence to a point latitude 17°30′ S, longitude 147°00′ E,
> thence to a point latitude 21°00′ S, longitude 152°55′ E,
> thence to a point latitude 24°30′ S, longitude 154°00′ E,
> thence to a point on the coast of Australia in
> latitude 24°42′ S, longitude 153°15′ E.

6   *International voyage* means a voyage from a country to which the present Convention applies to a port outside such country, or conversely.

7   *Person means* member of the crew and passengers.

8   *Anniversary date* means the day and the month of each year which will correspond to the date of expiry of the International Sewage Pollution Prevention Certificate.

Regulation 2
Application

1   The provisions of this Annex shall apply to the following ships engaged in international voyages:

.1   new ships of 400 gross tonnage and above; and

.2   new ships of less than 400 gross tonnage which are certified to carry more than 15 persons; and

.3   existing ships of 400 gross tonnage and above, five years after the date of entry into force of this Annex; and

.4   existing ships of less than 400 gross tonnage which are certified to carry more than 15 persons, five years after the date of entry into force of this Annex.

2   The Administration shall ensure that existing ships, according to subparagraphs 1.3 and 1.4 of this regulation, the keels of which are laid or which are of a similar stage of construction before 2 October 1983 shall be equipped, as far as practicable, to discharge sewage in accordance with the requirements of regulation 11 of the Annex.

Regulation 3

Exceptions

1   Regulation 11 of this Annex shall not apply to:

    .1   the discharge of sewage from a ship necessary for the purpose of securing the safety of a ship and those on board or saving life at sea; or

    .2   the discharge of sewage resulting from damage to a ship or its equipment if all reasonable precautions have been taken before and after the occurrence of the damage, for the purpose of preventing or minimizing the discharge.

Chapter 2 – Surveys and certification

Regulation 4
Surveys

1   Every ship which, in accordance with regulation 2, is required to comply with the provisions of this Annex shall be subject to the surveys specified below:

> .1   An initial survey before the ship is put in service or before the Certificate required under regulation 5 of this Annex is issued for the first time, which shall include a complete survey of its structure, equipment, systems, fittings, arrangements and material in so far as the ship is covered by this Annex. This survey shall be such as to ensure that the structure, equipment, systems, fittings, arrangements and materials fully comply with the applicable requirements of this Annex.

> .2   A renewal survey at intervals specified by the Administration, but not exceeding five years, except where regulation 8.2, 8.5, 8.6 or 8.7 of this Annex is applicable. The renewal survey shall be such as to ensure that the structure, equipment, systems, fittings, arrangements and materials fully comply with applicable requirements of this Annex.

> .3   An additional survey, either general or partial, according to the circumstances, shall be made after a repair resulting from investigations prescribed in paragraph 4 of this regulation, or whenever any important repairs or renewals are made. The survey shall be such as to ensure that the necessary repairs or renewals have been effectively made, that the materials and workmanship of such repairs or renewals are in all respects satisfactory and that the ship complies in all respects with the requirements of this Annex.

2   The Administration shall establish appropriate measures for ships which are not subject to the provisions of paragraph 1 of this regulation in order to ensure that the applicable provisions of this Annex are complied with.

3.1   Surveys of ships as regards the enforcement of the provisions of this Annex shall be carried out by officers of the Administration. The Administration may, however, entrust the surveys either to surveyors nominated for the purpose or to organizations recognized by it.

3.2   An Administration nominating surveyors or recognizing organizations to conduct surveys as set forth in subparagraph 3.1 of this paragraph shall, as a minimum, empower any nominated surveyor or recognized organization to:

    .1   require repairs to a ship; and

    .2   carry out surveys if requested by the appropriate authorities of a Port State.

The Administration shall notify the Organization of the specific responsibilities and conditions of the authority delegated to the nominated surveyors or recognized organizations, for circulation to Parties to the present Convention for the information of their officers.

3.3   When a nominated surveyor or recognized organization determines that the condition of the ship or its equipment does not correspond substantially with the particulars of the Certificate or is such that the ship is not fit to proceed to sea without presenting an unreasonable threat of harm to the marine environment, such surveyor or organization shall immediately ensure that corrective action is taken and shall in due course notify the Administration. If such corrective action is not taken, the Certificate should be withdrawn and the Administration shall be notified immediately and if the ship is in a port of another Party, the appropriate authorities of the Port State shall also be notified immediately. When an officer of the Administration, a nominated surveyor or recognized organization has notified the appropriate authorities of the Port State, the Government of the Port State concerned shall give such officer, surveyor or organization any necessary assistance to carry out their obligations under this regulation. When applicable, the Government of the Port State concerned shall take such steps as will ensure that the ship shall not sail until it can proceed to sea or leave the port for the purpose of proceeding to the nearest appropriate repair yard available without presenting an unreasonable threat of harm to the marine environment.

3.4   In every case, the Administration concerned shall fully guarantee the completeness and efficiency of the survey and shall undertake to ensure the necessary arrangements to satisfy this obligation.

4.1   The condition of the ship and its equipment shall be maintained to conform with the provisions of the present Convention to ensure that the ship in all respects will remain fit to proceed to sea without presenting an unreasonable threat of harm to the marine environment.

4.2   After any survey of the ship under paragraph 1 of this regulation has been completed, no change shall be made in the structure, equipment, systems, fittings, arrangements or materials covered by the survey, without the sanction of the Administration, except the direct replacement of such equipment and fittings.

4.3   Whenever an accident occurs to a ship or a defect is discovered which substantially affects the integrity of the ship or the efficiency or completeness of its equipment covered by this Annex, the master or owner of the ship shall report at the earliest opportunity to the Administration, the recognized organization or the nominated surveyor responsible for issuing the relevant Certificate, who shall cause investigations to be initiated to determine whether a survey as required by paragraph 1 of this regulation is necessary. If the ship is in a port of another Party, the master or owner shall also report immediately to the appropriate authorities of the Port State and the nominated surveyor or recognized organization shall ascertain that such report has been made.

Regulation 5
Issue or endorsement of Certificate

1   An international Sewage Pollution Prevention Certificate shall be issued, after an initial or renewal survey in accordance with the provisions of regulation 4 of this Annex, to any ship which is engaged in voyages to ports or offshore terminals under the jurisdiction of other Parties to the Convention. In the case of existing ships this requirement shall apply five years after the date of entry into force of this Annex.

2   Such Certificate shall be issued or endorsed either by the Administration or by any persons or organization* duly authorized by it. In every case, the Administration assumes full responsibility for the Certificate.

Regulation 6
Issue or endorsement of a Certificate by another Government

1   The Government of a Party to the Convention may, at the request of the Administration, cause a ship to be surveyed and, if satisfied that the provisions of this Annex are complied with, shall issue or authorize the issue of an International Sewage Pollution Prevention Certificate to the ship, and where appropriate, endorse or authorize the endorsement of that Certificate on the ship in accordance with this Annex.

2   A copy of the Certificate and a copy of the survey report shall be transmitted as soon as possible to the Administration requesting the survey.

3   A Certificate so issued shall contain a statement to the effect that it has been issued at the request of the Administration and it shall have the same

---

* Refer to the Guidelines for the authorization of organizations acting on behalf of the Administration, adopted by the Organization by resolution A.739(18), and the Specifications on the survey and certification functions of recognized organizations acting on behalf of the Administration, adopted by the Organization by resolution A.789(19).

force and receive the same recognition as the Certificate issued under regulation 5 of this Annex.

4 No International Sewage Pollution Prevention Certificate shall be issued to a ship which is entitled to fly the flag of a State which is not a Party.

Regulation 7
Form of Certificate

The International Sewage Pollution Prevention Certificate shall be drawn up in a form corresponding to the model given in the appendix to this Annex. If the language used is not English, French or Spanish, the text shall include a translation into one of these languages.

Regulation 8
Duration and validity of Certificate

1 An International Sewage Pollution Prevention Certificate shall be issued for a period specified by the Administration which shall not exceed five years.

2.1 Notwithstanding the requirements of paragraph 1 of this regulation, when the renewal survey is completed within three months before the expiry date of the existing Certificate, the new Certificate shall be valid from the date of completion of the renewal survey to a date not exceeding five years from the date of expiry of the existing Certificate.

2.2 When the renewal survey is completed after the expiry date of the existing Certificate, the new Certificate shall be valid from the date of completion of the renewal survey to a date not exceeding five years from the date of expiry of the existing Certificate.

2.3 When the renewal survey is completed more than three months before the expiry date of the existing Certificate, the new Certificate shall be valid from the date of completion of the renewal survey to a date not exceeding five years from the date of completion of the renewal survey.

3 If a Certificate is issued for a period of less than five years, the Administration may extend the validity of the Certificate beyond the expiry date to the maximum period specified in paragraph 1 of this regulation.

4 If a renewal survey has been completed and a new Certificate cannot be issued or placed on board the ship before the expiry date of the existing Certificate, the person or organization authorized by the Administration may endorse the existing Certificate and such a Certificate shall be accepted as valid for a further period which shall not exceed five months from the expiry date.

5   If a ship at the time when a Certificate expires is not in a port in which it is to be surveyed, the Administration may extend the period of validity of the Certificate but this extension shall be granted only for the purpose of allowing the ship to complete its voyage to the port in which it is to be surveyed and then only in cases where it appears proper and reasonable to do so. No Certificate shall be extended for a period longer than three months, and a ship to which an extension is granted shall not, on its arrival in the port in which it is to be surveyed, be entitled by virtue of such extension to leave that port without having a new Certificate. When the renewal survey is completed, the new Certificate shall be valid to a date not exceeding five years from the date of expiry of the existing Certificate before the extension was granted.

6   A Certificate issued to a ship engaged on short voyages which has not been extended under the foregoing provisions of this regulation may be extended by the Administration for a period of grace of up to one month from the date of expiry stated on it. When the renewal survey is completed, the new Certificate shall be valid to a date not exceeding five years from the date of expiry of the existing Certificate before the extension was granted.

7   In special circumstances, as determined by the Administration, a new Certificate need not be dated from the date of expiry of the existing Certificate as required by paragraph 2.2, 5 or 6 of this regulation. In these special circumstances, the new Certificate shall be valid to a date not exceeding five years from the date of completion of the renewal survey.

8   A Certificate issued under regulation 5 or 6 of this Annex shall cease to be valid in any of the following cases:

    .1   If the relevant surveys are not completed within the periods specified under regulation 4.1 of this Annex.

    .2   Upon transfer of the ship to the flag of another State. A new Certificate shall only be issued when the Government issuing the new Certificate is fully satisfied that the ship is in compliance with the requirements of regulations 4.4.1 and 4.4.2 of this Annex. In the case of a transfer between Parties, if requested within 3 months after the transfer has taken place, the Government of the Party whose flag the ship was formerly entitled to fly shall, as soon as possible, transmit to the Administration copies of the Certificate carried by the ship before the transfer and, if available, copies of the relevant survey reports.

Additional Information

Chapter 3 – Equipment and control of discharge

Regulation 9
Sewage systems

1   Every ship which, in accordance with regulation 2, is required to comply with the provisions of this Annex shall be equipped with one of the following sewage systems:

  .1   a sewage treatment plant which shall be of a type approved by the Administration, in compliance with the standards and test methods developed by the Organization,* or

  .2   a sewage comminuting and disinfecting system approved by the Administration. Such system shall be fitted with facilities to the satisfaction of the Administration, for the temporary storage of sewage when the ship is less than 3 nautical miles from the nearest land, or

  .3   a holding tank of the capacity to the satisfaction of the Administration for the retention of all sewage, having regard to the operation of the ship, the number of persons on board and other relevant factors. The holding tank shall be constructed to the satisfaction of the Administration and shall have a means to indicate visually the amount of its contents.

Regulation 10
Standard discharge connections

1   To enable pipes of reception facilities to be connected with the ship's discharge pipeline, both lines shall be fitted with a standard discharge connection in accordance with the following table:

---

* Reference is made to the international specifications for effluent standards, construction and testing of sewage treatment systems adopted by the Organization by resolution MEPC.2(VI) on 3 December 1976. For existing ships, national specifications are acceptable.

468

Standard dimensions of flanges for discharge connections

Description	Dimension
Outside diameter	210 mm
Inner diameter	According to pipe outside diameter
Bolt circle diameter	170 mm
Slots in flange	4 holes, 18 mm in diameter, equidistantly placed on a bolt circle of the above diameter, slotted to the flange periphery. The slot width to be 18 mm
Flange thickness	16 mm
Bolts and nuts: quantity and diameter	4, each of 16 mm in diameter and of suitable length

The flange is designed to accept pipes up to a maximum internal diameter of 100 mm and shall be of steel or other equivalent material having a flat face. This flange, together with a suitable gasket, shall be suitable for a service pressure of 6 kg/cm^2.

For ships having a moulded depth of 5 m and less, the inner diameter of the discharge connection may be 38 mm.

2   For ships in dedicated trades, i.e. passenger ferries, alternatively the ship's discharge pipeline may be fitted with a discharge connection which can be accepted by the Administration, such as quick-connection couplings.

Regulation 11
Discharge of sewage

1   Subject to the provisions of regulation 3 of this Annex, the discharge of sewage into the sea is prohibited, except when:

.1   the ship is discharging comminuted and disinfected sewage using a system approved by the Administration in accordance with regulation 9, paragraph 1.2 of this Annex at a distance of more than 3 nautical miles from the nearest land, or sewage which is not comminuted or disinfected at a distance of more than 12 nautical miles from the nearest land, provided that, in any case, the sewage that has been stored in holding tanks shall not be discharged instantaneously but at a moderate rate when the ship is *en route* and proceeding at not less than 4 knots; the rate of discharge shall be approved by the Administration based upon standards developed by the Organization; or

.2   the ship has in operation an approved sewage treatment plant which has been certified by the Administration to meet the operational requirements referred to in regulation 9, paragraph 1.1 of this Annex, and

.2.1   the test results of the plant are laid down in the ship's International Sewage Pollution Prevention Certificate; and

.2.2   additionally, the effluent shall not produce visible floating solids nor cause discoloration of the surrounding water.

2   The provisions of paragraph 1 shall not apply to ships operating in the waters under the jurisdiction of a State and visiting ships from other States while they are in these waters and are discharging sewage in accordance with such less stringent requirements as may be imposed by such State.

3   When the sewage is mixed with wastes or waste water covered by other Annexes of MARPOL 73/78, the requirements of those Annexes shall be complied with in addition to the requirements of this Annex.

Chapter 4 — Reception facilities

Regulation 12
Reception facilities

1   The Government of each Party to the Convention, which requires ships operating in waters under its jurisdiction and visiting ships while in its waters to comply with the requirements of regulation 11.1, undertakes to ensure the provision of facilities at ports and terminals of the reception of sewage, without causing delay to ships, adequate to meet the needs of the ships using them.

2   The Government of each Party shall notify the Organization, for transmission to the Contracting Governments concerned, of all cases where the facilities provided under this regulation are alleged to be inadequate.

Additional Information

Appendix to Annex IV

Form of Certificate

INTERNATIONAL SEWAGE POLLUTION PREVENTION CERTIFICATE

Issued under the provisions of the International Convention for the Prevention of Pollution from Ships, 1973, as modified by the Protocol of 1978 relating thereto, and as amended by resolution MEPC. . .(. . .), (hereinafter referred to as "the Convention") under the authority of the Government of:

. . . .

(full designation of the country)

by. . . . .

*(full designation of the competent person or organization
authorized under the provisions of the Convention)*

Particulars of ship[1]

Name of ship . . . .

Distinctive number or letters . . . .

Port of registry . . . .

Gross tonnage . . . .

Number of persons which the ship is certified to carry . . . .

IMO Number[2] . . . .

New/existing ship*

Date on which keel was laid or ship was at a similar stage of construction or, where applicable, date on which work for a conversion or an alteration or modification of a major character was commenced. . . . .

---

* Delete as appropriate.

THIS IS TO CERTIFY:

1. That the ship is equipped with a sewage treatment plant/comminuter/ holding tank and a discharge pipeline in compliance with regulations 9 and 10 of Annex IV of the Convention as follows:*

    1.1 Description of the sewage treatment plant*

    Type of sewage treatment plant . . . .

    Name of manufacturer . . . .

    The sewage treatment plant is certified by the Administration to meet the effluent standards as provided for in resolution MEPC.2(VI).

    1.2 Description of comminuter*

    Type of comminuter . . . .

    Name of manufacturer . . . .

    Standard of sewage after disinfection . . . .

    1.3 Description of holding tank equipment*

    Total capacity of the holding tank . . . . m^3

    Location . . . .

    1.4 A pipeline for the discharge of sewage to a reception facility, fitted with a standard connection.

2. The the ship has been surveyed in accordance with regulation 4 of Annex IV of the Convention.

3. That the survey shows that the structure, equipment, systems, fittings, arrangements and materials of the ship and the condition thereof are in all respects satisfactory and that the ship complies with the applicable requirements of Annex IV of the Convention.

This Certificate is valid until . . . .[3] subject to surveys in accordance with regulation 4 of Annex IV of the Convention.

Issued at. . . . .
(place of issue of Certificate)

. . . .
(date of issue)   *(signature of authorized official
                                        issuing the Certificate)*

(Seal or stamp of the authority, as appropriate)

---

* Delete as appropriate.

473

Additional Information

Endorsement to extend the Certificate if valid for less than 5 years where regulation 8.3 applies

The ship complies with the relevant provisions of the Convention, and this Certificate shall, in accordance with regulation 8.3 of Annex IV of the Convention, be accepted as valid until ....

Signed ....
(signature of authorized official)

Place ....

Date ....

(Seal or stamp of the authority, as appropriate)

Endorsement where the renewal survey has been completed and regulation 8.4 applies

The ship complies with the relevant provisions of the Convention, and this Certificate shall, in accordance with regulation 8.4 of Annex IV of the Convention, be accepted as valid until ....

Signed ....
(signature of authorized official)

Place ....

Date ....

(Seal or stamp of the authority, as appropriate)

Endorsement to extend the validity of the Certificate until reaching the port of survey or for a period of grace where regulation 8.5 or 8.6 applies

This Certificate shall, in accordance with regulation 8.5 or 8.6* of Annex IV of the Convention, be accepted as valid until ....

Signed ....
(signature of authorized official)

Place ....

Date ....

(Seal or stamp of the authority, as appropriate)

---

* Delete as appropriate.

[1] Alternatively, the particulars of the ship may be placed horizontally in boxes.

[2] In accordance with resolution A.600(15), IMO Ship Identification Number Scheme, this information may be included voluntarily.

[3] Insert the date of expiry as specified by the Administration in accordance with regulation 8.1 of Annex IV of the Convention. The day and the month of this date correspond to the anniversary date as defined in regulation 1.8 of Annex IV of the Convention.

7
Prospective amendments to Annex I

Resolution MEPC.95(46)

Amendments to the Annex of the Protocol of 1978 relating to the International Convention for the Prevention of Pollution from Ships, 1973

(Amendments to regulation 13G of Annex I to MARPOL 73/78 and to the Supplement to the IOPP Certificate)

adopted on 27 April 2001

THE MARINE ENVIRONMENT PROTECTION COMMITTEE,

RECALLING Article 38(a) of the Convention on the International Maritime Organization concerning the functions of the Marine Environment Protection Committee (the Committee) conferred upon it by international conventions for the prevention and control of marine pollution,

NOTING article 16 of the International Convention for the Prevention of Pollution from Ships, 1973 (hereinafter referred to as the "1973 Convention") and article VI of the Protocol of 1978 relating to the International Convention for the Prevention of Pollution from Ships, 1973 (hereinafter referred to as the "1978 Protocol") which together specify the amendment procedure of the 1978 Protocol and confer upon the appropriate body of the Organization the function of considering and adopting amendments to the 1973 Convention, as modified by the 1978 Protocol (MARPOL 73/78),

HAVING CONSIDERED the proposed amendments to regulation 13G of Annex I to MARPOL 73/38, which were approved by the forty-fifth session of the Committee and circulated in accordance with article 16(2)(a) of the 1973 Convention,

HAVING ALSO CONSIDERED the proposed amendments to the Supplement to the IOPP Certificate which are consequential amendments to the proposed amendments to regulation 13G of Annex I to MARPOL 73/78,

1.   ADOPTS, in accordance with article 16(2)(d) of the 1973 Convention, the amendments to regulation 13G of Annex I to MARPOL 73/78 and to

the Supplement to the IOPP Certificate, the text of which is set out at annex to the present resolution;

2.   DETERMINES, in accordance with article 16(2)(f)(iii) of the 1973 Convention, that the amendments shall be deemed to have been accepted on 1 March 2002 unless, prior to that date, not less than one-third of the Parties or Parties the combined merchant fleets of which constitute not less than 50 per cent of the gross tonnage of the world's merchant fleet have communicated to the Organization their objection to the amendments;

3.   INVITES the Parties to note that, in accordance with article 16(2)(g)(ii) of the 1973 Convention, the said amendments shall enter into force on 1 September 2002 upon their acceptance in accordance with paragraph 2 above;

4.   REQUESTS the Secretary-General, in conformity with article 16(2)(e) of the 1973 Convention, to transmit to all Parties to MARPOL 73/78 certified copies of the present resolution and the text of the amendments contained in the Annex; and

5.   REQUESTS FURTHER the Secretary-General to transmit to the Members of the Organization which are not Parties to MARPOL 73/78 copies of the present resolution and its annex.

Annex

Amendments to Annex I to MARPOL 73/78

1   *The existing text of regulation 13G is replaced by the following:*

"Regulation 13G
Prevention of oil pollution in the event of collision or stranding – Measures for existing tankers

(1)   This regulation shall:

    (a)   apply to oil tankers of 5,000 tons deadweight and above, which are contracted, the keels of which are laid, or which are delivered before the dates specified in regulation 13F(1) of this Annex; and

    (b)   not apply to oil tankers complying with regulation 13F of this Annex, which are contracted, the keels of which are laid, or are delivered before the dates specified in regulation 13F(1) of this Annex; and

(c) not apply to oil tankers covered by subparagraph (a) above which comply with regulation 13F(3)(a) and (b) or 13F(4) or 13F(5) of this Annex, except that the requirement for minimum distances between the cargo tank boundaries and the ship side and bottom plating need not be met in all respects. In that event, the side protection distances shall not be less than those specified in the International Bulk Chemical Code for type 2 cargo tank location and the bottom protection distances shall comply with regulation 13E(4)(b) of this Annex.

(2) For the purpose of this regulation:

(a) *Heavy diesel oil* means diesel oil other than those distillates of which more than 50% by volume distils at a temperature not exceeding 340°C when tested by the method acceptable to the Organization.[*]

(b) *Fuel oil* means heavy distillates or residues from crude oil or blends of such materials intended for use as a fuel for the production of heat or power of a quality equivalent to the specification acceptable to the Organization.[†]

(3) For the purpose of this regulation, oil tankers are divided into the following categories:

(a) *Category 1 oil tanker* means an oil tanker of 20,000 tons deadweight and above carrying crude oil, fuel oil, heavy diesel oil or lubricating oil as cargo, and of 30,000 tons deadweight and above carrying oil other than the above, which does not comply with the requirements for new oil tankers as defined in regulation 1(26) of this Annex;

(b) *Category 2 oil tanker* means an oil tanker of 20,000 tons deadweight and above carrying crude oil, fuel oil, heavy diesel oil or lubricating oil as cargo, and of 30,000 tons deadweight and above carrying oil other than the above, which complies with the requirements for new oil tankers as defined in regulation 1(26) of this Annex;

(c) *Category 3 oil tanker* means an oil tanker of 5,000 tons deadweight and above but less than that specified in subparagraph (a) or (b) of this paragraph.

(4) An oil tanker to which this regulation applies shall comply with the requirements of regulation 13F of this Annex not later than the anniversary of the date of delivery of the ship in the year specified in the following table:

---

[*] Refer to the American Society for Testing and Materials' Standard Test Method (Designation D86).

[†] Refer to the American Society for Testing and Materials' Specification for Number Four Fuel Oil (Designation D396) or heavier.

Category of oil tanker	Year
Category 1	2003 for ships delivered in 1973 or earlier 2004 for ships delivered in 1974 and 1975 2005* for ships delivered in 1976 and 1977 2006* for ships delivered in 1978, 1979 and 1980 2007* for ships delivered in 1981 or later
Category 2	2003 for ships delivered in 1973 or earlier 2004 for ships delivered in 1974 and 1975 2005 for ships delivered in 1976 and 1977 2006 for ships delivered in 1978 and 1979 2007 for ships delivered in 1980 and 1981 2008 for ships delivered in 1982 2009 for ships delivered in 1983 2010* for ships delivered in 1984 2011* for ships delivered in 1985 2012* for ships delivered in 1986 2013* for ships delivered in 1987 2014* for ships delivered in 1988 2015* for ships delivered in 1989 or later
Category 3	2003 for ships delivered in 1973 or earlier 2004 for ships delivered in 1974 and 1975 2005 for ships delivered in 1976 and 1977 2006 for ships delivered in 1978 and 1979 2007 for ships delivered in 1980 and 1981 2008 for ships delivered in 1982 2009 for ships delivered in 1983 2010 for ships delivered in 1984 2011 for ships delivered in 1985 2012 for ships delivered in 1986 2013 for ships delivered in 1987 2014 for ships delivered in 1988 2015 for ships delivered in 1989 or later

* Subject to compliance with the provisions of paragraph (7).

(5) Notwithstanding the provisions of paragraph (4) of this regulation:

    (a) in the case of a Category 2 or 3 oil tanker fitted with only double bottoms or double sides not used for the carriage of oil and extending to the entire cargo tank length or double hull spaces which are not used for the carriage of oil and extend to the entire cargo tank length, but which does not fulfil conditions for being exempted from the provisions of paragraph (1)(c) of this regulation, the Administration may allow continued operation of such a ship beyond the date specified in paragraph (4) of this regulation, provided that:

(i)   the ship was in service on 1 July 2001;

(ii)   the Administration is satisfied by verification of the official records that the ship complied with the conditions specified above;

(iii) the conditions of the ship specified above remain unchanged; and

(iv) such continued operation does not go beyond the date on which the ship reaches 25 years after the date of its delivery;

(b) in the case of a Category 2 or 3 oil tanker other than that referred to in subparagraph (a) of this paragraph which complies with the provisions of paragraph (6)(a) or (b) of this regulation, the Administration may allow continued operation of such a ship beyond the date specified in paragraph (4) of this regulation, provided that such continued operation shall not go beyond the anniversary of the date of delivery of the ship in 2017 or the date on which the ship reaches 25 years after the date of its delivery, whichever is the earlier date.

(6)   A Category 1 oil tanker of 25 years and over after the date of its delivery shall comply with either of the following provisions:

(a) wing tanks or double bottom spaces, not used for the carriage of oil and meeting the width and height requirements of regulation 13E(4), cover at least 30% of L_t, for the full depth of the ship on each side or at least 30% of the projected bottom shell area within the length L_t, where L_t is as defined in regulation 13E(2); or

(b) the tanker operates with hydrostatically balanced loading, taking into account the guidelines developed by the Organization.[*]

(7)   The Administration may allow continued operation of a Category 1 oil tanker beyond the anniversary of the date of delivery of the ship in 2005, and of a Category 2 oil tanker beyond the anniversary of the date of delivery of the ship in 2010, subject to compliance with the Condition Assessment Scheme adopted by the Marine Environment Protection Committee by resolution MEPC.94(46),[†] as may be amended, provided that such amendments shall be adopted, brought into force and take effect in accordance with the provisions of article 16 of the present Convention relating to amendment procedures applicable to an appendix to an Annex.

---

[*] Refer to the Guidelines for approval of alternative structural or operational arrangements, adopted by resolution MEPC.64(36); see appendix 8 to the Unified Interpretations of Annex I.

[†] See item 8 of this additional information.

(8) (a) The Administration of a State which allows the application of paragraph (5) of this regulation, or allows, suspends, withdraws or declines the application of paragraph (7) of this regulation, to a ship entitled to fly its flag shall forthwith communicate to the Organization for circulation to the Parties to the present Convention particulars thereof, for their information and appropriate action, if any.

      (b) A Party to the present Convention shall be entitled to deny entry of oil tankers operating in accordance with the provisions of paragraph (5) of this regulation into the ports or offshore terminals under its jurisdiction. In such cases, that Party shall communicate to the Organization for circulation to the Parties to the present Convention particulars thereof for their information."

Amendments to appendix II to Annex I to MARPOL 73/78

Amendments to the Supplement to the IOPP Certificate (Form B)

2 *The existing paragraph 5.8.4 is replaced by the following:*

"5.8.4 The ship is subject to regulation 13G and:

    .1 is required to comply with regulation 13F not later than . . . .   ☐

    .2 is so arranged that the following tanks or spaces are not used for the carriage of oil . . . .   ☐

    .3 is provided with the operational manual approved on .... in accordance with resolution MEPC.64(36)   ☐

    .4 is allowed to continue operation in accordance with regulation 13G(5)(a)   ☐

    .5 is allowed to continue operation in accordance with regulation 13G(5)(b)   ☐

    .4 is allowed to continue operation in accordance with regulation 13G(7)   ☐ "

8

Condition Assessment Scheme for amended regulation 13G* of Annex I

Resolution MEPC.94(46)

Condition Assessment Scheme

adopted on 27 April 2001

THE MARINE ENVIRONMENT PROTECTION COMMITTEE,

RECALLING Article 38(a) of the Convention on the International Maritime Organization concerning the functions of the Marine Environment Protection Committee (the Committee) conferred upon it by international conventions for the prevention and control of marine pollution,

RECALLING ALSO that, by resolution MEPC.52(32), the Committee adopted regulations 13F and 13G of Annex I to the Protocol of 1978 relating to the International Convention for the Prevention of Pollution from Ships, 1973 as amended (MARPOL 73/78), with a view to improving the requirements for the design and construction of oil tankers to prevent oil pollution in the event of collision or stranding,

HAVING ADOPTED, at its forty-sixth session, amendments to regulation 13G of Annex I to MARPOL 73/78 by resolution MEPC.95(46) to accelerate the phase-out of single-hull tankers in an effort to further enhance the protection of the marine environment,

NOTING that, in accordance with the revised regulation 13G of Annex I to MARPOL 73/78, an Administration may allow a Category 1 tanker to continue operating beyond the anniversary of the date of delivery of the ship in 2005 and a Category 2 tanker beyond the anniversary of the date of delivery of the ship in 2010, provided that the requirements of a Condition Assessment Scheme adopted by the Committee are complied with,

RECOGNIZING the need to provide the required Condition Assessment Scheme for the purposes of application of the revised regulation 13G of Annex I to MARPOL 73/78,

---

* See item 7 of this additional information.

HAVING CONSIDERED the draft Condition Assessment Scheme which was prepared by the MEPC Intersessional Working Group and further amended by the Committee at its forty-sixth session,

1.   ADOPTS the Condition Assessment Scheme, the text of which is set out at annex to the present resolution, with the understanding that the Model Survey Plan will be developed at MEPC 47 and will be made mandatory;

2.   REQUESTS the Secretary-General to transmit certified copies of the present resolution and the text of the Condition Assessment Scheme contained in the annex to all Parties to MARPOL 73/78;

3.   FURTHER REQUESTS the Secretary-General to transmit copies of the present resolution and its annex to Members of the Organization which are not Parties to MARPOL 73/78;

4.   INVITES the Maritime Safety Committee to note the Condition Assessment Scheme;

5.   URGES the Maritime Safety Committee to consider introducing and incorporating relevant elements and provisions of the Condition Assessment Scheme in the Guidelines on the Enhanced Programme of Inspections During Surveys of Bulk Carriers and Oil Tankers adopted by resolution A.744(18) as amended by resolution 2 of the 1997 SOLAS Conference, by resolution MSC.49(66) and by resolution MSC.105(73) when reviewing the Guidelines; and

6.   FURTHER URGES Parties to MARPOL 73/78 to:

  .1   transmit when a ship flying their flag is transferred under the flag of another Party to MARPOL 73/78, if they are requested by the latter Party to MARPOL 73/78 and for the purpose of ensuring the uniform and consistent implementation of the provisions of the Condition Assessment Scheme, copies of all documents and records relating to the assessment of the ship in question for compliance with the requirements of the Condition Assessment Scheme; and

  .2   accept, in the light of the fact that certain Category 1 oil tankers have to undergo the required CAS survey prior to 1 September 2002, valid Statements of Compliance issued pursuant to the provisions of the Condition Assessment Scheme following satisfactory completion of CAS surveys commenced prior to 1 September 2002.

Annex

Condition Assessment Scheme

1   PREAMBLE

1.1   The Condition Assessment Scheme (CAS) is intended to complement the requirements of Annex B of the Guidelines on the enhanced programme of inspections during surveys of bulk carriers and oil tankers (hereinafter called Enhanced Survey Programme), adopted by the Assembly of the International Maritime Organization by resolution A.744(18), as amended. The CAS is to verify that the structural condition of single-hull oil tankers at the time of survey is acceptable and, provided subsequent periodical surveys are satisfactorily completed and effective maintenance is carried out by the ship's operator, will continue to be acceptable for a continued period of operation, as indicated in the Statement of Compliance.

1.2   The requirements of the CAS include enhanced and transparent verification of the reported structural condition of the ship and verification that the documentary and survey procedures have been properly carried out and completed.

1.3   The Scheme requires that compliance with the CAS is assessed during the Enhanced Survey Programme of Inspections concurrent with intermediate or renewal surveys currently required by resolution A.744(18), as amended.

1.4   The CAS does not specify structural standards in excess of the provisions of other International Maritime Organization conventions, codes and recommendations.

1.5   The CAS has been developed on the basis of the requirements of resolution A.744(18), as amended, which were known* at the time of the adoption of the CAS. It is the intention to update the CAS as and when the need arises following amendments to resolution A.744(18), as amended.

2   PURPOSE

The purpose of the Condition Assessment Scheme is to provide an international standard to meet the requirements of regulation 13G(7) of Annex I of the International Convention for the Prevention of Pollution from Ships, 1973, as modified by the Protocol of 1978 relating thereto, as amended by resolution MEPC.95(46).

---

* Assembly resolution A.744(18) as amended by resolution 2 of the 1997 SOLAS Conference, by resolution MSC.49(66) and by resolution MSC.105(73).

483

3 DEFINITIONS

For the purpose of the CAS, unless expressly provided otherwise:

3.1 *MARPOL 73/78* means the Protocol of 1978 relating to the International Convention for the Prevention of Pollution from Ships, 1973, as amended.

3.2 *Regulation* means the regulations contained in Annex I of MARPOL 73/78.

3.3 *Resolution A.744(18), as amended* means the Guidelines on the Enhanced Programme of Inspections during Surveys of Bulk Carriers and Oil Tankers adopted by the Assembly of the International Maritime Organization by resolution A.744(18), as amended by resolution 2 of the 1997 SOLAS Conference and by resolutions MSC.49(66) and MSC.105(73).

3.4 *Recognized Organization (RO)* means an organization recognized by the Administration to perform the surveys in accordance with the provisions of regulation 4(3) of Annex I of MARPOL 73/78.*

3.5 *Administration* means the Government of the State as defined in article 2(5) of MARPOL 73/78.

3.6 *Category 1 oil tanker* means an oil tanker of 20,000 tons deadweight and above carrying crude oil, fuel oil, heavy diesel oil or lubricating oil as cargo, and of 30,000 tons deadweight and above carrying oil other than the above, which does not comply with the requirements for new oil tankers as defined in regulation 1(26) of Annex I of MARPOL 73/78.

3.7 *Category 2 oil tanker* means an oil tanker of 20,000 tons deadweight and above carrying crude oil, fuel oil, heavy diesel oil or lubricating oil as cargo, and of 30,000 tons deadweight and above carrying oil other than the above, which complies with the requirements for new oil tankers as defined in regulation 1(26) of Annex I of MARPOL 73/78.

3.8 *Company* means the owner of the ship or any other organization or person such as the manager or the bareboat charterer, who has assumed the responsibility for the operation of the ship from the owner of the ship and who, on assuming such responsibility, has agreed to take over all duties and responsibilities imposed by the International Safety Management (ISM) Code.

3.9 *Substantial corrosion* means an extent of corrosion such that the assessment of the corrosion pattern indicates wastage in excess of 75% of the allowable margins, but within acceptable limits.

---

* Under regulation XI/1 of SOLAS 74, as amended, resolutions A.739(18) and A.789(19) are applicable to Recognized Organizations.

3.10 *GOOD condition* means a coating condition with only minor spot rusting.

3.11 *Thickness Measurement (TM) Firm* means a qualified company certified by a RO in accordance with the principles stipulated in annex 7 to Annex B to resolution A.744(18), as amended.

3.12 *Critical structural areas* are locations which have been identified from calculations to require monitoring or from the service history of the subject ship or from similar or sister ships to be sensitive to cracking, buckling or corrosion which would impair the structural integrity of the ship.

3.13 *Suspect areas* are locations showing substantial corrosion and/or are considered by the attending surveyor to be prone to rapid wastage.

3.14 *Organization* means the International Maritime Organization.

4 GENERAL PROVISIONS

4.1 The Administration shall issue, or cause to be issued, detailed instructions to the RO which shall ensure that the CAS surveys are carried out in accordance with the provisions of sections 5 through 10 of this Scheme.

4.2 Nothing in this Scheme shall prevent an Administration from carrying out the CAS surveys itself, provided that such surveys are at least as effective as those prescribed in sections 5 through 10 in this Scheme.

4.3 The Administration shall require Category 1 and Category 2 oil tankers flying its flag to remain out of service during the periods referred to in paragraphs 5.1.1 and 5.1.2 respectively, until these oil tankers are issued with a valid Statement of Compliance.

5 APPLICATION, SCOPE AND TIMING

5.1 Application

The requirements of the CAS apply to:

.1 Category 1 oil tankers, as defined in section 3, where authorization is requested for continued service beyond the anniversary of the date of delivery of the ship in 2005, through to the date as specified in the schedule indicated for compliance with the double-hull requirements of regulation 13F, detailed in regulation 13G.

Additional Information

.2 Category 2 oil tankers, as defined in section 3, where authorization is requested for continued service beyond the anniversary of the date of delivery of the ship in 2010, through to the date as specified in schedule indicated for compliance with the double-hull requirements of regulation 13F, detailed in regulation 13G.

5.2 Scope of the CAS

The CAS shall apply to surveys of the hull structure in way of cargo tanks, pump-rooms, cofferdams, pipe tunnels, void spaces within the cargo area and all ballast tanks.

5.3 Timing

5.3.1 The first CAS survey shall be aligned to the Enhanced Survey Programme of Inspection and shall be carried out concurrent with the scheduled intermediate or renewal survey due prior to the anniversary of the date of delivery of the ship in 2005 for Category 1 oil tankers and prior to the anniversary of the date of delivery of the ship in 2010 for Category 2 oil tankers.

5.3.2 Any subsequent CAS surveys, required for the renewal of the Statement of Compliance, shall be carried out concurrently with the intermediate or renewal survey which has to be completed by the date of expiry of the Statement of Compliance.

5.3.3 Notwithstanding the above, the Company may, with the agreement of the Administration, opt to carry out the first CAS survey at a different time from that of the due survey referred to above, provided that all the requirements of the CAS are complied with.

6   SURVEY PLANNING REQUIREMENTS

6.1   Preparations for the CAS survey

6.1.1   *General procedures*

6.1.1.1   Early and detailed planning to identify areas of potential risk is a prerequisite for the successful and timely completion of the CAS. The following sequence of events shall be observed.

6.1.1.2   Notification from the Company to the Administration and to the RO of its intention to proceed with the CAS shall be submitted not less than 8 months prior to the planned commencement of the CAS survey.

6.1.1.3   Upon receipt of such notification the RO shall:

    .1   issue to the Company the Survey Planning Questionnaire (see appendix 2) not later than 7 months prior to the planned commencement of the CAS survey; and

    .2   advise the Company whether there have been any changes to the maximum acceptable structural corrosion diminution levels applicable to the ship.

6.1.1.4   The Company shall complete and return the Survey Planning Questionnaire to the RO not less than 5 months prior to the planned commencement of the CAS survey. A copy of the completed questionnaire shall be forwarded by the Company to the Administration.

6.1.1.5   The Survey Plan for the CAS shall be completed and submitted in signed order by the Company to the RO not less than 2 months prior to the planned commencement of the CAS survey. A copy of the Survey Plan for the CAS shall be forwarded by the Company to the Administration.

6.1.1.6   In special circumstances, such as re-activation from lay-up or unexpected events such as an extended stoppage period for hull or machinery damage, the Administration may, on a case-by-case basis, relax the timeframe, outlined in 6.1.1.2 to 6.1.1.5, for commencement of CAS procedures.

6.1.1.7   Such relaxation shall, at all times, be subject to the RO having sufficient time to complete the CAS survey and for the Administration to review the CAS Final Report and issue the Statement of Compliance prior to the dates referred to in 5.1.

6.1.2 *Survey Plan for the CAS*

6.1.2.1   The Survey Plan for the CAS shall be developed by the Company in co-operation with the RO. The Administration may participate in the development of the Survey Plan, if it deems necessary. The RO shall be fully satisfied that the Survey Plan complies with the requirements of 6.2.2 prior to the CAS survey being commenced. The CAS survey shall not commence unless and until the Survey Plan has been agreed.

6.1.2.2   The Survey Planning Questionnaire shall be drawn up based on the format set out in appendix 2.

6.2   Survey Plan documentation

6.2.1   In developing the Survey Plan, the following documentation shall be collected and reviewed with a view to identifying tanks, areas and structural elements to be examined:

    .1   basic ship information and survey status;

Additional Information

.2   main structural plans of cargo and ballast tanks (scantling drawings), including information regarding use of high-tensile steels (HTS);

.3   Condition Evaluation Report, according to annex 9 of Annex B of resolution A.744(18), as amended, and, where relevant, any previous CAS Final Reports;

.4   thickness measurement reports;

.5   relevant previous damage and repair history;

.6   relevant previous survey and inspection reports from both the RO and the Company;

.7   cargo and ballast history for the last 3 years, including carriage of cargo under heated conditions;

.8   details of the inert-gas plant and tank cleaning procedures as indicated in the Survey Planning Questionnaire;

.9   information and other relevant data regarding conversion or modification of the ship's cargo and ballast tanks since the time of construction;

.10   description and history of the coating and corrosion protection system (including anodes and previous class notations), if any;

.11   inspections by the Company's personnel during the last 3 years with reference to:

     .1   structural deterioration in general;

     .2   leakages in tank boundaries and piping;

     .3   condition of the coating and corrosion protection system (including anodes), if any;

.12   information regarding the relevant maintenance level during operation, including:

     .1   port State control reports of inspection containing hull-related deficiencies;

     .2   Safety Management System non-conformities relating to hull maintenance, including the associated corrective action(s); and

.13   any other information that will help identify suspect areas and critical structural areas.

6.2.2   The Survey Plan shall include relevant information so as to enable the successful and efficient execution of the CAS survey and shall set out the requirements with respect to close-up surveys and thickness measurements. The Survey Plan shall include:

     .1   basic ship information and particulars;

.2   main structural plans of cargo and ballast tanks (scantling drawings), including information regarding use of high-tensile steels (HTS);

.3   arrangement of tanks;

.4   list of tanks with information on their use, extent of coatings and corrosion protection systems;

.5   conditions for survey (e.g. information regarding tank cleaning, gas-freeing, ventilation, lighting, etc.);

.6   provisions and methods for access to structures;

.7   equipment for surveys;

.8   identification of tanks and areas for the close-up survey;

.9   identification of tanks for tank testing, as per annex 3 of Annex B of resolution A.744(18), as amended;

.10   identification of areas and sections for thickness measurement;

.11   identification of the Thickness Measurement (TM) Firm;

.12   damage experience related to the ship in question; and

.13   Critical structural areas and suspect areas, where relevant.

Additional Information

6.3   Documentation on board

6.3.1   The Company shall ensure that, in addition to the agreed Survey Plan, all other documents used in the development of the Survey Plan referred to in 6.2.1 are available on board at the time of the CAS survey.

6.3.2   Prior to the commencement of any part of the CAS survey, the attending surveyor(s) shall examine and ascertain the completeness of the on-board documentation and shall review its contents with a view to ensuring that the Survey Plan remains relevant.

7   CAS SURVEY REQUIREMENTS

7.1   General

7.1.1   Prior to the commencement of any part of the CAS survey, a meeting shall be held between the attending surveyor(s), the Company's representative(s) in attendance, the TM Firm operator (as applicable) and the master of the ship for the purpose of ascertaining that all the arrangements envisaged in the Survey Plan are in place, so as to ensure the safe and efficient execution of the survey work to be carried out.

7.1.2   The CAS survey shall be carried out by not less than two qualified exclusive surveyors of the RO. A qualified surveyor of the RO shall attend on board during the taking of the thickness measurements for the purpose of controlling the process.

7.1.3   The RO shall designate the surveyor(s) and any other personnel who will be engaged in the CAS survey of each vessel and shall keep records to this end. A qualified surveyor(s) shall have documented experience in carrying out intermediate or renewal surveys in accordance with the Enhanced Survey Programme of Inspection for tankers. In addition, all RO personnel to be assigned duties in connection with the CAS shall complete, prior to the assignment of such duties, an appropriate training and familiarization programme to enable the RO to ensure the consistent and uniform application of the CAS. The Administration shall require the RO to keep records of the qualifications and experience of the surveyors and of other personnel assigned to carry out work for the CAS. The Administration shall require the RO to monitor the performance of the personnel who have carried out or have been engaged in any CAS work and to keep records to this end.

7.1.4   When the CAS survey is split between survey stations, a list of the items examined and an indication of whether the CAS survey has been completed shall be made available to the attending surveyors at the next survey station prior to continuing the CAS survey.

7.1.5   Whenever the attending surveyors are of the opinion that repairs are required, each item to be repaired shall be identified in a numbered list. Whenever repairs are carried out, details of the repairs effected shall be reported by making specific reference to relevant items in the numbered list.

7.1.6   Whenever the attending surveyors are of the opinion that it is acceptable to defer hull repairs beyond the due date previously assigned, such a decision shall not be left to the sole discretion of the attending surveyors. The RO Headquarters shall be consulted in such circumstances and shall give specific approval to the recommended action.

7.1.7   The CAS survey is not complete unless all recommendations/conditions of class which relate to hull structures under review by the CAS survey have been rectified to the satisfaction of the RO.

7.2   Extent of overall and close-up surveys

7.2.1   *Overall survey*

An overall survey of all spaces set out in 5.2 shall be carried out at the CAS survey.

7.2.2   *Close-up survey*

The requirements for close-up surveys at the CAS survey are set out in the table below.

Table 7.2.2 – Close-up survey requirements

All web frame rings, in all ballast tanks (see note 1)
All web frame rings, in a cargo wing tank (see note 1)
A minimum of 30% of all web frame rings, in each remaining cargo wing tank (see note 1)
All transverse bulkheads, in all cargo and ballast tanks (see note 2)
A minimum of 30% of deck and bottom transverses, including adjacent structural members, in each cargo centre tank
Additional complete transverse web frame rings or deck and bottom transverse including adjacent structural members as considered necessary by the attending surveyor

Notes:

1   *Complete transverse web frame ring including adjacent structural members.*

2   *Complete transverse bulkhead, including girder and stiffener systems and adjacent members.*

7.2.3   The attending surveyors may extend the scope of the close-up survey as considered necessary, taking into account the Survey Plan, the condition of the spaces under survey, the condition of the corrosion-prevention system and also the following:

.1   any information that may be available on critical structural areas;

.2   tanks which have structures with reduced scantlings in association with a corrosion-prevention system approved by the RO.

7.2.4   For areas in tanks where coatings are found to be in GOOD condition, the extent of close-up surveys according to 7.2.2 may be specially considered by the RO. However, sufficient close-up surveys shall be carried out, in all cases, to confirm the actual average condition of the structure and to note the maximum observed diminution of the structure.

7.3   Extent of thickness measurements

7.3.1   The thickness measurements shall be recorded using the tables contained in appendix 2 of annex 10 of Annex B of resolution A.744(18), as amended. It is recommended that these records be kept in an electronic medium.

Additional Information

7.3.2　The thickness measurements shall be carried out either prior to or, to the maximum extent possible, concurrently with the close-up survey.

7.3.3　The minimum requirements for thickness measurements for the CAS surveys shall be those set out in the table below:

Table 7.3.3 – Thickness measurements requirements

1	Within the cargo area:
	.1　Each deck plate
	.2　Three transverse sections
	.3　Each bottom plate
2	Measurements of structural members subject to close-up survey according to 7.2.2, for general assessment and recording of corrosion pattern
3	Suspect areas
4	Selected wind and water strakes outside the cargo area
5	All wind and water strakes within the cargo area
6	Internal structure in the fore and aft peak tanks
7	All exposed main deck plates outside the cargo area and all exposed first tier superstructure deck plates

7.3.4　Where substantial corrosion is found, the extent of the thickness measurements shall be increased in accordance with annex 4 of Annex B of resolution A.744(18), as amended.

7.3.5　In addition, the thickness measurements may be extended as considered necessary by the attending surveyors.

7.3.6　For areas in tanks where coatings are found to be in GOOD condition, the extent of thickness measurements, according to paragraph 7.3.3, may be specially considered by the RO. However, sufficient thickness measurements shall be taken, in all cases, to confirm the actual average condition and the maximum observed diminution of the structure.

7.3.7　The thickness measurement to be taken shall be sufficient to enable the reserve strength calculations in accordance with annex 12 of Annex B of resolution A.744(18), as amended.

7.3.8　Transverse sections shall be chosen where the maximum diminutions are expected to occur or are revealed from deck plating thickness measurements. At least one transverse section shall include a ballast tank within 0.5L amidships.

8   ACCEPTANCE CRITERIA

The acceptance criteria for the CAS shall be those set out in resolution A.744(18), as amended.

9   CAS SURVEY REPORTS

9.1   A survey report shall be completed for the CAS survey. The report shall indicate the date, location (place), and where relevant, whether or not the CAS survey was carried out in dry-dock, afloat or at sea. When the CAS survey is split between different survey stations, a report shall be made for each portion of the CAS survey.

9.2   Survey records relating to the CAS survey, including actions taken, shall form an auditable documentary trail, which shall be made available to the Administration, if requested.

9.3   In addition, the following shall be included in each CAS survey report:

    .1   Extent of the survey:

        .1   identification of the spaces where an overall survey has been carried out;

        .2   identification of location, in each space, where a close-up survey has been carried out, together with the means of access used; and

        .3   identification of the spaces, and locations in each space, where thickness measurements have been carried out; and

    .2   Results of the survey:

        .1   extent and condition of coating in each space. Identification of spaces fitted with anodes and the overall condition of the anodes;

        .2   structural condition reporting for each space, which shall include information on the following, as applicable:

            .1   corrosion (location and type of corrosion, such as grooving, pitting, etc.);

            .2   cracks (location, description and extent);

            .3   buckling (location, description and extent);

            .4   indents (location, description and extent); and

            .5   areas of substantial corrosion; and

    .3   Actions taken with respect to findings:

        .1   details of repairs completed on structural members in identified spaces, including the repair method and extent; and

Additional Information

493

.2   list of items to be kept under observation for planning future inspections and surveys, including any thickness measurements.

9.4   Where no defects are found, this shall be stated in the report for each space.

9.5   The narrative report shall be supplemented by photographs showing the general condition of each space, including representative photographs or sketches of any of the above reported items.

9.6   The thickness measurement report shall be verified and endorsed by the attending surveyor.

9.7   The attending surveyors shall sign the CAS survey report.

10   CAS FINAL REPORT TO THE ADMINISTRATION

10.1 Review of the CAS by the RO

10.1.1   The RO Headquarters shall carry out a verification review of the CAS survey reports, the documents, photographs and other records relating to the CAS, as specified in section 9, for the purpose of ascertaining and confirming that the requirements of the CAS have been met.

10.1.2   The RO reviewing personnel shall not be engaged in any way whatsoever with the CAS survey under review.

10.2 CAS Final Report to the Administration

10.2.1   The RO shall prepare a CAS Final Report to the Administration upon completion of the CAS survey and following the review of the CAS survey reports by the RO's Headquarters, as specified in paragraph 10.1.1.

10.2.2   The CAS Final Report shall be submitted by the RO to the Administration without delay and in any case not later than 2 months prior to the date the ship is required to be issued with a Statement of Compliance.

10.2.3   The CAS Final Report shall, at least, include:

.1   the following general particulars:
Ship's name
IMO number
Flag State
Port of registry
Gross tonnage
Deadweight (metric tonnes)
Summer load line draught
Date of delivery
Category of ship

Date for compliance with regulation 13F
Company
Report identification reference

.2 a summary as to where, when, by whom and how the CAS survey was carried out;

.3 a statement identifying all survey documentation, including the Survey Plan;

.4 a statement as to the condition of the corrosion-prevention system(s) applied to the spaces;

.5 a statement identifying all thickness measurement reports;

.6 a summary of the findings of the overall surveys;

.7 a summary of the findings of the close-up surveys;

.8 a summary of the hull repairs carried out;

.9 an identification, together with the location, the extent and the condition, of all areas with substantial corrosion;

.10 a summary of the results of the evaluation of the thickness measurements, including identification of the areas and sections where thickness measurements were carried out;

.11 an evaluation of the structural strength of the vessel and an assessment of compliance with the acceptance criteria set out in section 8;

.12 a statement as to whether all the applicable requirements of the CAS have been met;

.13 a recommendation to the Administration as to whether the ship should be allowed to continue operating until the date envisaged in regulation 13G for compliance with the requirements of regulation 13F or for the period of validity of the CAS, if earlier; and

.14 conclusions.

11   VERIFICATION OF THE CAS BY THE ADMINISTRATION

11.1   In addition to any instructions the Administration may have issued to the RO authorized to carry out surveys under the Enhanced Survey Programme on its behalf, the Administration shall issue instructions to the RO and to Companies operating Category 1 and Category 2 oil tankers flying its flag, so that the Administration is able to monitor the performance of and verify compliance with the CAS.

Additional Information

11.2 The Administration, for the purpose of ensuring uniform and consistent implementation of the CAS, shall establish, at least, procedures through which it will:

.1 give effect to the requirements of the CAS;

.2 monitor the CAS work the RO is carrying out on its behalf;

.3 review the CAS Final Report;

.4 review cases of ships which have been submitted for the CAS re-assessment; and

.5 issue the Statement of Compliance.

11.3 The Administration shall review the CAS Final Report prior to the issue of the Statement of Compliance, shall record and document the findings and conclusions of the review and its decision as to the acceptance or rejection of the CAS Final Report and shall produce a Review Record.

11.4 The Administration shall ensure that any persons assigned to monitor the execution of the CAS or to review a CAS Final Report:

.1 are adequately qualified and experienced to the satisfaction of the Administration;

.2 are under the direct control of the Administration; and

.3 have no connection whatsoever with the RO which carried out the CAS survey under review.

12 RE-ASSESSMENT OF SHIPS FOLLOWING FAILURE TO MEET THE REQUIREMENTS OF THE CAS

12.1 A ship which, in the opinion of the Administration, has failed to meet the requirements of the CAS may be submitted for the CAS re-assessment. In such a case the grounds on which the Administration declined the issue of a Statement of Compliance to the ship shall be addressed and dealt with and the remedial actions shall, thereafter, be reviewed for the purpose of ascertaining whether the requirements of the CAS have been complied with.

12.2 Such re-assessment, as a rule, shall be carried out by the RO and by the Administration who carried out the previous CAS.

12.3 If a ship which has failed to obtain a Statement of Compliance changes flag, the new Administration shall, in accordance with the provisions of regulation 8(3), request the previous Administration to transmit to them copies of the CAS documentation relating to that ship for the purpose of ascertaining whether the grounds on the basis of which the previous Administration declined the issue to the ship of a Statement of Compliance are dealt with and that the CAS is implemented in a consistent and uniform manner.

12.4 As a rule, the CAS re-assessment shall be carried out as soon as possible and in any case, subject to the provisions of paragraph 5.3, not later than 6 months following the date on which the Administration has made the decision to decline the issue of a Statement of Compliance to the ship.

13 STATEMENT OF COMPLIANCE

13.1 The Administration shall, in accordance with its procedures, issue to each ship which completes the CAS to the satisfaction of the Administration, a Statement of Compliance.

13.2 The Statement of Compliance shall be drawn up in the official language of the issuing Administration in a form corresponding to the model given in appendix 1. If the language used is neither English, French or Spanish, the text shall include a translation into one of these languages.

13.3 The original of the Statement of Compliance shall be placed on board the ship as a supplement to the ship's International Oil Pollution Prevention Certificate.

13.4 In addition, a copy of the CAS Final Report which was reviewed by the Administration for the issue of the Statement of Compliance and a copy of the Review Record, specified in paragraph 11.3, shall be placed on board to accompany the Statement of Compliance.

13.5 A certified copy of the Statement of Compliance and a copy of the Review Record, specified in paragraph 11.3, shall be forwarded by the Administration to the RO and shall be kept together with the CAS Final Report.

13.6 The Statement of Compliance shall be valid, following the completion of the CAS survey, until the earlier date of either:

   .1   the earlier date by which the ship is required to complete:

      .1   an intermediate survey, in accordance with regulation 4(1)(c); or

      .2   a renewal survey, in accordance with regulation 4(1)(b);

   or

   .2   the date by which the vessel is required, in accordance with regulation 13G, to comply with the requirements of regulation 13F.

13.7 If the Statement of Compliance expires prior to the date by which the ship is required, in accordance with regulation 13G, to comply with the requirements of regulation 13F, that ship, in order to continue operating

Additional Information

after the expiry of its Statement of Compliance, shall carry out a renewal CAS survey in accordance with the requirements of sections 5 to 10.

13.8 The Administration may consider and declare that the Statement of Compliance of a ship remains valid and in full force and effect if:

.1 the ship is transferred to a RO other than the one that submitted the CAS Final Report that was reviewed and accepted for the issue of the Statement of Compliance; or

.2 the ship is operated by a Company other than the one that was operating the ship at the time of the completion of the CAS survey

provided the period of validity and the terms and conditions for the issue of the Statement of Compliance in question remain those adopted by the Administration at the time of the issue of the Statement of Compliance.

13.9 If a ship with a valid Statement of Compliance is transferred to the flag of another Party, the new Administration may issue to that ship a new Statement of Compliance on the basis of the Statement of Compliance issued by the previous Administration, provided that the new Administration:

.1 requests and receives from the previous Administration, in accordance with regulation 8(3), copies of all the CAS documentation relating to that ship which the previous Administration has used for the issue or renewal and the maintenance of the validity of the Statement of Compliance the ship was issued with at the time of the transfer;

.2 establishes that the RO which submitted the CAS Final Reports to the previous Administration is an RO authorized to act on its behalf;

.3 reviews the documentation referred to in subparagraph .1 and is satisfied that the requirements of the CAS are met; and

.4 limits the period and the terms and conditions of validity of the Statement of Compliance to be issued to those established by the previous Administration.

13.10 The Administration shall:

.1 suspend and/or withdraw the Statement of Compliance of a ship if it no longer complies with the requirements of the CAS; and

.2 withdraw the Statement of Compliance of a ship if it is no longer entitled to fly its flag.

14 COMMUNICATION OF INFORMATION TO THE ORGANIZATION

14.1 The Administration shall communicate to the Organization:

.1 particulars of the Statements of Compliance issued;

.2 details of the suspension or withdrawal of the Statements of Compliance issued; and

.3 particulars of the ships to which it has declined the issue of a Statement of Compliance and reasons thereof.

14.2 The Organization shall circulate the aforementioned information to all Parties to MARPOL 73/78 and shall maintain an electronic database containing the aforesaid information, accessible only to Parties to MARPOL 73/78.

Additional Information

Appendix 1

Form of Statement of Compliance

STATEMENT OF COMPLIANCE

Issued under the provisions of the Condition Assessment Scheme (CAS) adopted by the Organization by resolution MEPC.94(46) under the authority of the Government of:

....

(full designation of the country)

Particulars of ship

Name of ship ....

Distinctive number or letters ....

Port of registry ....

Gross tonnage ....

Deadweight of ship (metric tons) ....

IMO number ....

Category of tanker ....

THIS IS TO CERTIFY:

1. That the ship has been surveyed in accordance with the requirements of CAS (resolution MEPC.94(46));

2. That the survey showed that the structural condition of the ship is in all respects satisfactory and the ship complied with the requirements of the CAS.

This Statement of Compliance is valid until ....

Issued at ....

(place of issue)

....

(date of issue)   *(signature of duly authorized official*
                                         issuing the Statement)

(Seal or stamp of the authority,
as appropriate)

Appendix 2

Survey Planning Questionnaire

The following information will enable the Company, in co-operation with the RO, to develop a Survey Plan complying with the requirements of the CAS.

It is essential that the Company provides, when completing the present questionnaire, up-to-date information.

The present questionnaire, when completed, shall provide all information and material required by the CAS.

Particulars

Ship's name:
IMO number:
Flag State:
Port of registry:
Gross tonnage:
Deadweight (metric tonnes):
Summer load line draught:
Date of delivery:
Category of ship:
Date for compliance with regulation 13F:
Company:
Report identification reference:

Information on access provision for close-up surveys and thickness measurement:

The Company is requested to indicate, in the table below, the means of access to the structures subject to close-up survey and thickness measurement.

A *close-up* survey is an examination where the details of structural components are within the close visual inspection range of the attending surveyor, i.e. preferably within reach of hand.

Additional Information

Space \ Access	Temporary staging	Rafts	Ladders	Direct access	Other means (please specify)
Forepeak					
Wing tanks — Under deck					
Wing tanks — Side shell					
Wing tanks — Bottom transverse					
Wing tanks — Longitudinal bulkhead					
Wing tanks — Transverse bulkhead					
Centre tanks — Under deck					
Centre tanks — Bottom transverse					
Centre tanks — Transverse bulkhead					

Tank cleaning procedures:
Indicate the frequency of the tank washing, especially in way of uncoated tanks:
Washing medium used: Crude oil: Yes/No Heated seawater: Yes/No Other medium (specify):

Inert gas system installed: Yes/No
Indicate average oxygen content during inerting:
Details of use of the inert gas plant:

Cargo history for the last 3 years together with indication as to whether cargo was heated

Ballast history for the last 3 years

Additional Information

Inspections by the Company

Using a format similar to that of the table below (which is given as an example), the Company should provide details of the results of their inspections, for the last 3 years – in accordance with the requirements of resolution A.744(18), as amended, and of the CAS – on all CARGO and BALLAST tanks and VOID spaces within the cargo area.

Spaces (include frame numbers and **p** or **s**)	Corrosion protection (1)	Coating extent (2)	Coating condition (3)	Structural deterioration (4)	Tank history (5)
Cargo centre tanks					
Cargo wing tanks					
Slop tanks					
Ballast tanks*					
Aft peak					
Forepeak					
Miscellaneous spaces:					

* Indicate tanks which are used for oil/ballast.

Notes:
(1) HC = hard coating; SC = soft coating;
    A = anodes; NP = no protection
(2) U = upper part; M = middle part;
    L = lower part; C = complete
(3) G = good; F = fair; P = poor; RC = recoated
(4) N = no findings recorded;
    Y = findings recorded, description of findings is to be attached to the questionnaire
(5) D & R = Damage & Repair; L = Leakages;
    CV = Conversion; CPS = Corrosion protection system (reports to be attached)

Company: ....

Name/signature: ....

Date: ....

Reports of port State control inspections

List the reports of port State control inspections containing hull-related deficiencies and relevant information on the deficiencies:

Safety Management System

List non-conformities related to hull maintenance, including the associated corrective actions:

Name of the Thickness Measurement (TM) Firm

9

Resolution MEPC.82(43)

Guidelines for monitoring the world-wide average sulphur content of residual fuel oils supplied for use on board ships

adopted on 1 July 1999

THE MARINE ENVIRONMENT PROTECTION COMMITTEE,

RECALLING Article 38(a) of the Convention on the International Maritime Organization concerning the function of the Committee conferred upon it by international conventions for the prevention and control of marine pollution,

BEING AWARE that the Conference of Parties to MARPOL 73/78 was held in September 1997 and that the Conference adopted the Protocol of 1997 to amend the International Convention for the Prevention of Pollution from Ships, 1973, as modified by the Protocol of 1978 relating thereto, which sets out in its annex the new Annex VI, Regulations for the Prevention of Air Pollution from Ships, and eight conference resolutions including resolution 4, which provides for the development of guidelines for monitoring the world-wide average sulphur content of residual fuels oil supplied for use on board ships,

RECOGNIZING regulation 14 of Annex VI to MARPOL 73/78 which requires Parties to Annex VI of MARPOL 73/78 to monitor world-wide average sulphur content of residual fuel oil supplied for use on board ships taking into account guidelines to be developed by the Organization,

1.   ADOPTS the Guidelines for Monitoring the World-wide Average Sulphur Content of Residual Fuel Oils Supplied for Use on Board Ships as set out in the annex to the present resolution;

2.   URGES Member Governments and interested organizations to make available the resources and expertise necessary for the implementation of these guidelines.

Annex

Guidelines for monitoring the world-wide average sulphur content of residual fuel oils supplied for use on board ships

Additional Information

Preface

1   The primary objective of the guidelines is to establish an agreed method to monitor the average sulphur content of residual fuel oils supplied for use on board ships. A further objective of the guidelines is to re-open the discussion in MEPC on measures to reduce SO_x emissions from ships, should the average sulphur level in fuels, calculated on the basis of these guidelines, show a sustained increase.

Introduction

2   The basis for these guidelines is provided in regulation 14(2) of Annex VI of MARPOL 73/78 and in Conference Resolution 4 (in MP/CONF.3/ 35), on monitoring the world-wide average sulphur content of residual fuel oil supplied for use on board ships. Among the emissions addressed by Annex VI are emissions resulting from the combustion of fuels containing sulphur. An upper limit for the sulphur content of fuels was set and it was further decided to monitor the average sulphur content of fuel.

It is estimated that independent testing companies undertake up to 50,000 tests annually, which cover between 25 and 35% of all deliveries. From the data gathered by these testing services, the current average figures for the sulphur content of residual fuels can be derived. These figures are publicised regularly and are currently in the order of 3% by mass.

Definitions

3   For the purpose of these guidelines the following definitions shall apply:

(1)   *Residual fuel:*
Fuel oil for combustion purposes delivered to and used on board ships with a kinematic viscosity at 100°C greater than or equal to 10.0 centistokes.[*]

---

[*] Reference is made to ISO Standard 8217, 1996.

(2)   *Provider of sampling and testing services:*
A company that, on a commercial basis, provides testing and sampling services of bunker fuels delivered to ships for the purpose of assessing quality parameters of these fuels, including the sulphur content.

(3)   *Reference value A_w:*
The value of the world-wide average sulphur content in residual fuel oils supplied for use on board ships, based on the first three years of data collected and as determined on the basis of paragraphs 4 and 5 of these guidelines.

4   Monitoring and calculation of yearly and three-year rolling average

4.1   *Monitoring*

Monitoring shall be based on calculation of average sulphur content of residual fuels on the basis of sampling and testing by independent testing services. Every year the average sulphur content of residual fuels shall be calculated. After three years the reference value for monitoring will be set as described in paragraph 5.

4.2   *Calculation of yearly average*

At the basis of monitoring is the calculation, on an annual basis, of the average sulphur content of residual fuel.

The calculation of the average sulphur content is executed as follows:

For a certain calendar year, the sulphur contents of the samples analysed (one sample for each delivery of which the sulphur content is determined by fuel oil analysis) are recorded. The sulphur contents of the samples analysed are added up and divided by the number of samples. The outcome of that division is the average sulphur content of residual fuel for that year.

As a basis for well informed decisions, a graphical representation of the distribution of the global sulphur content in residual fuels in terms of the % sulphur in increments of 0.5% sulphur plotted against the quantity of fuel associated with each incremental sulphur content range shall be made available by 31 January of each year.

The mathematical formula for the method of calculation described is given in appendix 1 to these guidelines.

4.3   Three-year rolling average

A three-year rolling average shall be calculated as follows:

$$A_{cr} = (A_{c1} + A_{c2} + A_{c3})/3$$

in which:

A_{cr} = rolling average sulphur content of all deliveries tested over a 3-year period

A_{c1}, A_{c2}, A_{c3} = individual average sulphur contents of all deliveries tested for each year under consideration

A_{cr} is to be recalculated each year by adding the latest figure for A_c and deleting the oldest.

Setting of the reference value

5   The reference value of the world-wide average sulphur content of residual fuel oils supplied for use on board ships shall be A_w, where $A_w = A_{cr}$ as calculated in January of the year following the first three years in which data were collected on the basis of these guidelines. A_w shall be expressed as a percentage.

Agenda setting of consideration of measures to reduce SO$_x$ emissions

6   If, in any given year following the setting of the reference value, A_{cr} exceeds A_w by a number equal to or greater than 0.2%, the Marine Environment Protection Committee shall consider the need for further measures to reduce SO$_x$ emissions from ships, so as to decide whether it should be considered a high-priority item for the Committee. MEPC shall continually review this excess value (now 0.2%) once the reference value has been set.

Providers of sampling and testing services

7   For the purpose of the initial five years operational period, there are presently three providers of sampling and testing services under these guidelines.

Any additional providers of sampling and testing services shall be subject to the following criteria:

.1   They shall preferably be IACS members, but shall in any way be subject to the approval of the Marine Environment Protection Committee, which shall apply these criteria.

.2   They shall be provided with a technical and managerial staff of qualified professionals providing adequate geographical coverage and local representation to ensure quality services in a timely manner.

.3   They shall provide services governed by a documented Code of Ethics.

.4   They shall be independent as regards to commercial interest in the outcome of monitoring.

.5   They shall implement and maintain an internationally recognized quality system, certified by an independent auditing body, which ensures reproducibility and repeatability of services which are internally audited, monitored and carried out under controlled conditions.

.6   They shall take a significant number of samples on an annual basis for the purpose of globally monitoring average sulphur content of residual fuels.

Standardized method of calculation

8   Each of the providers of sampling and testing services shall provide the necessary information for the calculation of the average sulphur content of the residual fuels to the Secretariat of IMO or another agreed third party on the basis of a mutually agreed format, approved by MEPC. This party will process the information and will provide the outcome in the agreed format to MEPC. From the viewpoint of competitive positions, the information involved shall be considered sensitive. Therefore the third party involved shall treat such information as indicated by any party involved as confidential, without prejudice to the information required by the Committee for the purposes of monitoring and related decision making.

Financial arrangements

9   The costs of monitoring consist of an initial amount for setting up monitoring by the providers listed in paragraph 7 and an annual fee to these providers for the provision of the update.

The costs of monitoring shall, for a 5-year operational period, be borne on a voluntary basis by the Member States listed in appendix 2 of these guidelines. The experience gained shall be evaluated by the end of the fourth year of operation. On that basis, the Marine Environment Protection Committee, in consultation with the Secretary-General of IMO, is invited to consider more permanent financial arrangements to cover the costs of the agreed monitoring system.

Additional Information

Appendix 1
Calculation of average sulphur content

Note: wherever "all deliveries" are mentioned, this is meant to refer to all deliveries sampled and tested for sulphur and being taken into account for the purpose of monitoring.

Calculation not weighted for quantity

$$A_{cj} = \frac{\displaystyle\sum_{i=1,2,3,\ldots N} a_i}{N_j}$$

in which:

A_{cj} = the average sulphur content of all deliveries sampled world-wide in year j

a_i = the sulphur content of individual sample for delivery i

N_j = total number of samples taken in year j

Additional Information

Appendix 2
List of countries bearing the costs of monitoring for a five-year period

Denmark

Finland

Netherlands

Norway

Sweden

United Kingdom